3XA2I

ex-lib

Multicomputers and Image Processing

Algorithms and Programs

Notes and Reports
in
Computer Science and Applied Mathematics

Editor
Werner Rheinboldt
University of Pittsburgh

Multicomputers and Image Processing

Algorithms and Programs

Edited by

Kendall Preston, Jr.

Department of Electrical Engineering
Carnegie-Mellon University
and Department of Radiation Health
University of Pittsburgh
Pittsburgh, Pennsylvania

Leonard Uhr

Department of Computer Sciences
University of Wisconsin, Madison
Madison, Wisconsin

1982

ACADEMIC PRESS
A Subsidiary of Harcourt Brace Jovanovich, Publishers

New York   London
Paris   San Diego   San Francisco   São Paulo   Sydney   Tokyo   Toronto

Academic Press Rapid Manuscript Reproduction

ACADEMIC PRESS, INC.
111 Fifth Avenue, New York, New York 10003

United Kingdom Edition published by
ACADEMIC PRESS, INC. (LONDON) LTD.
24/28 Oval Road, London NW1 7DX

Library of Congress Cataloging in Publication Data
Main entry under title:

Multicomputers and image processing.

(Notes and reports in computer science and applied
mathematics)
Papers from a workshop held in Madison, Wis., May 27-
30, 1981.
Includes index.
1. Image processing--Digital techniques--Congresses.
2. Parallel processing (Electronic computers)--
Congresses. 3. Computer networks--Congresses.
I. Preston, Kendall. II. Uhr, Leonard, Merrick.
III. Series.
TA1632.M84   621.36'7   82-1623
ISBN 0-12-564480-9   AACR2

PRINTED IN THE UNITED STATES OF AMERICA

82 83 84 85   9 8 7 6 5 4 3 2 1

Contents

Contributors

Numbers in parentheses indicate the pages on which the authors' contributions begin.

Dan Antonsson (31), *Department of Electrical Engineering, Linkoeping University, S-581 83 Linkoeping, Sweden*

Jean-Luc Basille (99), *Laboratoire C. E. R. F. I. A., Université Paul Sabatier, Toulouse, France*

Faye A. Briggs (319), *School of Electrical Engineering, Purdue University, West Lafayette, Indiana 47907*

Serge Castan (99), *Laboratoire C. E. R. F. I. A., Université Paul Sabatier, Toulouse, France*

Ronald S. Curtis (307), *Department of Computer Science, State University of New York at Buffalo, Amherst, New York 14226*

Renato De Mori (193), *Istituto di Scienze dell' Informazione, Università di Torino, Torino, Italy*

Bernard Delres (99), *Laboratoire C. E. R. F. I. A., Université Paul Sabatier, Toulouse, France*

Robert J. Douglass (207), *Department of Applied Mathematics and Computer Science, University of Virginia, Charlottesville, Virginia*

M. J. B. Duff (261), *Department of Physics and Astronomy, University College London, England*

Charles R. Dyer (409, 453), *Department of Information Engineering, University of Illinois at Chicago Circle, Chicago, Illinois 60607*

Glenn S. Fowler (431), *Department of Computer Science and Department of Electrical Engineering, Virginia Polytechnic Institute and State University, Blacksburg, Virginia 24060*

ix

Ariel J. Frank (307), *Department of Computer Science, State University of New York at Buffalo, Amherst, New York 14226*

K. S. Fu (319), *School of Electrical Engineering, Purdue University, West Lafayette, Indiana 47907*

Peter Gemmar (87), *Research Institute for Information Processing and Pattern Recognition, Karlsruhe, Federal Republic of Germany*

Barry K. Gilbert (385), *Biodynamics Research Unit, Department of Physiology and Biophysics, Mayo Foundation, Rochester, Minnesota*

Attilio Giordana (193), *Istituto di Scienze dell' Informazione, Universita di Torino, Torino, Italy*

Goesta H. Granlund (19), *Picture Processing Laboratory, Linkoeping University, S-581 83 Linkoeping, Sweden*

F. Gail Gray (431), *Department of Computer Science and Department of Electrical Engineering, Virginia Polytechnic Institute and State University, Blacksburg, Virginia 24060*

Björn Gudmundsson (31, 231), *Department of Electrical Engineering, Linkoeping University, S-581 83 Linkoeping, Sweden*

Concettina Guerra (221), *Istituto di Automatica, University of Rome, Italy*

Shin-ichi Hanaki (343), *C&C Systems Research Laboratories, Nippon Electric Co., Ltd., Kawasaki-city, Japan*

Pat Hanrahan (179), *Department of Computer Sciences, University of Wisconsin, Madison, Wisconsin 53706*

Robert M. Haralick (431), *Department of Computer Science and Department of Electrical Engineering, Virginia Polytechnic Institute and State University, Blacksburg, Virginia 24060*

Kai Hwang (319), *School of Electrical Engineering, Purdue University, West Lafayette, Indiana 47907*

Taizo Iijima (361), *Department of Computer Science, Tokyo Institute of Technology, Tokyo, Japan*

Yasushi Inamoto (353), *Fujitsu Laboratories Ltd., Kawasaki, Japan*

Mitsuo Ishii (353), *Fujitsu Laboratories Ltd., Kawasaki, Japan*

Ernest Kent (453), *Integrated Systems Laboratory, University of Illinois at Chicago Circle, Chicago, Illinois 60607*

Thomas M. Kinter (385), *Biodynamics Research Unit, Department of Physiology and Biophysics, Mayo Foundation, Rochester, Minnesota 55901*

Loren M. Krueger (385), *Biodynamics Research Unit, Department of Physiology and Biophysics, Mayo Foundation, Rochester, Minnesota 55901*

Björn Kruse (31, 125), *Department of Electrical Engineering, Linkoeping University, S-581 83 Linkoeping, Sweden*

Ichiro Kubota (125), *Institute of Industrial Science, University of Tokyo, Roppongi, Tokyo, Japan*

H. T. Kung (373), *Department of Computer Science, Carnegie-Mellon University, Pittsburgh, Pennsylvania 15213*

Pietro Laface (193), *C. E. N. S. — Istituto di Elettrotecnica Generale, Politecnico di Torino, Torino, Italy*

Christian Lantuéjoul (111), *Centre de Géostatistique et de Morphologie Mathématique, Ecole des Mines de Paris, Fontainebleau, France*

Jean-Yves Latil (99), *Laboratoire C. E. R. F. I. A., Université Paul Sabatier, Toulouse, France*

Stefano Levialdi (221), *Istituto Scienze dell'Informazione, University of Bari, Bari, Italy*

Martin D. Levine (149), *Department of Electrical Engineering, McGill University, Montreal, Quebec, Canada*

Kenneth Lundgren (19), *Picture Processing Laboratory, Linkoeping University, S-581 83 Linkoeping, Sweden*

Hiroyuki Matsuura (361), *Department of Computer Science, Tokyo Institute of Technology, Tokyo, Japan*

W. M. McCormack (431), *Department of Computer Science and Department of Electrical Engineering, Virginia Polytechnic Institute and State University, Blacksburg, Virginia 24060*

Bruce H. McCormick (453), *Integrated Systems Laboratory, University of Illinois at Chicago Circle, Chicago, Illinois 60607*

Ahmed Nazif (149), *Department of Electrical Engineering, McGill University, Montreal, Quebec, Canada*

Hidemitsu Ogawa (361), *Department of Computer Science, Tokyo Institute of Technology, Tokyo, Japan*

Morio Onoe (125), *Istitute of Industrial Science, University of Tokyo, Roppongi, Tokyo, Japan*

James Piper (161), *Medical Research Council, Western General Hospital, Edinburgh, Scotland*

J. L. Potter (275), *Digital Technology, Goodyear Aerospace Corporation, Akron, Ohio*

Kendall Preston, Jr. (135), *Department of Electrical Engineering, Carnegie-Mellon University, and Department of Radiation Health, University of Pittsburgh, Pittsburgh, Pennsylvania 15213*

Anthony P. Reeves (7), *School of Electrical Engineering, Purdue University, West Lafayette, Indiana 47907*

Azriel Rosenfeld (253), *Computer Vision Laboratory, Computer Science Center, University of Maryland, College Park, Maryland 20740*

Denis Rutovitz (161), *Medical Research Council Science, Western General Hospital, Edinburgh, Scotland*

Makoto Sato (361), *Department of Computer Science, Tokyo Institute of Technology, Tokyo, Japan*

Larry Schmitt (179), *Department of Computer Sciences, University of Wisconsin, Madison, Wisconsin 53706*

Howard Jay Siegel (241, 331), *Laboratory for Applications of Remote Sensing, and School of Electrical Engineering, Purdue University, West Lafayette, Indiana 47907*

Leah J. Siegel (241), *Laboratory for Applications of Remote Sensing, and School of Electrical Engineering, Purdue University, West Lafayette, Indiana 47907*

Bradley W. Smith (331), *Laboratory for Applications of Remote Sensing, and School of Electrical Engineering, Purdue University, West Lafayette, Indiana 47907*

S. W. Song (373), *Department of Computer Science, Carnegie-Mellon University, Pittsburgh, Pennsylvania 15213*

Stanley R. Sternberg (291), *Environmental Research Institute of Michigan, and University of Michigan, Ann Arbor, Michigan 48106*

James P. Strong (47), *Goddard Space Flight Center, Greenbelt, Maryland 20770*

Philip H. Swain (241, 331), *Laboratory for Applications of Remote Sensing, and School of Electrical Engineering, Purdue University, West Lafayette, Indiana 47907*

Steven L. Tanimoto (421), *Department of Computer Science, University of Washington, Seattle, Washington 98105*

Tsutomu Temma (343), *C&C Systems Research Laboratories, Nippon Electric Co., Ltd., Kawasaki-city, Japan*

Joseph G. Tront (431), *Department of Computer Science and Department of Electrical Engineering, Virginia Polytechnic Institute and State University, Blacksburg, Virginia 24060*

Leonard Uhr (1, 179), *Department of Computer Sciences, University of Wisconsin, Madison, Wisconsin 53706*

Larry D. Wittie (307), *Department of Computer Science, State University of New York at Buffalo, Amherst, New York*

Preface

Introduction: Fitting Algorithms to Multicomputer Architectures

This book is the second of a set presenting papers from a series of workshops exploring the large and powerful new multicomputer arrays and networks that are today just beginning to be built.

Three major topics are being examined in these workshops: (a) the architectures of multicomputer arrays and networks, with emphasis on those used for image processing; (b) higher level programming languages that aid and encourage the programmer in formulating parallel programs; (c) algorithms and programs that exploit the potentially enormous increases in speed that arrays and networks can offer.

The workshops were deliberately kept small and quite informal, in order to establish a congenial environment for lively and thoughtful discussions and examinations of problems of mutual interest. Many of the workers who are building multicomputer arrays and networks and designing higher level languages, and whose primary focus is on image processing, have participated in these workshops. Many of the leading researchers in image processing, pattern recognition, and perception systems, chosen because they have been taking parallel approaches, have also participated. The great interest and enthusiasm of the participants suggests that they have found these workshops worthwhile. The quality of the formal papers and of the research they describe, as attested to by these first two volumes, suggests that these meetings are beginning to form an important focal point for the researchers developing these new kinds of multicomputers.

The first two meetings, held in Great Windsor Park, near London, in 1979, and on Ischia, in the Bay of Naples, in 1980, focused on first languages and second architectures. The first volume in this set, "Languages and Architectures for Image Processing," edited by Michael Duff and Stefano Levialdi, and published in 1981

xiii

by Academic Press, London, contains papers from these two meetings. The papers collected in the present volume are from the third meeting, held in Madison, Wisconsin, May 27-30, 1981.

At the Madison meeting algorithms and programs were emphasized. But each workshop has explored all of the closely related aspects of developing algorithms and programs for new multicomputer architectures, and each of the two volumes contains chapters on each of the major topics. Indeed, a number of papers address several or all of these issues.

This reflects the need to explore this problem in an interrelated way. Algorithms suggest architectures that can most efficiently execute them, and languages in which they can most fluently be expressed, and programmed. Architectures and languages can encourage, or discourage, the efficient programming of algorithms.

The Organization and Contents of This Book

The papers in the present volume are organized to reflect the three major aspects of the problem: user algorithms and programs, higher level languages, and multicomputer architectures (see the Introduction).

User Algorithms and Programs for Arrays and Networks

Architectures provide us with the tool. Languages allow us to express ourselves and use the tool. But our ultimate purpose is to solve problems, in our case image processing and perceptual problems. It is the algorithms, and the programs that embody and execute these algorithms, that finally make the problem-oriented user happy, and justify all the hard work and frequent pain inherent in developing complex new multicomputer systems.

The development of algorithms with the extremely high degree of parallelism that the new highly parallel arrays and other kinds of networks can handle is an unusually difficult, at times almost mysterious, task. Almost all of our previous experience has been with serial computers, that is with computers that do only one thing at a time. Almost all of the algorithms developed to date are serial; and our programs are entirely serial in that they map algorithms onto a serial computer. In sharp contrast, we know that many problems have great degrees of parallelness, even though it is rarely known precisely how great. We are just beginning to attack what will become an increasingly important research problem, the development of parallel methods for solving problems with parallel multicomputers.

Several of the papers in this volume present relatively specific algorithms for specific applications; others present and examine larger structures of algorithms; others examine whole programs.

Examinations of a Variety of Image Processing Algorithms

Anthony Reeves examines basic algorithms for binary array processors (and gives a good succinct description of such systems). The algorithms described handle a wide variety of processes, from image registration and image enhancement to motion detection and object description.

Kenneth Lundgren and Goesta Granlund describe a very general operator that is the basis of and is embodied in Granlund's GOP computer (described briefly in this paper, and more extensively in the first volume, edited by Duff and Levialdi). They also show how this operator is used for contour detection.

Björn Kruse, Björn Gudmundsson, and Dan Antonsson describe the parallel filter processor (FIP) that is the local-neighborhood processing subsystem of PICAP II.

James Strong gives an overview of the very large and fast MPP array that is now being built by Goodyear-Aerospace for NASA-GODDARD, and examines a wide range of basic algorithms.

Peter Gemmar, in addition to describing the FLIP architecture, shows how it can be used for image correlation and object tracking algorithms.

Jean-Luc Basille, Serge Castan, Bernard Delres, and Jean-Yves Latil present a propagation algorithm and examine its implementation on the projected SYMPATI system, a two-level multicomputer with the lower level SIMD (single instruction-multiple data stream) and the upper level MIMD (multiple instruction-multiple data stream) processors.

Christian Lantuejoul describes geodesic segmentation techniques for transforming complex images, and examines their application to several interesting practical problems.

Mario Onoe and Ichiro Kubota describe a new algorithm for implementing the two-dimensional Fourier transform that eliminates the need for matrix transposition. This algorithm is of great significance in reducing synthetic aperture radar holograms to images.

Kendall Preston analyzes the maxmin propagation transform. He further shows how it may be carried out on a two-dimensional cellular logic machine such as the PHP now running at the joint Carnegie–Mellon University–University of Pittsburgh Image Processing Laboratory.

H. J. Siegel, Philip Swain, and Bradley Smith compare their proposed PASM multicomputer with the CDC Flexible Processor on a variety of basic image processing algorithms.

Perception Programs That Embody Complex Sets of Algorithms

The specific image processing algorithms must be combined into a very large and complex total system in order to actually recognize objects and describe whole scenes. This is an extremely difficult problem, when highly variable and unconstrained real-world images are used, and researchers are only beginning to develop such systems.

Martin Levine and Ahmed Nazif examine parallel and serial techniques for model-based segmentation in multilevel vision systems.

Denis Rutovitz and James Piper examine the problem of chromosome analysis, and how it might be attacked with systems that use a parallel array like CLIP and/or a serial computer.

Leonard Uhr, Larry Schmitt, and Pat Hanrahan describe a complex type of program that embodies a set of lower level and higher level algorithms for effecting visual perception taking a cone/pyramid multilayered converging approach. They then compare different array, network, and pipeline structures with respect to such a structure of algorithms.

Renato De Mori and Attilio Giordana describe a set of complex high-level algorithms for the closely related and comparably difficult problem of recognition of continuous spoken speech. These algorithms form a complete program for this task, one that would most suitably be executed on a network of the sort described in their paper.

General Issues and Problems of Algorithm Development

Robert Douglass examines the perception of occluded objects from the psychological point of view. He also proposes parallel algorithms and appropriate network architectures for handling this difficult and important problem.

Concettina Guerra and Stefano Levialdi survey major models for effecting local operations at a large number of locations in parallel, and explore techniques for describing these models and for expressing algorithms for computer systems built to embody these models.

Björn Kruse and Björn Gudmundsson briefly describe the PICAP system of specialized processors and pipelines, and examine the variety of types of local and global parallelism that it embodies.

Leah Siegel, H. J. Siegel, and Philip Swain examine and compare a variety of different criteria for evaluating the performance of parallel algorithms.

Azriel Rosenfeld reviews work on cellular bounded automata, both one-dimensional and two-dimensional. He also examines extensions to these near-neighbor systems that give them a more general interconnection topology, regular connections at increasing distances, and pyramid connections through successive converging layers.

Michael Duff examines the set of interrelated issues of computer architecture, programming language, and algorithm structure. He works through several interesting examples of algorithms to explore the need for and value of having the programmer be aware of and work with the architecture's structure.

Higher Level Programming Languages for Arrays and for Networks

We have been especially fortunate that virtually every researcher of whom we are aware working on higher level languages for image processing and perceptual tasks using arrays and networks has attended at least one in this series of workshops.

Most of these languages are described in the first volume of this series (edited by Duff and Levialdi). These include the ALGOL-based PIXAL developed by Stefano Levialdi and his associates, Zenon Kulpa's PICASSO system, Björn Gudmundsson's higher level language for PICAP, the language L being developed by Adolfo Guzman and his associates, Leonard Uhr's parallel extensions to PASCAL, and Robert Douglass's proposal for MAC, the first language of which we are aware that attempts to handle both SIMD and MIMD processes.

The present volume extends this list with papers on two important new languages.

Jerry Potter describes the new Fortran-based language being developed for the very large MPP. This language, because of the great potential power of the MPP, and the preponderance of Fortran programmers in the real world, is likely to have a major practical impact.

Stanley Sternberg describes his unusually elegant set-theory-based language for expressing the structural image processing and perceptual operations effected by his Cytocomputer and other cellular-array-motivated architectures.

The Architectural Structure of Arrays, Networks, and Pipelines

Several multicomputer architectures are described in this book. Often the description is combined with an examination of languages or algorithms for that computer (in which case we will refer to that chapter).

The major computer architectures that have been developed can be viewed and categorized from several points of view. Are they serial or parallel, general-purpose, specialized (but still general-purpose) or special-purpose? Does each processor work independently in MIMD mode; or do all processors execute the same instruction, in SIMD mode; or is there a pipeline, each processor continuing to execute the same instruction, but passing results on to the next processor in MISD assembly-line fashion? What kind of architecture does the individual processor have (e. g., 1-bit or 32-bit), and what interconnection topology exists between processors (e.g., bus, ring, cross-point switch, star, n-cube, array, tree, pyramid)? The following set of papers examine several major types of new architectures that are being developed for large parallel-serial multicomputer systems.

Very Large Parallel SIMD *Arrays of 1-Bit Computers*

Several very large arrays, typically organized in two-dimensional grids with each computer linked directly to its four, eight, or (when a hexagonal design is used) six nearest-neighbors have recently been built, or designed.

In order to build such large arrays, each computer has been made as simple as possible, usually executing 1-bit processes on 1-bit words. Longer strings and larger numbers are therefore processed in bit-serial fashion, as the price of achieving the highly parallel processing over the large array.

These arrays are also given a single controller, so that all computers are executing the same instruction at each moment in time, This makes them "SIMD" (single instruction-multiple data stream) multicomputers.

They are often called "special-purpose;" but since each of their thousands of computers is itself general-purpose (albeit small) that is not really the case. They might more aptly be called "specialized" (and one can argue that any computer is more or less specialized).

These arrays include the 96 by 96 CLIP4 built by Michael Duff and his associates and running at University College London, the 64 by 64 DAP built by Stewart Reddaway and his associates and sold by ICL England, and the planned 128 by 128 MPP, developed by James Strong and others at NASA-Goddard and being designed and built by Kenneth Batcher, Jerry Potter, and others at Goodyear-Aerospace. CLIP4 and DAP are described in the first volume of this series, while the MPP is described in this volume, by Jerry Potter and by James Strong.

Multicomputer MIMD *Networks, Fixed and Reconfigurable*

A number of networks have been designed, where, typically, more powerful computers are used, interconnected in a wide variety of ways, and not only in a near-neighbor array. These networks are typically designed with a much smaller number of each much more powerful 8-bit, 16-bit, 32-bit, or 64-bit computers, each with its own controller.

Today only two or three such networks with more than 50 computers actually appear to be running. These include Cm* (built at Carnegie-Mellon University) and the system described by Manara and Stringa in the first volume of this series, which is being used in about a dozen Italian and French Post Offices to recognize portions of addresses on letters.

Most of these networks are being designed with the hope they will handle a general mix of programs, rather than with any particular classes of applications in mind. This is turning out to be a very difficult task, and very few of these systems have yet reached the point where the design is ready for actual building.

Among the most promising and interesting of these is Larry Wittie's MICRONET (described with co-authors Ronald Curtis and Ariel Frank). It has 16

LSI-11s in its first prototype version, but it is being designed looking toward future networks of thousands of computers.

Banks of switches can be added, between processors and processors or between processors and memories, so that a network can be reconfiqured under program control.

Peter Gemmar has designed and built a reconfigurable "flexible image processing system" (FLIP) with 16 computers that is optimized to handle correlations and other operations frequently used in image processing systems (see also the first volume of this series).

Two large reconfigurable systems of much more powerful computers are being developed by Faye Briggs, Kai Hwang, and K. S. Fu, and by H. J. Siegel and his associates. Both of these very ambitious systems are envisioned to use up to 1024 computers. Although they are being designed to handle a wide mix of problems, the focus is on image processing and other perceptual tasks.

Specialized and Special-Purpose Image Processing Architectures

A variety of more or less specialized networks and image processing systems are being developed.

Shin-ichi Hanaki and Tsutomo Temma describe a multicomputer system designed to handle image processing with a data-flow, template/data-driven approach. They show how the "buttlerfly" perfect shuffle, to compute fast fourier transforms, can be implemented on a ring of computers built to implement their data-flow approach.

Mitsuo Ishii and Yasushi Inamoto describe a new system for image processing and computer aided design graphics that has an unusually large memory.

Makoto Sato, Hiroyuki Matsuura, Hidemitsu Ogawa, and Taizo Iijima describe an architecture for microcomputers that both puts a reasonably large number of them (32) on a ring and also links them all directly to a "father" node.

Two special-purpose systems of unusual interest are described in this volume: H. T. Kung's systolic array for convolutions and Barry Gilbert's very powerful and amazingly fast multicomputer for real-time three-dimensional x-ray CAT scanning.

A systolic array is a very efficient special-purpose system that has been designed to embody a particular algorithm in hardware (meaning off-the-shelf IC chips today, and specially designed VLSI chips tomorrow). Kung and Song describe a VLSI implementation of the first systolic array (for convolution) that has been designed for an image processing task. It is of special interest because an IC version is now being built by TRW and, to the extent a market develops, will probably be produced commercially (this was described at the Madison workshop by Kung and R. L. Picard, but that paper is not included in this volume).

Gilbert's system to handle three-dimensional CAT scans in real time is now being completed. It may well turn out to be one of the fastest convolvers and one of the most powerful computers ever built. It will push ECL chips to subnanosecond clock times, and is expected to execute more than 3 billion floating-point instructions per second.

Pipelines

Pipelines of computers can be built, and data pumped through them, much as material flows through factory assembly lines.

Björn Kruse's PICAP systems combine this kind of pipeline with a variety of other more or less specialized processors and general-purpose serial computers, into a total MIMD system that has been carefully tuned for efficient image processing and pattern recognition.

Stanley Sternberg's Cytocomputer, with over 100 computers in the pipe, has extended the pipeline to its greatest length to date. There are plans for new Cytocomputers that will be built with each processor on its own VLSI chip. These could have even longer pipes, when desired.

Proposals for Future Three-Dimensional Networks and Arrays

Three-dimensional networks and arrays are just beginning to be seriously examined and designed. Here we have been especially fortunate that four of the most interesting systems were described at the Madison workshop and are included in this volume.

Charles Dyer has been exploring the design of pyramid computers side-by-side with his closely related work on algorithms for cellular arrays and pyramids.

Steven Tanimoto is designing, and hoping to fabricate, a VLSI chip that will be used to build a pyramid that, roughly, starts with a CLIP4-like array at its base. Tanimoto examines the programming of such a system.

Robert Haralick and his co-workers (W. M. McCormack, Gail Gray, Joseph Trout, and Glenn Fowler) examine architectural issues in using multicomputer structures to handle combinatorial problems.

Bruce McCormick was the architect of ILLIAC-III, one of the very first, and still one of the most interesting, of the 1-bit arrays (unfortunately demolished by fire just before it would have been completed). In this book McCormick, with coauthors Ernest Kent and Charles Dyer, gives an intriguing picture of a three-dimensional system.

This volume is probably best read together with the first volume (edited by Duff and Levialdi). For the first time, so far as we are aware, they give the reader access to papers on virtually all of the major new array, network, and special-purpose architectures emphasizing image processing, and to the major higher level languages for these new multicomputers, along with a wide variety of algorithms and programs.

INTRODUCTION: TOWARD VERY LARGE MULTI-COMPUTERS

Leonard Uhr

Department of Computer Sciences
University Of Wisconsin
Madison, Wisconsin

We are just beginning to attack the many problems of designing
parallel algorithms and parallel multi-computer architectures.
The advent of LSI and VLSI (Very Large Scale Integration) chips
each of which contains thousands, or in 5 to 10 years millions, of
devices equivalent to transistors or logic gates for the first
time makes possible the design and fabrication of enormously large
yet relatively cheap parallel-serial multi-computer arrays, pipelines, and other types of networks.

Continuing progress on a number of extremely important problems, such as image processing, pattern recognition, scene
description, and other aspects of the perceptual process, demand
the development of these large and powerful multi-computers. For
these problems need enormous amounts of computing power. [There
are a number of other very large problems, of modelling 3-
dimensional masses of matter, for example to design airplanes and
missiles, to predict the weather, and to study large-scale
phenomena like earthquakes, storms, and neuronal firings in the
brain (brainstorms). Often, but not always, these appear to con-
front the computer architect with a similar set of demands. But
since this book focuses on image processing and related perceptual
tasks these are merely mentioned.]

Typical estimates suggest that to recognize and describe the
objects in a single image, many billions of instructions are needed. And it is often of vital importance that an image be processed within a very short "real-time" interval. For example,
each of the successive images in a moving scene, whether input
from a movie camera, TV camera or some other sensing device, must
be processed within 20 to 50 milliseconds.

MULTICOMPUTERS AND IMAGE PROCESSING
ALGORITHMS AND PROGRAMS

1

MAPPING PROGRAM GRAPH INTO MULTI-COMPUTER GRAPH

Our underlieing purpose is to solve problems, which means to transform information from the initial specification of the problem to the required set of answers.

It is illuminating to conceptualize this process as one of mapping:

A) the graph determined by the flow of information through the processes that transform that information from input problem to output solution onto

B) the graph of physical processors, that is the actual multi-computer network that effects those processes on information input to it.

Flow charts and data flow graphs are often used to conceptualize a program. We can think of each instance of these as the program graph. Multi-computer architects are exploring a wide variety of graphs in their search for network structures onto which programs can be mapped so that they will execute fast and efficiently.

We thus have a graph-matching problem. If we could express our algorithms by coding data-flow graphs where each node could be equated with a processor, and each edge that indicated the flow of information from one process to the next could be equated with a physical link (a wire or bus of wires) between the two physical processors that actually executed the two successive processes, we would be able to map the program directly onto the network, with great efficiency.

In effect we would have two graphs (the program graph and the multi-computer graph) that were isomorphic and could therefore be mapped onto one another perfectly. We could now consider maximizing information-flow within and between nodes. We could attempt to balance the load across nodes, and to increase the parallel bandwidth of the network, as desired, to increase and speed up the flow of information as it is transformed from posed problem to achieved solution.

Networks, Reconfigurable and Fixed

This suggests having a "general reconfiguring capability" that would take the set of computers in the network and restructure them by changing the links between them to exactly mirror the program graph. This is the goal of reconfigurable computers like STARAN, PASM and PUMPS. But reconfiguring switches appear to be very expensive, and general reconfiguring has major problems.

Another alternative is to use some particular graph structure, but one that appears to be relatively "general" enough so that most programs can be mapped into them, or into sub-graphs that they contain, with relatively small distortions. A network of computers is best modelled as a graph whose vertices are computers (or specific computer resources like processors, memories, input-output devices), and whose edges are data paths linking these resources. This means that the variety of possible networks is enormous, since, potentially, it includes all possible graphs. Here people have suggested and are exploring a great variety of different structures, including cross-point switches, busses, rings, stars, clusters of clusters, lenses, n-cubes, lattices, pyramids, trees and augmented trees. But relatively few networks have actually been built, and very few firm comparisons can be made. [See Thurber, 1976, Kuck, 1978, Stone, 1980, Baer, 1980, and Uhr, 1982 for descriptions and examinations of some of these systems.]

More or Less Specialized Computers

An alternate approach attempts to develop what I will call a "specialized" network structure that at least to some extent mirrors the structures used by the program(s) that network will execute. Whereas the re-configurable network will attempt to mirror the program by throwing switches, this kind of specialized network will be built with a typical program structure as its anatomy.

There can be many degrees of specialization.

Two-dimensional arrays of computers are probably the best example of powerful specialized systems, since they are clearly useful when processing large multi-dimensional arrays of data (e.g., a 512 by 512 TV image, or a matrix of numbers representing a 2-dimensional or 3-dimensional mass of matter, as for wind tunnel or weather computations).

Pyramids of computers, where several successively smaller arrays are linked together, or where the buds in a tree of computers are stitched to an array, appear to a number of researchers to offer worthwhile specializations that move them closer to data structures and information flows appropriate for image processing and other aspects of perception.

In addition, a wide variety of architectures that are more or less specialized to image processing have been built, or designed.

Special-Purpose Processors that Embody a Procedure in Silicon

Another possibility is to build a "special-purpose" system, one that is not "general-purpose" in the sense that it can be programmed to execute any possible program, but rather has been built to execute some particular program, or some particular class of programs. Such a system can be given the exact structure of the program or the procedure it is designed and built to execute. The program can actually be built into the succession of gates and wires, rather than being stored in memory to be fetched and decoded by the controller.

This also has several variants: occasionally a whole program can be embodied in this way into a special-purpose computer. More typically, some particular instruction or procedure that is repeatedly used, e.g., floating point multiplication, or convolutions, can be cast in silicon and wires. Note that such special-purpose processors will usually be linked to the bus of a larger general-purpose computer. Today this larger computer is almost always a conventional single-CPU serial Von Neumann computer. But there are interesting possibilities that are just beginning to be explored for combining a number of different special-purpose, specialized and more general computer nodes into larger multi-computer networks.

Serial Computers (Single-Node Graphs)

Finally, one can take the traditional approach, of mapping the program graph onto a conventional single-processor computer. Such a conventional computer is represented by the very simple graph that contains one node (and no links). It matches and executes the program graph much as one single ant would move along a path that traversed all nodes and edges.

The programming of this matching process is usually relatively simple, since it is a matter of traversing all parts of the graph, in a reasonably efficient order. But such a serial process is inevitable slow. And a number of inefficiencies are introduced, in the need to backtrack and store intermediate results, and to index through, and test for borders in, arrays and other sets of regularly structured information that could, potentially, be handled in parallel.

Most people today think of the serial computer as very general and efficient. A graph with one node has minimal structure (except within that node), and can easily be used to trace along any other graph. But this is bought at the price of extremely slow serial processing, either a very large random access memory or input-output slowdowns, and the various aspects of the Von Neumann bottleneck.

POTENTIAL IMPROVEMENTS IN SPEED AND EFFICIENCY OF MULTI-COMPUTERS

We traditionally consider a computer's efficiency in terms of the percent of the time that its processor is being used. But we typically build a serial computer with 2,000 to 20,000 gates in its CPU and 2,000,000 to 80,000,000 gates in its high speed main memory. Almost all of these gates are sitting around almost all of the time doing nothing, slaves designed to provide the CPU immediately with whatever information it might need.

Multi-computer and multi-processor networks offer the possibility of large numbers of active processors simultaneously transforming information as it flows through them.

In the extreme, an array of computers exactly the size of the array of information they are processing can eliminate the appreciable overhead of indexing and testing for borders, as well as executing transformations on all cells in the array at once.

A pipeline of gates, processors, computers, arrays or networks can eliminate the need to continually load new program instructions and store intermediate results (since each stage in the pipe repeatedly executes the same instruction on a continuing sequence of data, and the output from one stage is immediately gobbled up as the input to the next stage).

Perhaps the most interesting alternatives are those that combine several, or all, of the above possibilities.

THE POTENTIAL ENORMOUS SIZE OF FUTURE MULTI-COMPUTER NETWORKS

Typically, a computer system is considered "large" (that is, relatively expensive and difficult to build) when it contains 10,000 or more components. But today each component can be a chip with 100,000 or more gates. Reasonably firm projections (based upon a continuing doubling every 14 to 18 months for the past 20 years of the number of devices that can be fabricated successfully on a single chip) suggest that we can think in terms of 1,000,000 gates per chip by 1986-1996, and even 10,000,000 gates per chip by 1990-2001.

The variety of multi-computer architectures that can be built with 10,000 or so chips of such size and potential complexity is mind boggling.

Today's 32-bit computers typically have roughly 10,000 gates in their CPUs, and use from 1 to 8 gates to store each bit of memory. The 1-bit processors used in the multi-computer arrays have only 50 to 500 gates each, and only 32 bits (CLIP4) to 4,000 bits (DAP) of memory per processor. An 8-bit or 16-bit processor has 2,000 to 5,000 gates. A specialized processor that forms one stage in a larger pipeline or systolic array might have as few as 5 or 10 gates.

We can begin to consider a 32x32 or even a 64x64 array of 1-bit processors (albeit limited in input-output to the chip, and with roughly 1,000 bits of memory per processor). A Cray-N super-computer will need less than one chip (and, probably, several additional memory chips).

The multi-computer architect will soon have enormous quantities of relatively maleable raw materials with which to build networks of more or less specialized, general, or/and special-purpose nodes, interconnected in any desired graph structure, or, if preferred, reconfigurable to a variety of possible structures.

The programmer/algorithm developer will have similar opportunites to structure the graph of processes through which information flows and by which that information is transformed. Both architect and programmer can play a major role in modifying and improv-

ing the structure of the multi-computer network and the structure
of the information-flow program graph, so that they can be better
mapped onto one another with relative ease.

The problem of mapping one graph onto another is, when defined
in a general way, simply the exponentially hard graph isomorphism
problem. Sub-graph isomorphism is, in the general case, NP-
Complete. We are therefore confronted with problems that are
quite intractable for graphs with thousands or millions of nodes,
except when very special sub-classes of graphs are very carefully
chosen and matched. We can view our exploration of parallel algorithms and multi-computer architectures as one of choosing, modifying and conditioning information-flow graphs and multi-computer
graphs so that they can be mapped onto one another.

REFERENCES

Baer, J-L. (1980). "Computer Systems Architecture", Computer
    Sci. Press, Potomac, Md.
Kuck, D.J. (1978). "The Structure of Computers and Computation:
    Vol. 1", Wiley, New York.
Stone, H.S. (1980). "Introduction to Computer Architecture",
    SRA, Chicago.
Thurber, K.J. (1976). "Large Scale Computer Architecture", Hay-
    den, Rochelle Park, N.J.
Uhr, L. (1982). Computer Arrays and Networks: Algorithm-
    Structured Parallel Architectures, Academic Press, New York.

PARALLEL ALGORITHMS FOR REAL-TIME IMAGE PROCESSING

Anthony P. Reeves

School of Electrical Engineering
Purdue University
West Lafayette, Indiana 47907

I. INTRODUCTION

The very high computation speed necessary for the analysis of real-time TV video imagery may be achieved with emerging parallel processors. In this paper algorithms suitable for implementation on highly parallel binary array processors (BAP's) are described. These algorithms deal with the low level image processing operations on a sequence of video images, i.e. image registration, image enhancement, moving object detection and object description extraction.

The image registration is necessary to compensate for motion of the TV camera between adjacent frames. Noise on the raw TV data may cause processing problems. Since adjacent images change by only small amounts several images, when registered, may be combined to reduce noise effects. Furthermore, the small differences between registered images can be used to detect moving objects. Finally, once objects have been detected a shape description may need to be extracted for further analysis and classification.

A typical scenario for real-time video imagery may involve images in the order of 512 x 512 8-bit pixels which are obtained at the rate of 30/second. Therefore, there are about 33 m.s. available to process completely an image of 262,144 pixels. Execution times have been estimated for implementing the described algorithms on an extended MPP BAP; these times are within the above 33 m.s. bound.

Anthony P. Reeves

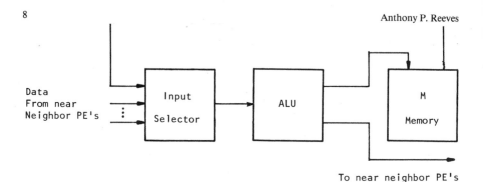

Fig. 1. BAP PE organization

II. BINARY ARRAY PROCESSORS

Binary Array Processors (Reeves, 1980a) operate in the single instruction steam-multiple data stream (SIMD) mode with a matrix of identical processing elements (PE's). The whole image or a consecutive block of the image is distributed through the PE's and processed in parallel. All data paths within a PE are only one bit wide and each PE is connected to PE's adjacent to it.

The main features of the BAP scheme result from the bit-serial architecture of the PE's and near-neighbor interconnection scheme. The bit-serial architecture allows flexible data formats and makes the BAP very efficient with respect to memory and processing resource utilization. Many image processing algorithms require that data within local areas of each pixel is to be combined; the near neighbor interconnection scheme enables these algorithms to be efficiently implemented.

The main functional units which characterize most BAP PE architectures are shown in Fig. 1. All data paths are 1-bit wide and data is stored in a 1-bit wide memory M. Data selection is achieved by the input select unit which can obtain data from the memory M or from near neighbor PE's. Data processing is achieved with a simple ALU which can process two or more single-bit operands per instruction. The input select unit and ALU may contain several 1-bit registers for holding temporary data values. Some PE designs involve a 1-bit mask register which, when set, inhibits the PE from processing data.

The data interconnections between PE's are shown in Fig. 2. Most PE architectures have interconnections with the 4 nearest neighbor PE's. Some BAP designs also have interconnections with the 4 next nearest neighbor PE's as indicated by the broken lines in Fig. 2. Hexagonal PE interconnections have also been considered and in several BAP's either a square or a hexagonal interconnection scheme may be selected by a user programmable

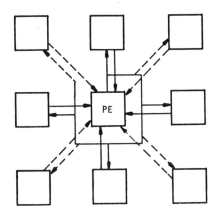

Fig. 2. BAP near neighbor interconnections

option. The input select unit can select data from any memory location within its own PE or any adjacent-connected-PE memory location.

Several large scale BAP systems have been built and designed, including CLIP4 (Duff 1976), DAP (Readdaway 1979) and the Massively Parallel Processor (MPP) (Batcher 1980). A general diagram of a large scale BAP is shown in Fig. 3. the data pro-

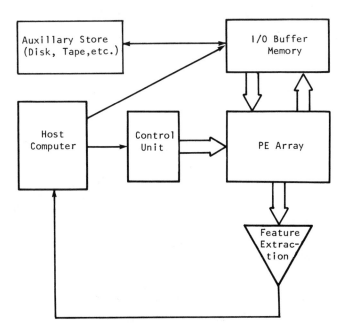

Fig. 3. BAP system organization

cessing is achieved by a matrix of PE's which simultaneously process a submatrix of an image. The total image is processed as a
sequence of these submatrices. Data is input to and output from
the PE array via the I/O buffer memory which communicates the
data to image peripherals and conventional computer bulk storage
devices. With the current LSI systems each bit plane is input
along one edge of the PE array one column on each clock cycle.
Each row of the PE array acts as a shift register. When the
complete bit plane has been input it is stored in the local
memory in one clock cycle. Bit-serial instructions to the PE
array are issued by a single high speed microprogrammed controller. The whole system synchronization is maintained by a
conventional host computer which issues macro instructions to
the controller. Some feature information may be extracted from
the PE array by the global information extraction mechanism.

For example, the MPP involves a 128 x 128 matrix of 16384
PE's. Instructions, involving a memory fetch and an operation,
will be executed in a single 100 n.s. clock cycle. If the data
paths for a BAP are restricted to 1-bit width and all operands
are maintained in the local memory then the minimum number of
memory cycles for basic operations may be determined. The number of memory cycles for N-bit-integer operands is given in
Table 1.
Most simple operations require the same amount of time as for
addition. Multiplication requires N additions and therefore requires N times the addition cost. In future VLSI BAP designs
additional PE hardware could be used for multiplication (Reeves
1980b) which would require the minimum number of memory cycles
(4N).

Table 1.   Estimated execution times for basic operations on a
           BAP for N-bit integer operands.

Operation	Memory Cycles	MPP (time in μs) 128 x 128	512 x 512
Logical	3	0.3	4.8
Addition	3N	0.3N	4.8N
Multiplication	$3N^2$ $(4N)^*$	$0.1N^2$	$1.6N^2$
Translate one Pixal position	2N	0.2N	4.8N
Array sum (typical value A=8)	A+N	$0.8+0.1N^{**}$	$0.8+1.6N^{**}$

[*]possible with additional PE hardware, see Reeves (1980b).
[**]not implemented on the current MPP design.

Execution times for the MPP are also given in table 1. The multiplication is achieved in N^2 instructions by storing one of the operands in a multi-bit register in the PE ALU. A 512 x 512 image is processed in 16 blocks. Operations therefore require 16 times the 128 x 128 times except for translation which is slightly slower because of the time needed to transfer data between adjacent blocks.

The MPP has only a simple OR global feature extraction mechanism which returns a one if any bit in a bit plane is set. Counting the bits in a bit plane is an important operation in the algorithms to be described. Special bit counting hardware has been designed which can count a bit-plane each clock cycle but involves an 8 cycle pipeline delay (Reeves 1980c). In table 1 the time required to sum the elements in an image using this mechanism is given; adding this to the MPP would involve less than a 10% increase in PE hardware.

From table 1 it can be seen that all basic operations given, except multiplication, require execution times proportional to the operand length N. The multiplication could be executed in a time proportional to N if substantial additional logic is added to each PE.

III. IMAGE REGISTRATION

A. Grey Level Registration

The problem of image registration is to determine the transformation between two related images. If the transformation is a simple translation in the x-y plane then two images may be registered by sliding one image over the other until a minimum in a similarity measure between them is found. Potential similarity measures include the correlation coefficient, correlation function and the sum of the absolute differences (Svedlow et al., 1978).

For the application of sequential video imagery the difference between grey levels of registered adjacent frames should be very small; a suitable dissimilarity measure is the mean squared error (MSE) which, for two images f(x,y) and g(x,y) is defined by

$$MSE = C \sum_{i=1}^{X} \sum_{j=1}^{Y} ((f(i,j) - g(i,j))^2$$

Where C is a normalizing constant.

As one image is slid over the other the area of overlap between them will change; this could cause normalization problems since we wish to compare the error values. One solution to this problem is to use a consistent subset of the image $f(x,y)$ for all correlations. For a maximum displacement of r elements between the two images all elements of $f(x,y)$ within r elements of the edge must be discounted from the correlation computations. A measure for relative difference may be defined as follows.

$$V(k,l) = \sum_{i=r+1}^{X-r} \sum_{j=r+1}^{Y-r} (f(i,j) - g(i+k,j+l))^2$$

where k is the horizontal displacement and l is the vertical displacement between g and f. C is a constant value for all $V(k,l)$ therefore the minimum value of V indicates the best correlation.

In general, it is reasonable to assume that the displacement between adjacent images will not be more than 10 pixels in any direction. Therefore, the matrix V to be computed will contain 21 x 21 = 441 elements. For very fast motion the 10 pixel limit may be exceeded, however for such exceptional cases it would probably be possible to predict, externally, the displacement between images to within 10 pixels. The algorithm for implementing this algorithm on a BAP is described below:

1. Starting with $g(x,y)$ located where registration is expected, $g(x,y)$ is shifted in a spiral search pattern (which requires only one pixel translation at a time).
2. The squared difference between $f(x,y)$ and $g(x,y)$ is computed.
3. $V(k,l)$ is obtained by computing the array sum of the result from step 2. A binary mask inhibits elements near the edge of the matrix from being considered.
4. The location and value of the minimum element of V to date is maintained, if the most recent value $V(k,l)$ is lower than this value the retained minimum is updated.
5. Steps 1 through 4 are repeated until all values of V have been computed. The retained location of the minimum value of V indicates the displacement of the images for best registration.

B. Binary Image Registration

If the images are only binary valued the expression for $V(k,l)$ reduces to

$$V(k,l) = \sum_{i=r+1}^{X-r} \sum_{i=j+1}^{Y-r} f(i,j) \oplus g(i+k,j+l)$$

This function may be computed much more rapidly on a BAP than the grey level function. However, the new image must first be reduced to a binary image. This is achieved in two stages; first, impulse noise is removed by a median filter then the binary image is computed using a adaptive thresholding algorithm.

1. Pseudo Median Filtering. This is achieved in two stages. First, each pixel is replaced by the median of itself and its two horizontal adjacent near neighbors. Then each pixel of the resulting image is replaced by the median of itself and the two vertically adjacent pixels to it. This algorithm is much simpler to compute than a true 3x3 median and it ensures that the final result will be within 1 ranked value of the true median. The pseudo median operation was also useful for preprocessing the image data before using the grey level registration algorithm.

2. Max-min Thresholding. The binary image is computed from the grey level image by the adaptive thresholding algorithm defined by

$$B(i,j) = 2g(i,j) > lmax(g(i,j)) + lmin(g(i,j))$$

The lmax and lmin functions compute the maximum and minimum respectively of the local neighborhood of $g(i,j)$. In practice a local neighborhood of 15 x 15 pixels was used; lmax and lmin each required 30 shift operations and 8 compare-and-select operations to implement.

IV. IMAGE BACKGROUND ENHANCEMENT

Once a set of registered imges has been obtained we can combine them to enhance the stationary part of the image background. Three enhancement algorithms are considered below.

A. Mean Enhancement.

The background may be enhanced by computing the mean of the last M frames. An efficient algorithm is to maintain the M previous images and the mean of these images. A new enhanced image may be computed from the old one by subtracting the oldest image from it and adding the new one to it. A value of M = 8 has been found to give good results.

B. Median Enhancement.

This algorithm is similar to the mean enhancement except that
the median is computed. Updating for a new image involves M
compare-and-exchange operations to unsort the oldest image from
an ordered set of M images and a further M operations to sort
the new one into the ordered set. The cost in storage and computation is higher than for the mean algorithm but slightly
better results have been observed.

C. Mean Update Enhancement

For the mean update algorithm the new image is added to a
weighted multiple of the previously enhanced image. In this
case the enhanced image is a weighted mean of previous images
with the largest emphasis on the most recent image. This algorithm requires the least computation but is not as good as the
others.

An example of registration and mean enhancement is shown in
Figs. 4-7. The data was obtained by digitizing a set of sequential film frames; each image is 256 x 240 pixels. Fig. 4 shows
an image of a missile being launched fig 5 is the image ocurring
7 frames after Fig. 4; both the missile and the camera have
moved significantly. The binary image used for registration is
shown in Fig. 6 and a mean enhanced version of all 8 registered
frames is shown in Fig. 7. Much of the noise has been removed
from the background but the missile is smeared due to its motion.

V. MOVING OBJECT DETECTION

A simple example of moving object detection is shown in Fig.
8 which is the difference image between Fig. 5 and Fig. 7. Both
the missile and a scratch in the film show as moving objects.
The image in Fig. 5 may be processed to remove the isolated
noise points and extract the regions which contain moving objects.   More sophisticated methods of detecting moving objects
are being considered such as locating maxima in the variance of
the last 8 registered frames.

VI. OBJECT DESCRIPTION EXTRACTION

In this section the extraction of a description of a segmented object, in terms of moments, from the BAP will be considered.

Fig. 4. First frame of a
sequence

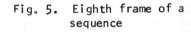

Fig. 5. Eighth frame of a
sequence

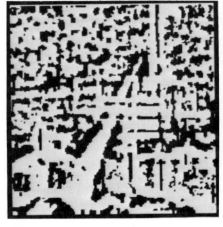

Fig. 6. Binary image of Fig. 5

Fig. 7. Mean of 8 registered
frames

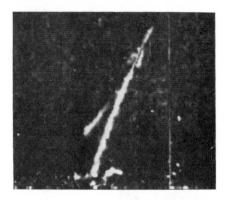

Fig. 8. Difference image of Fig. 5 with Fig. 7

It is assumed that a potential object of interest has been segmented from the background such that an image exists which contains only the object pixels and is zero elsewhere. The method
of segmentation is highly application dependent and is not considered here.

A segment of an image may be represented by a set of moments.
The moment of order (p+q) for a segment $f(x,y)$ is defined by

$$m_{p,q} = \sum_x \sum_y x^p y^q \, f(x,y)$$

A complete moment set (CMS) of order n consists of all the
moments of order n and lower. Simple operations are defined for
a CMS which correspond to translation, rotation and scale change
of the image segment (Reeves and Rostampour, 1980). Also, simple operations between CMS's are defined which correspond to the
addition, subtraction and concatenation of the image segments
they represent.

If the order n of the CMS is sufficiently large then the CMS
exactly represents the image segment. In practice, a value of n
in the range 3-15 is typical, depending upon the accuracy of
representation required for the application (the value of n=3
has received the most attention).

The algorithm for generating a CMS from a segmented object on
a BAP is as follows: First, generate two matrices in the PE array; one contains the x location of its PE and the other contains the y location. Second, generate each moment value in an
ascending order sequence. The data to be summed for each moment
may be obtained by multiplying the data for a lower order moment
by either the x or y matrix. For example the moment $M_{i,j}$ is

computed from the array sum of $\mathrm{D} = x^i y^j f(x,y)$, the moment $M_{i+1,j}$ can then be computed from the array sum of x.p. Therefore, each moment may be generated with a single multiplication and an array sum. Unfortunately, the dynamic range of the moments increases very rapidly with their order. Some simple dynamic scaling of the data may be necessary to ensure that each moment is represented with a reasonable number of bits. A CMS of order n involves $(n+1)(n+2)/2$ moment values; this is also the number of multiplications and array sums needed to generate them. When a segmented object spans more than one subimage then a CMS for each subimage can be computed separately; these CMS's may then be simply combined to form a single composite CMS for this object.

VII. ALGORITHM EXECUTION TIME

The estimated execution times of all the algorithms on an extended MPP with a fast bit counter are given in Table 2. These estimates are simply based on the figures given in Table 1, it is expected that lower times could be achieved for all these algorithms with careful assembly programming.

With the exception of grey registration and order 8 moment extraction on a 512 x 512 image the algorithms can be implemented in real time within the 33 m.s. bound. The binary registration has given results as good as the grey registration for the data we have used. Also, in typical applications it is expected that objects of interest will be much smaller than 512 x 512 pixels and therefore much faster to extract.

Table 2.   Upper bound execution time of algorithms on the extended MPP for 8-bit image data.

Algorithm	MPP (time in μs)	
	128 x 128	512 x 512
Grey registration (441 points)	5,645	85,025
Binary registration (441 points)	618	5,292
P-median filter	26	461
Max-min threshold	139	2,982
Mean of 8 Enhance	117	2,765
Median of 8 Enhance	276	6,196
Mean update Enhance	25	525
Moving object detection	3	39
Object description extraction		
(order = 3, precision = 20 bits)	428	6,728
(order = 8, precision = 20 bits)	1,926	302,760

VII. SUMMARY

A set of algorithms has been described for real-time video processing on a highly parallel processor. These algorithms demonstrate that a BAP such as the MPP can achieve a substantial amount of low-level image processing on full size TV video imagery in real-time. Also, algorithms may be modified to match the architecture of a BAP e.g., the binary registration algorithm, and thereby be more efficient than conventional algorithms. In many practical applications it is not necessary to process every 512 x 512 frame in real-time. In such cases a much smaller, cheaper BAP than MPP can be used. Such a BAP is the BASE-8 processor which may be considered as an add-on device to a conventional minicomputer (Reeves and Rindfuss 1979).

REFERENCES

Batcher, K.E. (1980), "Design of a Massively Parallel Processor," IEEE Trans. on Computers, Vol. C-29, No. 9, 836-840.

Duff, M.J.B. (1976), "CLIP4 A Large Scale Integrated Circuit Array Parallel Processor," 3rd International Joint Conference on Pattern Recognition, 728-732.

Readdaway, S.F. (1979), "The DAP Approach," Infotech State of the Art Report on Supercomputers, Vol. 2,

Reeves, A.P. (1980a), "A Systematically Designed Binary Array Processor," IEEE Trans. on Computers, Vol. C-29, No. 4, 278-287.

Reeves, A.P. (1980b), "The Anatomy of VLSI Binary Array Processors," Workshop on New Computer Architectures and Image Processing, Ischia, Italy.

Reeves, A.P. (1980c), "On Efficient Global Feature Extraction Methods for Parallel Processing," Computer Graphics and Image Processing, Vol. 14, 159-169.

Reeves, A.P. and Rindfuss, R. (1979), "The BASE 8 Binary Array Processor," Proceedings of the IEEE Conference on Pattern Recognition and Image Processing, 250-255.

Reeves A.P. and Rostampour, A. (1981), "Shape Analysis of Segmented Objects Using Moments," Proceedings of the 1981 IEEE Conference on Pattern Recognition and Image Processing.

Svedlow M., McGillem, C.D. and Anuta, P.E. (1978), "Image Registration: Similarity Measure and Preprocessing Method Comparisons," IEEE Trans. on Aerospace and Elctronic Systems, Vol. AES-14, No. 1, 141-150.

IMAGE PROCESSING APPLICATIONS ENABLED BY MIMD PROCESSOR STRUCTURES

Kenneth Lundgren
Goesta H. Granlund

Picture Processing Laboratory
Linkoeping University
Linkoeping, Sweden

INTRODUCTION

Most real life images contain a great deal of ambiguous information, such as noisy lines and edges, texture, color, etc, which requires large processing capabilities. A number of special hardware processors have been developed, but so far they have mainly dealt with binary images. In order to fully use the information available in gray scale and color images for image analysis, new methods have to be developed. An important question is the one of representation of image information. An operator has been defined which describes and detects structure as opposed to uniformity, whatever structure implies at a certain level. The operator has proved very useful for description of texture and for detection of faint and noisy lines and edges. The information representation used allows very effective procedures for filtering. Higher level contextual features of the image can be used to guide the processing at lower levels in a feed-back mode. A hardware processor, the GOP image processor, has been developed which implements this operator, as well as most other picture processing operations suggested.

The processor is divided into two parts; Part I, a fast, parallel, pipelined processor, where image and mask data are combined. Part II, a fast, serial processor with a high degree of flexibility, takes care of data out from Part I. The two parts are independently microprogrammable, and algorithms can easily be specified through use of medium level and high level interactive languages. This architecture of the processor makes it in effect a MIMD (Multiple Instruction strean - Multiple Data stream) parallel

MULTICOMPUTERS AND IMAGE PROCESSING
ALGORITHMS AND PROGRAMS

19

machine with an extreme flexibility to be reconfigured and reprogrammed for new tasks.

The processor can be connected to any system for picture processing where it speeds up the processing by a factor of 400-2000
depending upon the situation.

THE GENERAL OPERATOR CONCEPT

The General Operator builds upon two fundamental assumptions
concerning representation of image information.

1. Image information can be described as locally one-dimensional
    structures.

2. Directional information of structures is extremely important,
    and it has to be integrated in processing.

These assumptions have important consequences for the definition of operations on image information. We will not go into a
discussion of the relevance of these assumptions as they are outside the scope of this paper. These matters have been discussed in
more detail earlier (Granlund, 1978).

The effect of the operator is to produce a complex value for a
local region with two components:

1. A magnitude reflecting the amount of variation within the
    window, e.g. step size of an edge.

2. An angle determined by the direction in which we find the
    largest magnitude component according to 1).

Without going into the details of the representation let us
mention that with the convention for directionality used, the output from a transformation of a disc will appear as in Figure 1.

The operator, of a certain size, say 15*15 elements, scans the
input image step by step.

In the computation of the amount of variation within the image
region, the image content is matched with a combination of edge
and line detectors for a number of different directions. In
practice, the number of orientations of the filters is restricted
to 4 in the range $(0,\pi)$. The magnitude, $B(x,y)$, and direction,
$\theta(x,y)$, of the output image at the point (x,y) are derived from
the input image $F(x,y)$ in the following way. First the image $F(x,y)$
is convolved with the set of 4 line and edge filters, $L_i(x,y)$ and

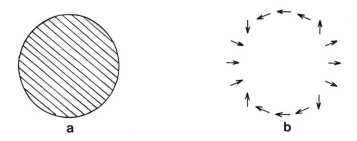

Fig. 1. Image of disc (a) with its transform (b).

$E_i(x,y)$ respectively, to give the magnitude in the ith direction, $B_i(x,y)$:

$$B_i(x,y) = \sqrt{(S_i^2(x,y) + C_i^2(x,y))/V(x,y)} \qquad i=1,2,3,4 \qquad (1)$$

where

$$S_i(x,y) = F(x,y)*E_i(x,y)$$
$$C_i(x,y) = F(x,y)*L_i(x,y) \qquad\qquad (2)$$

and

$$V(x,y) = [\sum_{i=1} S_i^2(x,y) + C_1^2(x,y)]^\beta \qquad 0\leq\beta\leq1 \qquad (3)$$

In eqn (2), * denotes convolution. In eqn (3) the exponent β is used to control the degree to which absolute ($\beta=0$) or relative ($\beta=1$) magnitude is important. The overall magnitude, $B(x,y)$ is then estimated by adding the squared differences of the components in orthogonal directions:

$$B(x,y) = [B_1(x,y) - B_3(x,y)]^2 + [B_2(x,y) - B_4(x,y)]^2 \qquad (4)$$

and the direction $\theta(x,y)$ is estimated using

$$\sin 2\theta(x,y) = [B_2(x,y) - B_4(x,y)]/B(x,y) \qquad (5)$$
$$\cos 2\theta(x,y) = [B_1(x,y) - B_3(x,y)]/B(x,y) \qquad (6)$$

Note that the estimation of $\theta(x,y)$ is expressed in the form of eqns (5) and (6) to avoid the degeneracy associated with the inverse trigonometric functions. Thus if all the energy lies along one of the directions, $B(x,y) = B_2(x,y)$, say then $\theta = \frac{\pi}{4}$.

One important feature of the operator is that it can be used repeatedly upon earlier transforms to detect more global properties in the image, e.g. boundaries between textures. An example of this, using the GOP processor, is given in Fig. 2, which displays a sequence of two transforms of an image of tweed from Brodatz book on textures.

In Fig. 2a we have the original image with two different texture regions. Fig. 2b shows the first transform which gives a description of the textures in terms of variation content and directionality. The second transform, Fig. 2c, gives the border between the textures.

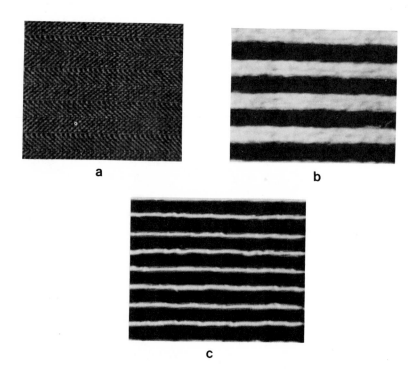

a

b

c

Fig. 2. Result of two transformations on tweed. (a) Original image; (b) First order transform; (c) Second order transformation.

THE GOP IMAGE PROCESSOR

The GOP (General Operator Processor) image processor has been developed to implement the General Operator approach, as well as most other image processing operations.

Part I of the processor is a reconfigurable pipelined parallel processor, where data from the image segment memory and weights from the mask memory are combined in four parallel pipelines. Part II of the processor has an entirely different architecture. After processing in part I, the amount of information is reduced considerably. Now a high degree of flexibility is required to combine intermediary results derived by part I. These combinations are usually highly nonlinear operations, determined from one point to the other by some particular transform of the image to process. Part II is consequently a serial, special purpose processor. A block-diagram of the processor is shown in Fig. 3.

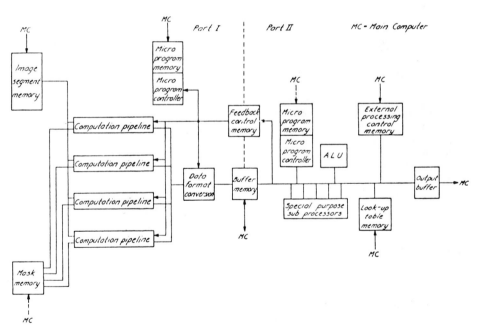

Fig. 3. Block diagram of GOP image processor.

The image segment memory has a capacity of 64 kbyte. It can be restructured to fit the current processing situation, such that any number of input images of any size can be involved in processing simultaneously and operators with a size up to 180 x 180

pixels can be used. Software in the external computer determines
the allowable length of the image segment, which depends upon the
number of images and the mask size. Usually the image segment has
a length equal to the original image, and a width equal to the
size of a neighborhood. Data is moved to the processor one line
at a time. There it substitutes the oldest line in a "rolling"
fashion.

The mask data is stored in a memory of size 16K words of 24
bits. This storage can be restructured in a number of ways. The
normal configuration is that the mask memory is divided into four
sections, one section providing each pipeline with weight coefficients. In parallel with the first section is another memory of
4K words of 14 bits. The content of this later memory points to
the image segment memory selecting data points to be processed.
This means that points can be picked arbitrarily within the 64
kbyte image segment memory to form a neighborhood subregion of up
to 4096 points sampled in any order or arrangement. This allows
e.g. masks of different sizes to be used on different input image
planes. The mask memory can be organized to contain up to 4096
different masks with any distribution of size within the limits
of the mask memory itself. These masks can be freely combined in up
to 1024 different mask sets. Which one of these mask sets to use
can be determined by part II, e.g. in response to image data intended to control the processing.

In order to allow fast computation, part I of the processor
uses fixed point arithmetic. However, great care has been taken
not to cause errors due to overflow or underflow. Consequently,
at the end of the pipeline there is a dynamic range of 48 bits.
Before data enters part II of the processor it is converted to
either of three data modes: 16 bit fixed-point data, logarithmic
data or floating-point data. The data mode can be selected with
respect to demands on accuracy and speed in the computation in
part II. The communication from part I of the processor to part
II is done over a dual memory of 2·4K words of 16 bits. Part I
can write into one half of this memory at the same time as part
II reads from the other half.

The central units in part II are a microprogram controller
and an arithmetic logic unit. A fast microprogram memory (access
time 55 nsec) of size 2K words of 64 bits (expandable to 4K words
of 80 bits) gives a cycle time of 150 nsec for the processor. All
units communicate over a 16-bit bus. A special work memory of 2K
words (expandable), with possibilities of indirect addressing
facilitates programming. In order to obtain fast processing of
complicated algorithms, a 16K word memory area can be used for
look-up tables. A memory area of 4K words is available for external processing control. In this memory lines from up to 8 images
can be stored to control the processing point by point. On the
main bus are attached a number of special purpose processing units.

They perform operations such as fast multiplication, scaling and shifting (up to 16 steps in one cycle), floating point operations, etc. The computation within the processor can be performed using fixed-point 16-bit representation, logarithmic 16-bit representation or floating-point representation, or any desired mix of these during a particular procedure.

Part I is controlled by part II regarding what operations to perform, but does not interfere during the computations set up for a neighborhood. However, the configuration of the pipeline can be changed after the computation, and an entirely different configuration can be set up instantaneously for a different type of computation on the same (or different) neighborhood. This allows maximal flexibility in conjunction with high speed. In normal processing the pipelines remain in the same mode for the whole image. Parts I and II run simultaneously at maximum speed with data exchanged over the twin buffer.

SOFTWARE

In order to obtain an easily workable system with a processor as flexible as the GOP, it is necessary to have a good software system. For that reason an extensive, interactive program system has been developed. The goal has been to provide program routines for most commonly occurring processing tasks, as well as to provide an attractive environment for the researcher who wants to investigate new algorithms and develop his own programs. The program system is built around three levels of languages. The intention has been that the program system should be easily transportable between different computers.

EXAMPLE OF PROCESSING

As one example of the many uses of the GOP image processor we will look at how it can be used for contour detection and masking in a feedback mode. The architecture of the processor allows us to make a processing in a hierarchical way with feedback from one level to another. A set of images can be used to control the operations to be performed from one neighborhood to another. See Fig. 4.

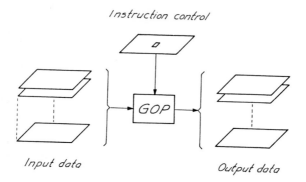

Fig. 4. Input-Output structure of GOP processor.

In this example a structure, as illustrated in Fig. 5, is used.

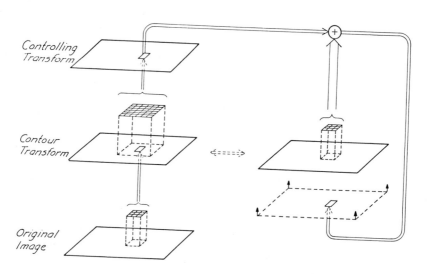

Fig. 5. Feedback structure for contour detection and masking.

The original image is processed with the General Operator producing a transform containing essentially contours. In general the first order transform is sufficient for this purpose. In the case where different regions are defined at least partly by different textures, it may be necessary to produce a second order transform which will then give borders between different textures. The first and second order transform can then be combined which will give optimal definition of existing boundaries.

From the contour transform a controlling transform is produced. The controlling transform is a function of the contour transform. In the simplest form it may consist of the contour transform or a low pass filtered version of the contour transform. The controlling transform thus displays the direction of the dominant variation within the neighborhood.

The contour transform is now brought over to the iteration loop. A specially designed operator is used here, which performs a nonlinear differentiation in one direction and an integration in the perpendicular direction. The directions for differentiation and integration are determined for every point by the controlling transform.

Fig. 6 shows an example of the processing that can be performed. Fig. 6a shows the original and 6b is the contour transform. The iteratively processed image appears in 6c.

We can see that there is a fairly efficient thinning of the main contours. Another important property of the processing is the one of masking, implying that weak contours are suppressed next to strong ones.

ACKNOWLEDGMENTS

This research was supported by The National Swedish Board for Technical Development. The authors also want to express their appreciation of the enthusiastic work by the GOP group.

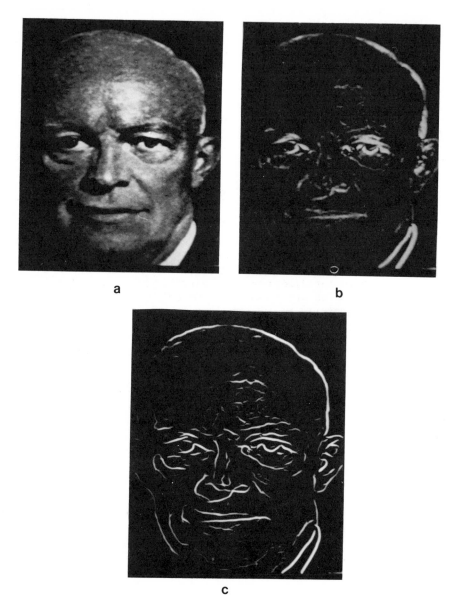

Fig. 6. Example of processing in contour detection and masking.

REFERENCES

Briggs, F.A., Fu, K.S., Hwang, K. and Patel, J.H. (1979). "PM[4] - A Reconfigurable Multiprocessor System for Pattern Recognition and Image Processing." National Computer Conference. AFIPS-Conference Proceedings, Vol. 48.

Duff, M.J.B. (1978). "Review of the CLIP Image Processing System." National Computer Conference 1978, AFIPS Conference Proceedings, Vol. 47.

Granlund, G.H. (1978). "An Architecture of a Picture Processor Using a Parallel General Operator". Proc. from the Fourth International Joint Conference on Pattern Recognition, Kyoto, Japan.

Granlund, G.H. (1978). "In Search of a General Picture Processing Operator". Computer Graphics and Image Processing, 2, 155-173.

Granlund, G.H., Hedlund, M. "Feedback Structures for Image-Content Dependent Filtering". Internal Report No. LiTH-ISY-I-0398, Department of Electrical Engineering, Linkoeping University, Sweden.

Hedlund, M., Granlund, G.H., and Knutsson, H. (1981). "Image Filtering and Relaxation Procedures Using Hierarchical Models". Proc. from the Second Scandinavian Conference on Image Analysis, Espoo 15, Finland.

Knutsson, H., Wilson, R. (1981). "Anisotropic Filtering Operations for Image Enhancement and their Relation to the Visual System". Submitted to IEEE Comp. Soc. Conf. on Pattern Recognition and Image Processing, Dallas, Texas.

Knutsson, H., Wilson, R., and Granlund, G.H. (1981). "An-isotropic Filtering Controlled by Image Content". Proc. from the Second Scandinavian Conference on Image Analysis, Espoo 15, Finland.

Kruse, B. (1980). "System Architecture for Image Analysis. Structured Computer Vision". Eds S. Tanimoto and A. Klinger, Academic Press, 169-212.

Lougheed, R.M., McCubbrey, D.L., and Sternberg, S.R., (1980). "Cytocomputers: Architectures for Parallel Image Processing". Environmental Research Institute of Michigan, Ann Arbor, Michigan.

Preston, K. Jr., Duff, M.J.B., Levialdi, S., Norgren, P.E., and Toriwaki, J-I (1979). "Basics of Cellular Logic with Some Applications in Medical Image Processing". Proceedings IEEE, Vol. 67, No. 5, pp. 826-856.

Rosenfeld, A., et al. (1976). "Scene Labelling by Relaxation Operations". IEEE Tr. Systems, Man & Cybernetics, 420-439.

PICAP AND RELATIONAL NEIGHBORHOOD
PROCESSING IN FIP

Björn Kruse
Björn Gudmundsson
Dan Antonsson

Department of Electrical Engineering
Linköping University
Linköping, Sweden

The PICAP II parallel picture processing system is a research vehicle in the area of high speed analysis of images. The main features of the system are a high speed bus, (40 Mbyte/sec), a large random access memory (4 Mbyte) and powerful processors (10^8 instructions/sec). In this paper an overview of the system will be given together with a more specific presentation of the logical computing power of one of the processors in the system, FIP. A brief introduction to relational neighborhood processing is given in order to show an example of FIP capabilities. The basic hardware structure of PICAP II including FIP has been built and tested (June 1980) and the completion of the system is now in progress).

1.   INTRODUCTION

Parallel algorithms are becomming increasingly popular in picture processing as more difficult problems are being studied. Research and development of such algorithms are often hampered by limitation in both processing speed and memory space. To overcome these problems there are essentially two approaches. One way to achieve increased speed is to use a vast number of simple processing elements in a regular structure [21,16,4,19,1] with main memory evenly

distributed among the processors. The large number of processing
elements more than compensates for their simplicity. In the other
approach relatively few but powerful modules are connected together [10,8,6,5,17,7,20] to form a system of concurrently operating, not necessarily identical, processors. As contrasted with
the former scheme, main memory does not form an integrated part of
the processors themselves. Instead it is a central resource shared
among the different processors. In this paper a system of the latter
kind will be described.

The PICAP II system [11,9] is a research vehicle for parallel
picture processing developed at the Picture Processing Laboratory,
Linköping University. The unique design of PICAP II is centered
around a high speed bus which connects the large memory - 4 Mbytes
of high speed interleaved memory banks - to a varity of processors
[2,3]. The basic processor for parallel processing is FIP, the
filter processor, which performs 10^8 elementary operations/sec on
8-bit data. The background for the design of FIP is several years
of experience with the PICAP I system [12,13,14] in numerous application programs. In this paper the nonlinear processing capabilities (relational neighborhood processing) of FIP will be described.

A. PICAP II Picture Processing System

PICAP II is an entire general purpose picture processing system including high speed processing hardware for various purposes,
a large memory for image storage and a fairly large general purpose host computer, see figure 1. The purpose of building such a
system is that real-life picture processing problems of moderate
and great complexity are very time-consuming to study, and not
the least, to develop algorithms for. The requirements for high
speed processing and large memory size are therefore high even in
the development stage and not only in production.

The architecture of PICAP II is built around a high speed bus,
as can be seen in figure 1. The bus itself transfers 32-bit data
to and from memory and the processors. The memory is divided into
sixteen identical and independent banks, each having a capacity of
256 Kbytes. This means that information may be read and stored using interleaving which gives a high transfer speed. Since the memory is interleaved sixteen ways the potential speed-improvement-
factor would be sixteen. However, the present design of the bus
does only allow for 40 Mbytes/sec which corresponds to a time-slot
of 100 ns. The details of priority and anticollision logic involved are described in [2]. The memory modules employ correction
of 1-bit errors and detection of 2-bit errors to ensure safe operation.

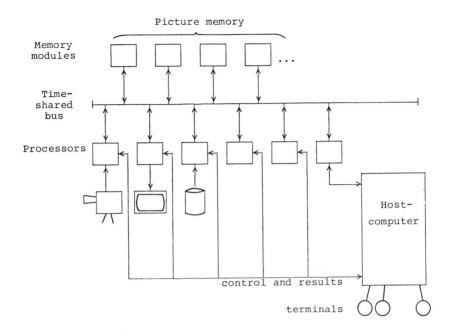

Fig. 1. System architecture of PICAP II.

Memory allocation of images is such that an image is primarily viewed as a one-dimensional array of data, since the memory space is linear. As the basic interleaved speed of the memory modules is four times the actual speed of the bus, it is possible to have the same fast access also in two dimensions provided the rowlengths of the images are restricted to values that for example do not cause adjacent pixels in a column to belong to the same module, see figure 2. The conditions of allowing fast access is readily created simply by filling a suitable number of blank pixels. Along the bus there are sixteen processor slots equipped with the necessary logic to allow for simple interfacing. The interface logic also creates a uniform electrical load on the bus, regardless of whether a processor is inserted in the corresponding slot or not, which is essential for high speed.

Each processor is designed for a certain domain of operation in order to create a general purpose picture processing system. The host computer controls all activities and issues commands to the processors describing the operand(s), the type of operation and

of course the destination of results. As a whole the system operates as a MIMD machine. The list of processors that are in operation, currently under design or planning are:

- Video-input processor
- Display processor
- Processor for graphical overlays
- Filter processor
- Logical neighborhood processor
- Segmentation processor

The word *processor* is in our vocabulary loosely used to mean
a device that executes a given task on its own. For example, the
video-input processor (VIP), which one normally perhaps wouldn't
call a processor, executes the task of selecting a specified window
within the camera field of vision, digitizing the pixels inside the
window and correcting them for shading abberations. The digitization is done at video rate and the image is stored directly into
main memory.

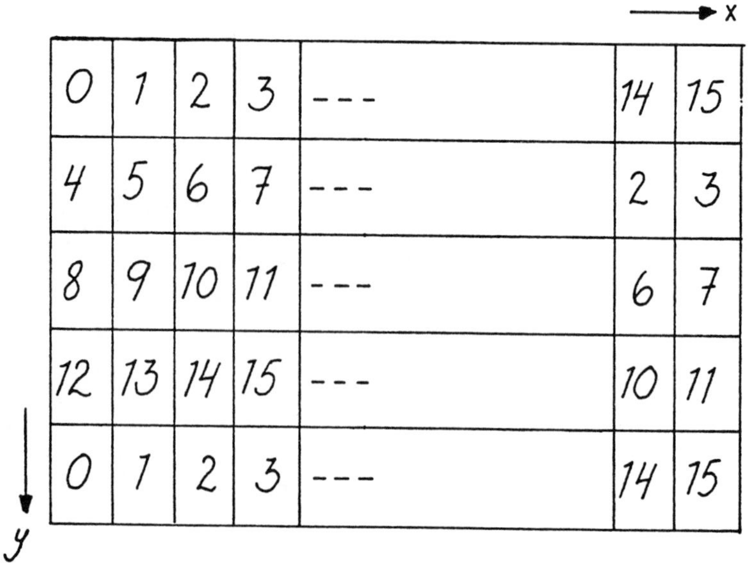

Fig. 2. Example of bank assignment for fast access in
two dimensions.

The display processor (DIP) operates according to a given display-file. The display-file itself is assembled by the host computer and transferred to the display processor as part of its command. The display-file code contains all the necessary information as to the location of the images and the windows that are to be presented on the monitor screen. The image is refreshed and remains on the screen until a new command is given. The use of a display-file to control the display gives direct access to the image information itself without the need of a separate display area which of course conserves memory space.

Contrary to the display processor the processor for graphical overlays (GRP) operates directly on memory. Characters, lines and areas can be painted as pseudo-colour overlays on the image information.

The filter processor (FIP) performs both linear and non-linear operations on gray-level images. Since the next section presents the logical part of FIP and the linear part has been described in [15] it suffices here to state that it differs from the logical neighborhood processor which will operate on 3x3 binary neighborhoods using table-look-up methods. According to an unpublished investigation one may save some time or hardware in using a sequential approach to the evaluation of neighborhood contents [18]. As will be seen in the next section the logical unit operates sequentially. However, compared to table-look-up methods which are trivial on 3x3 binary neighborhoods, not much can be gained.

The processors so far presented have processed the information in image form creating a new image as result. In pattern recognition applications these transformations yield only intermediate results and need to be further processed to allow for example for object identification and scene description. The segmentation processor (SEG) performs essential tasks for that purpose. From an image, in which the pixels are given labes indication their class belonging, SEG produces   a map showing the separate connected areas of uniform labels. Furthermore a list is produced giving the class and some measurements of all the different areas, optionally together with the object border in chain-link form.

The capacity of the bus is essential in the type of architecture that PICAP II represents. Sharing the memory as a common resource, which is natural in a general purpose system, requires high transfer rates. The bus-speed in PICAP II corresponds to the video-rate of 4 TV-channels allowing for simultaneous input, output and processing in real time. The distributed processing power

allows for a high degree of overlapping operation which of course
makes the high speed possible. The features of PICAP II are very
well suited for both algorithm development operation, which is typically a multiuser situation, and production.

2. THE FILTER PROCESSOR (FIP)

Before the relational operators are described a brief presentation of the FIP structure is necessary. The overall PICAP II architecture is of type MIMD but some of the individual processors,
among them FIP, are in themselves of type SIMD. In FIP, four subprocessors (P_1-P_4) operate in parallel according to a common microprogram, accessing pixels from a shared local memory. To further
increase the processor performance the subprocessors are pipelined.
A comparison with a conventional computer instruction set shows
that FIP executes in excess of 10^8 basic instructions per second.
An overview of FIP is shown in figure 3. A more detailed description is given in [15].

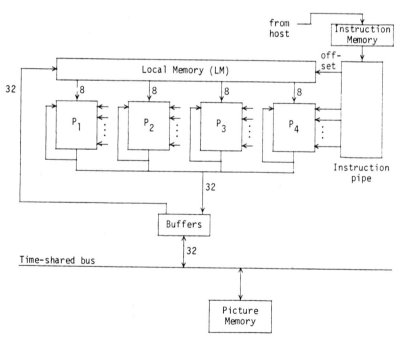

Fig. 3.   FIP

The local memory (LM) stores a horizontal strip from a picture giving access to all neighborhoods along one line, (see figure 4). The operator moves along the line and when ready a new
line is entered into LM while the result is stored back in picture memory. For 3-D-neighborhood operators a set of strips from
the involved pictures are stored such that corresponding lines
are adjacent to each other. The size of LM is 32 kbyte which for
example corresponds to a 64x512 neighborhood operator on a 512x
512 picture. The accessing mechanism that enables the four subprocessors to retrieve data from LM obviously requires a partitioning of LM into four separate modules. The accessed pixels are
horizontally adjacent, one for each module and they are supplied
one to each subprocessor after appropriate flipping. A base register holds the location of the current neighborhood position.
The base address is incremented by four when four neighborhoods have
been processed.

The subprocessors of FIP are identically controlled by a common microprogram. The microinstructions are fed through an instruction pipe from which appropriate control signals are derived (see
figure 5). A pixel flowing through the pipelined processor is subjected to the following transformation:

$$ACK^+ = F_2(ACK, F_1(\omega, MAP(p)))$$

p: pixel from LM or a register

ω: constand (immediate operand) from instruction

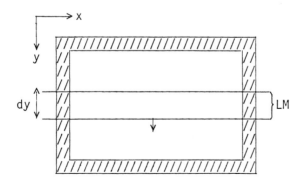

Fig. 4. Local memory contents.

MAP($\cdot$): pixel to pixel mapping

F_1,F_2: Linear and non-linear functions

ACK: accumulator in Arithmetic Logic Unit (ALU)

In the case of for example a linear filter (convolution), F_1 is multiplication and F_2 is addition. Since the resulting pixel can be stored back to a register, it is possible to compose very powerful compund operators involving both linear and non-linear functions.

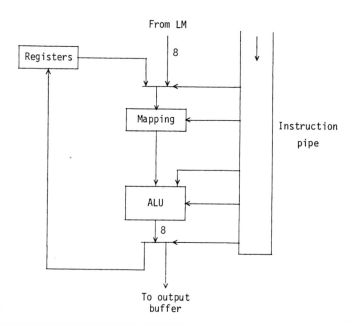

Fig. 5. Subprocessor.

3. RELATIONAL NEIGHBORHOOD PROCESSING

Relations have long been used in PICAP I to define logical neighborhood operations [14]. Equal to, greater than are typical relations for PICAP I. They are defined over a 3x3 neighborhood whose elements are related to instruction-supplied constants. The outcomes over the neighborhood are checked and when they all hold a pixel-transformation occurs. In the following we will describe relational neighborhood processing intended to serve as an example of the processing power of FIP:s logical unit.

In this context we will not describe so much why, but more how the transformations are done. Let it be enough to state that through the relations each neighborhood is given a representation that re- veals the microstructure. A very natural, and also simple, way of representing the microstructure is to relate the neighborhood ele- ments to the centre pixel.

$$R: \quad q \rightarrow r$$

Where $R=\{R_k:k=1,2,\ldots,n\}$ is a set of relations that relates the neighborhood element q_o to its n neighbors. Furthermore

$$r = < r^1, r^2, \ldots, n^n >$$

where

$$r^k = q_o R_k q_k$$

is the outcome of the relation $\#$ k. To simplify the description we will use the following relation

$$R_k: \quad q_o < q_k - T \quad k=1,2,\ldots,8$$

where the neigborhood for example is the set of eight nearest neigbors.

q_4	q_3	q_2
q_5	q_0	q_1
q_6	q_7	q_8

The relations $q_o R_k q_k$ simply state that the first order approximations q_k-q_o of the gradients components in certain directions are greater than T. The value of T may be derived from the image itself by examination of the different relation code frequencies. Clearly the directionality of the data is retained by the transformation. For example the outcome

r^4	r^3	r^2		0	1	1
r^5	X	r^1	=	0	X	1
r^6	r^7	r^8		0	0	1

indicates a neighborhood in which there is a positive gradient in the two o'clock direction.

Surprisingly much of the original information is left after this seemingly crude operator. This is evidenced by figure 6 in which the original image is shown together with a reconstruction from its relational representation. The reconstruction is not perfect and probably not even a true solution to the system of $2\cdot10^6$ simultaneous inequalitites. However, evident from the images in figure 6, the important information from a perceptional point of view is retained.

4. THE LOGICAL UNIT OF FIP

A schematic of the logical unit of FIP is shown in figure 7. The unit is part of the ALU-pipeline as shown in figure 5 and shares the instruction-pipe with the arithmetical unit. Data-paths not important in this context have been omitted. The principal parts are the comparator (COMP), the three-input combinatorical net (COMB), the binary LIFO stack, the two eight-bit shift- and count-registers and the M-register that serve as output to the buffer. It should be clear from the preceding that there are four sets of hardware of this kind in FIP, the four logical sub-processors.

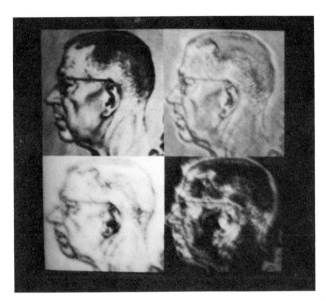

Fig. 6. a) Original b) Reconstruced image c) and d) estimations of maximum and minimum values.

Input to the logical subprocessor is from LM or register, not shown in figure 7, and from the instruction-pipe (immediate constant). The two 8-bit inputs are compared in the comparator and according to the test, greater than, equal to or less than etc., a binary signal is generated. The previous binary values, stored in the stack, may be combined with the comparison outcome in the combinatorical unit to generate a new binary value that in its turn may be pushed on to the stack. This way, complex functions of relations can easily be programmed. There is virtually no restriction on neighborhood size or complexity of the operator. The stack-top element which is identical to the combinatoric output delayed one unit in time, may either be stored as the result of an evaluation or used as a condition for controlling other activities in the unit. This applies to the M-register, the counter and the shift register which may be conditionally loaded, incremented and shifted respectively. In the counter, for example, the number of occurrencies over the neighborhood of one or several features may be computed. This is an important type of operation for example in texture analysis. The conditional shifting of a zero or a one into the shift-register can be used for creating binary feature vectors for example as microstructure representation which we will see below.

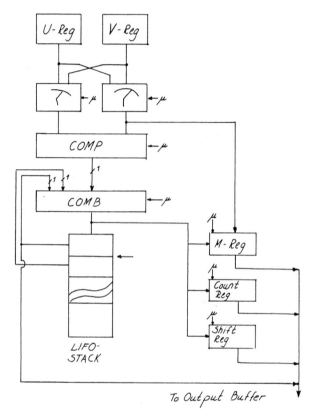

Fig. 7. The logical unit in FIP.

The expression for the processing in the pipelined logical
unit is not as straight-forward to write down as the one for the
arithmetical unit. However, if we restrict ourselves to the binary
logical output the following is a slightly simplified expression
for the processing steps involved.

$$ST^+ \quad = L(ST,ST1,R(\omega,MAP(p)))$$

p: pixel from LM or a register

ω: constant from instruction pipe

MAP(·):   pixel to pixel mapping

L: a logical function of three variables

R: a relation

ST,ST1: the two top elements of the stack.

In the simple case of template matching the relation is of type "equal to" and the logical function is an AND of R($\cdot$) and ST disregarding ST1. An initial setting of ST is also required. This covers virtually all instructions of logical type in PICAP I [6,13]. Microprogramming definition for template matching:

OP	CONST	REL	ADDRESS	MAP
PUSH	1	EQ	A(0,1)	0
AND	1	EQ	A(0,0)	0
AND	1	EQ	A(1,1)	0
AND	0	EQ	A(1,0)	0
		$\vdots$		
ST	–	–	–	–
END	–	–	–	–

The result of the operation is a logical "1" only if the pixels all are equal to the template defined by the constants.

Template

$$\begin{array}{|ccc|} \hline . & 0 & 1 \\ . & 1 & 1 \\ . & . & . \\ \hline \end{array}$$

The operator definition for the relation operator is as follows in terms of simplified microinstructions.

OP	CONST	REL	ADDRESS	MAP	
LD	–	–	A(0,0)	0	
ADD	T	–	R1	–	
ST	–	–	R1	–	$R1 \leftarrow Q_0 + T$
LDV	–	–	R1	0	
SHFT	–	LT	A(0,1)	0	$R1 < Q1 \Rightarrow$ shift in 1 else 0
SHFT	–	LT	A(1,1)	0	$R1 < Q2 \Rightarrow$ shift in 1 else 0
SHFT	–	LT	A(1,0)	0	.
			:		.
			:		.
SHFT	–	LT	A(1,-1)	0	
STSR	–	–	BUF	–	Store shift register
END	–	–	–	–	contents

The neighborhood centre pixel Q_0 is first loaded and incremented by T. Since this is done in the arithmetical part of FIP the result has to be intermediately stored in a register, in this case R1, before it can be fetched and used in the logical unit. Then the result of adding T to Q_0 is compared to the surrounding pixels in succession and the outcomes are shifted into the shift register to form the desired relation vector. This completes the operator.

5. CONCLUSIONS

The available space does not permit any detailed presentation of the PICAP II system. However, we hope its fundamental features of high speed parallel processing and multiprocessor structure have been made clear. The high performance of 10^8 basic operations per second of FIP has but fragmentarily been described. For example in this paper only two-dimensional data has been considered in the examples. There are of course no such restrictions in the hardware and complex compound operators can be applied to one-, two- or three-dimensional data with equal ease.

The basic hardware structure of PICAP II including FIP has been built and tested and the completion of the system is now in progress. (June 1980).

ACKNOWLEDGMENTS

A large system such as PICAP II has many contributors to both its design and construction. The authors would like to acknowledge the contributions by C.V. Kameswara Rao (evaluation of pixel parallelism), Tomas Ohlsson, Tomas Hedblom and Arne Linge (the memory and the time-shared bus), Peter Lord and Torbjörn Eriksson (software) and many other members of the PICAP group.

REFERENCES

1. Batcher, K., The Massively Parallel Processor (MPP) System, Goodyear Aerospace Co. Akron, Ohio.
2. Danielsson, P.E., The Vth Int. Conf. on Pattern Recognition, (1980).
3. Danielsson, P.E., Kruse, B., and Gudmundsson, B., Workshop on Picture Data Description and Management, Asilomar, (1980).
4. Duff, M.J.B., Int. Joint Conf. on Pattern Recognition, Coronado, Calif., (1976).
5. Gerritzen, F.A., and Aardema, L.G., First Scandinavian Conf. on Image Analysis, (1980).
6. Golay, M.J.E, IEEE Tr. Comp., Vol C-18, p. 1007, (1971).
7. Granlund, G., First European Signal Proc. Conf., Lausanne, (1980).
8. Gray, B.S., Inf. Int. Inc., Los Angeles, Calif., (1972).
9. Gudmundsson, B., The Vth Int. Conf. on Pattern Recognition, Miami, (1980).
10. Kruse, B., IEEE Tr. Comp., Vol C-22, 12, p. 1075, (1973).
11. Kruse, B., Danielsson, P.E., Gudmundsson, B., in Special Comp. Architectures for Pattern Processing" (Fu, K.S. and Ichikawa, T, eds.), CRC Press Inc., (1981).
12. Kruse, B., Third Int. Joint Conf. on Pattern Recognition, p. 875, (1976).
13. Kruse, B., Proc. National Comp. Conf., Vol 47, p. 1015, Irvine, (1978).
14. Kruse, B., in "Structured Computer Vision", (Tanimoto and Klinger, eds.), Academic Press, (1980).
15. Kruse, B., Gudmundsson, B., Antonsson, D., The Vth Int. Conf. on Pattern Recognition, Miami, (1980).
16. McGormick, B.H., IEEE Tr Electr. Comp., Vol EC-12, No. 3, p. 791, (1963).
17. Mori, K., Proc. AFIPS, Vol 47, pp. 1025-1031, (1978).

18. Private communication with Kameswara Rao, C.V.
19. Reddaway, *in* "DAP - Distributed Array Procesor", Academic Press, (1975).
20. Sternberg, S., *Third Int. IEEE COMPSAC, Chicago,* (1979).
21. Unger, S.H., *Proc. IRE, p. 1744,* (1958).

BASIC IMAGE PROCESSING ALGORITHMS

ON THE

MASSIVELY PARALLEL PROCESSOR

James P. Strong

Goddard Space Flight Center

Greenbelt, Maryland

1. Introduction

The Massively Parallel Processor (MPP) [1,2] is being
developed by the Goddard Space Flight Center and
fabricated by the Goodyear Aerospace Corporation. It
is an array of 16,384 processing elements arranged in a
128x128 array. The MPP is an SIMD machine in the sense
that each processing element performs the same
instruction simultaneously on different data. In image
processing, the data consists of elements of the image.

The primary purpose for developing the MPP is to
perform the image processing tasks necessary to analyze
NASA's imagery from remote sensing satellites at speeds
commensurate with the rate at which the imagery is
obtained. In order to establish the feasibility of
meeting this requirement, basic benchmark image
processing tasks have been established. These tasks
are:

    1. Maximum Liklihood Classification

    2. Maximum Value Detection

    3. Cross Correlation

    4. Image Resampling

    5. Rotation

MULTICOMPUTERS AND IMAGE PROCESSING
ALGORITHMS AND PROGRAMS

This paper discusses the techniques developed by
Goodyear for performing these basic image processing
tasks. To obtain maximum speed, Goodyear has matched
the algorithms for these tasks closely to the hardware
design of the MPP. Therefore, to aid in understanding
these algorithms, the basic design of the MPP will now
be presented.

2. MPP DESIGN

     2.1 System Configuration

          The configuration of the Massively Parallel
Processor is shown in Figure 1. The Array Unit in this
configuration consists of 16,384 processing elements
each performing the same instruction simultaneously.
The Array Unit is connected to a Staging Buffer and an
Array Control Unit. The Staging Buffer acts as the
data interface between the Array Unit and input devices
such as disk drives and tapes. The Array Control Unit
generates the instructions for the Array Unit. It also
performs all operations on scalar variables. The
Companion Processor is the interface between programmer
and the MPP. It provides software and hardware for
generating, testing, running and storing programs.

     2.2 Array Unit Design

          The design of the array unit is shown
schematically in Figure 2.

In this diagram the basic data entity is a 128x128
element binary image [3]. All buses are in a 128x128
array format transferring 128x128 element binary images
or bit planes from memory to registers or from one
register to other registers. All devices shown in the
diagram operate on or store bit planes.

There are two bitplane processing devices in the array
unit. One is the logic device which performs logic
functions on the two bitplanes at its input. One of
its inputs is the output from the P-register. The
other comes from which ever device is outputing to the
bus. The result is stored back into the P-register.
For example, a typical logic operation of "OR"ing two
bitplanes would be given by the following sequence of
operations:

          1. Load P-register (through logic device)
with bitplane from Memory Location 1

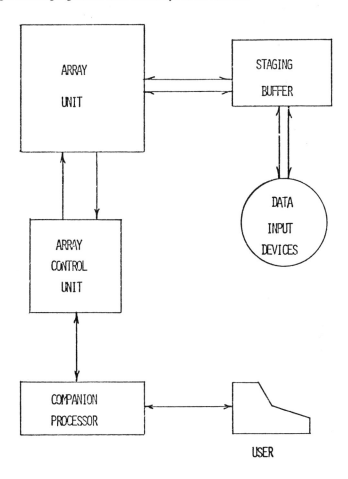

Fig. 1. Architecture of the Massively Parallel Processor.

    2. "OR" with bitplane from Memory Location 2

    3. Store resulting bitplane in P-register into Memory Location 3.

The second bitplane processing device in the array is the Adder. The Adder sums the bitplanes stored in the A-register, the P-register and the C-register. The resulting sum bitplane is stored in the B-register and the carry bitplane replaces the previous contents of the C-register. The use of the P-register as an input

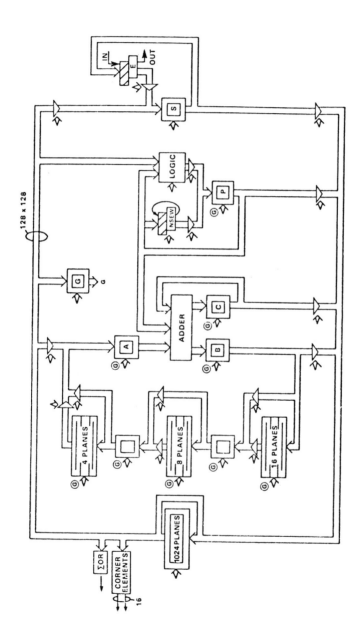

Fig. 2. Architecture of the Array Unit showing the basic data entity as a 128 x 128 bitplane.

facilitates inversion of one input to the adder which is required for subtraction. Addition and subtraction are performed in bit sequential fashion each step being performed in the manner described above for logic operations. At each step, the carry register makes the carry bitplane available for the addition of the next most significant bitplanes in the next step.

In order to translate images vertically or horizontally, the translation device labeled NSEW is used. The arrow going from the output to the input of the device in Figure 2 indicates the possibility for cylindrical or "wraparound" translations where the top row is connected to the bottom row or the left edge is connected to the right edge.

The Array Unit has a shift register stack which is basically a stack of bitplanes. With a shift instruction, bitplane data at each level of the stack is transfered to the next higher level. The length of the shift register (or height of the stack) can be varied by gating the data bus around various sections of the stack. The shift register stack is used in multiplication operations, floating point operations, and as temporary storage. Its use is a major factor in Goodyear's algorithms for cross correlation, rotation and resampling.

The MPP can perform masked operations in which some elements of the bitplanes are operated upon and others are not. Masked operations are controlled by a "mask" bitplane stored in the G-register. Where this bitplane has value 1 masked instructions are performed. Where it has value zero, masked instructions are not performed. Devices whose operations can be controlled by the mask are indicated in Figure 2 with a G. Masked shifts in the shift register stack are used frequently in the rotation and resampling tasks.

Bitplane data is input and output one column at a time from the array by way of the S-register. Each time a new column is input, the data in the S-register is cycled through the horizontal translator attached to the S-register moving it over one column to make room for the next column. On input, the bitplane is stored after 128 columns are input. On output, data is first loaded into the S-register and is completely output after 128 columns have been output. With this arrangement, it is possible for bitplanes to be input and output simultaneously.

Most image processing tasks require some quantative
information about the images (areas, types of
classifications, etc.). To facilitate this information
transfer from the Array Unit to the Array Control Unit,
the Array Unit has two information output devices. One
device OR's all elements of the bitplane on the bus.
Using this device, if any element in the bitplane is
non zero, a logical 1 is output to the ACU. This "Sum-Or"
device has the label Σ OR in Figure 2. The other
device outputs values from 16 elements in the bitplane.
These elements are located in the lower right hand
corner of the sixteen 32x32 non overlapping
sub-bitplanes making up a 128x128 element bitplane.

While the Array Unit does the actual image data
manipulation, the Array Control Unit performs all the
other tasks in the image processing programs. To
understand how image processing programs are run on the
MPP, the conceptual design of the Array Control Unit
and the philosophy of the programming of the MPP will
now be presented.

2.3   Array Control Unit

        As shown in Figure 3, the ACU is made up of 3
sections. The Processing Element Control Unit (PECU or
PE Control Unit) broadcasts instructions to all
elements in the array. The I/O Control Unit (IOCU)
controls the transfer of bitplanes between the Staging
Buffer and the Array Unit. The Main Control Unit
performs all tasks in an image processing program not
dealing directly with image data. It initiates all
image manipulating tasks done by the Array Unit and
initiates all data input and output operations between
the Array Unit and the Staging Buffer.

2.3.1   Processing Element Control Unit

        The PE Control Unit must not only
generate instructions for the Array Unit, must compute
Array Unit memory addresses and compute and act upon
any conditions for program branching in addition to
setting various command lines. To perform all these
operations efficiently, the PECU has a large 64 bit
instruction and eight index registers any of which can
be coupled to the address lines to the Array Unit
memory. The 64 bit instruction allows simultaneous
modification of all index registers and the testing of
the status of any register for branching conditions.

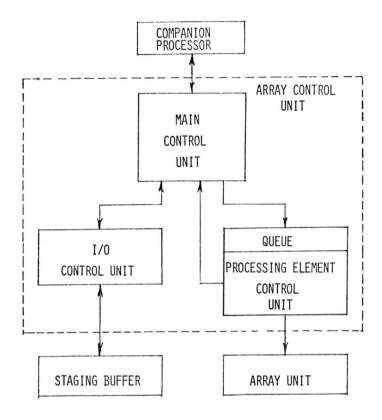

Fig. 3. Architecture of the Array Control Unit.

Generally, programs in the PECU perform short "subroutines" which cause the Array Unit to perform such operations as image addition, multiplication and translation. Image processing tasks are performed as a sequence of "subroutine calls" from the Main Control Unit. The interface between the PECU and the Main Control Unit is a queue for storing a sequence of subroutine calls. As each subroutine is finished by the PECU, the next is begun.

2.3.2 I/O Control Unit

Image data is seldom stored in bit
plane format on mass storage media such as tape or
disks. The MPP's staging buffer has the capability of
converting data stored in standard word format into
bitplane format. It is the task of the I/O Control
Unit to generate the control signals to the Staging
Buffer for the desired format and to generate the
signals to transfer the data into (or out of) the Array
Unit. This unit like the PE Control Unit operates by
performing subroutines.

2.3.3 Main Control Unit

For an image processing program, the Main Control Unit
performs much like a serial computer. Its task is to
perform all non-image manipulating tasks and to
generate image manipulating subroutine calls by placing
the parameters to be passed and the address of the
subroutine into the queue betwen the Main Control Unit
and the PECU. Typically, the subroutines performed by
the PECU require more time than it takes for the MCU to
generate the required parameters so that the queue will
always be filled. It is only when image data or
information from the Array Unit is required for a
branching decision that the queue empties. The PECU
subroutine which transfers the data from the Array Unit
to the Main Control Unit must be completed before the
Main Control Unit program can proceed with the
branching operations and generate the next subroutine
call.

2.4 Programming Languages for the MPP

The PECU, IOCU, and Main Control Unit can be programmed
in assembly language. There will be a library of
Primative PECU Programs for performing basic basic
arithmetic types of operations on images. Using this
library, most image processing programs will require
only a Main Control assembly language program. A high
level language is presently being developed for the MPP
at Purdue University. This language is a parallel
extension of Pascal. When implemented, the Main
Control Unit will perform the parts of the high level
program matched to serial processing. At the same
time, the PE Control Unit will cause the Array Unit to
perform those portions which deal directly with image
data.

3. BENCHMARK TASKS FOR IMAGE PROCESSING

The tasks chosen to demonstrate the efficiency with which the MPP can perform image processing tasks can be divided into those tasks requiring few or no data translations and those which require many. The maximum likelihood classification task and maximum value task fit the first group. The cross correlation, resampling, and rotation tasks fit the second. This distinction becomes important when processing large images of interest to NASA. The tasks falling in the first group are easily applied to large images when using a parallel array of processors. Those falling into the second group are more difficult to perform with a parallel array. (The efficient processing of large images using the MPP is very much task dependent and is the subject of present studies at Goddard.). For the benchmark tasks it is assumed that the images consist of 128x128 elements. The following sections describe the algorithms and techniques developed by Goodyear [4] to be used on the MPP to perform the five benchmark tasks.

3.1 Maximum Likelihood Classification

The increase in performance of the maximum likelihood classification algorithm obtained with the MPP over that obtained with conventional computers is primarily because of the number of picture elements which can be processed simultaneously by the MPP. The input data for this algorithm consists of images from 5 spectral bands, each ten bits in brightness resolution. Outputs are 10 binary images corresponding to the 10 possible classifications. A "1" at an element in the ith output binary image assigns that element to the ith classification.

The algorithm evaluates the likelihood equation below for each classification:

$$G^i = C^i - \sum^{H^i_{j,k}} (X_j - M^i_j) \quad (X_k - M^i_k) h^i_{j,k}$$

In this equation, the superscript i indicates the class, and the subscripts j and k indicate spectral bands. The other variables are given below along with their assumed accuracy.

G^i -   Likelihood function image for i^{th} class (11 bits)

C^i -   Constant corresponding to i^{th} class (10 bits)

X_j, X_k -   Images in j^{th} and k^{th} spectral bands respectively (10 bits)

M^i_j, M^i_k -   constants representing the mean values of brightness expected in each spectral band for elements in i^{th} class (10 bits)

$h^i_{j,k}$ -   constants representing the correlation between spectral bands j and k respectively for elements in the ith class (20 bits)

In evaluating the equation, the scalar mean value constants are subtracted from each spectral band image first and stored as images XP_n for n = 1 to 5. Then, for each value of j, inner product images, calculated as image XP_k times constant $h^i_{j,k}$, are formed and summed for values of k ranging from 1 to j. The resulting sum image is multiplied by image XP_j. The resulting product image is added to an outer sum image each time j is incremented. After the j=5 step, the constant C^i is added to the final outer sum image forming G^i.

An element in the image is assigned to classification q if its value in G^q is greater than in all other G^i's. In order to conserve memory space in the MPP, the classification of elements is updated after each likelihood function calculation. When each G^i image is calculated, it is compared at each element to the maximum of all previous values at each element. At the elements where G^i is greater than the maximum, the element is tentatively assigned to the ith classification by placing a 1 at that element in the ith binary output image and placing a zero in all other binary output images. Where G^i is less than the maximum, the classification is not changed. After all ten iterations of the computation of G^i and the classification update are completed, the output binary images indicate the classification of all elements of the input image.

3.2   Maximum Value Algorithm

The maximum value algorithm might typically be used after a cross correlation is performed to locate the x-y position of the maximum of the function. When defining the requirements for the maximum value task, it was assumed that it would be applied to the output of the benchmark cross correlation task. Thus the input image for this task has an accumulated wordlength of 21 bits.

The maximum value algorithm places the maximum value and the x-y coordinate of its location into the memory of the ACU Main Control Unit. It is assumed that all values of the input image are positive.

To facilitate the finding of the location of the maximum, Goodyear's algorithm first "tags" each array element with its x-y row and column coordinate. This is done by creating an "x-coordinate" array and a "y-coordinate" array. In the "x-coordinate" array, each element has the value corresponding to its column number. For instance the elements in column 1 have the value 1, those in column 2 have the value 2 and so on. The "y-coordinate" array is similar except that the elements have the value corresponding to their row number. These two arrays are appended to the least significant bitplanes of the cross correlation function such that a new 35 bit function is created given by:

$$A = F \times 16384 + 128 \mathrm{x} X + Y$$

where F is the input image, X is the x-coordinate array and Y is the y-coordinate array. Note that A will be a maximum wherever F is a maximum. Also, if F has two or more elements at the same maximum value, A will have only one corresponding to the maximum value of $128 \mathrm{x} X + Y$.

The algorithm determines the maximum value of A by examining each bit plane starting with the most significant and workding down. Here the sum-or device described in section 2.2 on the Array Unit is used to determine if any elements in the bitplanes contain a 1. Starting with the most significant bit plane, the test is repeated on each decending value bitplane until the first bitplane containing a 1 is found. (Note that the output of the sum-or device in these steps will be the bit value of the maximum at each significance level.

When the sum-or output is a 1, the elements containing a one are candidates for the maximum value. From this point in the algorithm, all further decending valued bitplanes are "AND"ed with the results of the previous step. For instance, if two elements had a 1 in the previous bitplane above, and only one had a 1 in the next least significant bitplane, one would be greater than the other and the AND function would eliminate the smaller one. The result of the AND function is then ANDed with the next least significant bitplane. In the case where two elements had a 1 in the previous bitplane and both had zeros in the next significant bitplane, the output of the sum-or device would be zero (again indicating the bit value of the maximum). In this case, both elements are still equal. The previous results containing the two 1's is then used to examine the next significant bitplane. When the algorithm has iterated through the 35 bits, the sum-or devices will have output the maximum value and its x and y coordinates to the ACU. Figure 4 shows a flow diagram of this algorithm and an illustrative example for 3 elements in 5 bit words.

3.3   Cross Correlation Task

The cross correlation task requires the cross correlation a 128x128 input image consisting of 6 bit data values with a 20x20 reference image also of 6 bit data. The algorithm proposed by Goodyear performs this task in 3 phases. In the first phase an image consisting of the least significant two bit planes of the input image is cross correlated with the reference image. In the second phase, an image consisting of the next two significant bitplanes of the input image is cross correlated with the reference image. The resulting cross correlation function is multiplied by four and added to the first. Finally, the process is repeated with the most significant two bitplanes of the input image. The resulting cross correlation function is multiplied by 16 and added to the previous sum. The final result is the cross correlation function required. The reason for dividing the task into three phases is that a very efficient scalar multiplication algorithm can be implemented.

3.3.1   Scalar Multiplication on the MPP

In scalar operations on the MPP, all elements of an image or array of data are operated upon by the same number. This number is located in the ACU

EXAMPLE: For 3 Elements, Each 5 Bits Long —

ELEMENT	MSB				LSB
1	0	1	0	1	0
2	0	1	0	0	0
3	0	0	1	0	0

Status of Variables During 5 Steps

	A[I] ELEMENTS 1 2 3			P ELEMENTS 1 2 3			B ELEMENTS 1 2 3			MAXVAL(I)
Step 1	0	0	0							0
Step 2	1	1	0							1
Step 3	0	0	1	1	1	0	0	0	0	0
Step 4	1	0	0	1	1	0	1	0	0	1
Step 5	0	0	0	1	0	0	0	0	0	0

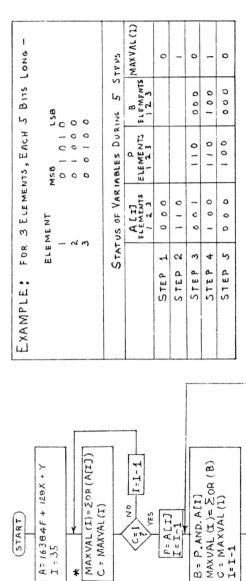

* MAXVAL (1-7) ⇒ Y-coordinate of Maximum
MAXVAL (8-15) ⇒ X. coordinate of Maximum
MAXVAL (16-35) ⇒ Maximum Value in Array F

Fig. 4. Flow diagram for the extraction of the maximum value in an image. A five bit example demonstrates the results at each step of the algorithm.

memory. In scalar multiplication, the values of the
bits in the scalar multiplier are used by the ACU to
control the updating of the partial sum in the Array
Unit which when all bits in the multiplier have been
considered will be the product. The partial sum is
updated only for bits in the multiplier whose values
are non zero. If the multiplier can be converted into
a format with fewer non zero bits, the scalar
multiplican can proceed faster.

One method of accomplishing this is to introduce the
concept of negative values in a binary number. For
instance, the binary number for 15 contains four 1's.
If negative values could be used in binary digits as
well as positive ones, 15 would be written as 1 0 0 -1
where -1 is a negative one. The meaning is that 15 = 1
x 2^4 - 1 x 2^0. This number has only two non zero
coefficients. Multiplication by this representation of
15 involves only two updates of the partial sum instead
of 4. Subtraction from the partial sum requires no
more time than addition.

Consider the multiplication of a two bit binary number
$a_1 a_0$ by the binary number for 5 (101). The result is
$a_1 a_0 a_1 a_0$. Note that each bit of the result is one or
the other of the bits of the multiplicand $a_1 a_0$ and that
no additions had to be performed to obtain this
product. This will be true so long as the multiplier
does not have two adjacent non zero bits. It can be
shown that any binary number can be converted to meet
this criteria when allowed negative values. Figure 5
shows an example where sequentially all adjacent pairs
or groups of 1's are converted to a 1 followed by at
least one zero followed by a negative 1. Note that the
resulting variable has at most only one more bit.

```
 1 0 1 1 1 0 1 0 1 1 0 1   = 2048 + 512 + 256 + 128 + 32 + 8 + 4 + 1
 1 0 1 1 1 0 1 1 0-1 0 1   = 2048 + 512 + 256 + 128 + 32 + 16 - 4 + 1
 1 0 1 1 1 1 0-1 0-1 0 1   = 2048 + 512 + 256 + 128 + 64 - 16 - 4 - 1
 1 1 0 0 0-1 0-1 0-1 0 1   = 2048 + 1024 - 64 - 16 - 4 + 1
1 0-1 0 0 0-1 0-1 0-1 0 1  = 4096 - 1024 - 64 - 4 - 1
```

Fig. 5. Breaking down a binary number into an equivalent one
having the property that each non-zero valued bit is separated by
at least one zero valued bit.

While the inclusion of negative values in the binary representation of a number allows the ifficient multiplication of two bit data, it imposes a subtraction step not necessary when all bits are positive. For instance the result of multiplying the two bit number a_1a_0 by 5 required creating the produce $a_1a_0a_1a_0$ by manipulating bits (bitplanes in the MPP) a_0 and a_1. However, multiplying by 3 (11 = 1 0 -1) requires subtracting $a_1 \times 2^1 + a_0 \times 2^0$ from $a_1 \times 2^3 + a_0 \times 2^2$. This subtraction step can be eliminated by using a two's complement representation of the subtrahend. The negative of the value $a_1 \times 2^1 + a_0 \times 2^0$ is equal to $\bar{a}_1 \times 2^1$ + $\bar{a}_0 \times 2^0 + 1 - 2^2$ where $\bar{a}_0$ and $\bar{a}_1$ are the complements of a_0 and a_1. The product can therefore be written as $a_1a_0\bar{a}_1\bar{a}_0 + (1-2^2)$. Thus the multiplication of two bits, or bitplanes in the MPP, is reduced to manipulating and complementing the bitplanes and subtracting a constant from the result. The product is obtained by placing bitplanes a_1 and a_0 in the nth and $n+1^{st}$ bitplane position respecitvely of the product for every positive one in the nth bit of the multiplier and placing $\bar{a}_0$ and $\bar{a}_1$ in the mth and $m+1^{st}$ bitplane position respectively for each negative one in the mth position of the multiplier. The value of the constant is given by:

$$C = \sum_{i=1}^{n} (2^{i+1} + 2^i) \times (\max(-x_i,0))$$

where x_i is the ith bit of the multiplier (x_i=1,0,or -1).

In performing the cross correlation function between a reference image and an image consisting of two bitplanes of the input image, a summation is performed where each term of the sum is the product of the reference image element $q(i,j)$ and the two bitplane image translated i elements in the x direction, and j elements in the y direction. For this cross correlation task, the sumation is performed for i and j varying between 0 and 19 corresponding to the 20x20 reference image. Using the above multiplication technique, each product corresponding to $q(i,j)$, is too large by the constant $C(i,j)$ when considering only

James P. Strong

While the inclusion of negative values in the binary
representation of a number allows efficient
multiplication of two bit data, it imposes a
subtraction step not necessary when all bits are
positive. For instance the result of multiplying $a_1 a_0$
by 5 required creating the product $a_1 a_0 a_1 a_0$ by
manipulating bits (bitplanes in the MPP) a_0 and a_1.
However, multiplying by 3 (11 = 1 0 -1) requires
subtracting $a_1 \times 2^1 + a_0 \times 2^0$ from $a_1 \times 2^3 + a_0 \times 2^2$.
This subtraction step can be eliminated by using a
two's complement representation of the subtrahend. The
value $a_1 \times 2^1 + a_0 \times 2^0$ is equal to $\bar{a}_1 \times 2^1 + \bar{a}_0 \times 2^0 + 1$
$- 2^2$ where $\bar{a}_0$ and $\bar{a}_1$ are the complements of a_0 and
a_1. The product can therefore be written as $a_1 a_0 \bar{a}_1 \bar{a}_0$
$+(1 - 2^2)$. Thus the multiplication of two bits, or
bitplanes in the MPP, is reduced to manipulating and
complementing the bitplanes and subtracting a constant
from the result. The product is obtained by placing
bitplanes a_1 and a_0 in the nth and $n+1^{st}$ bitplane
position respectively of the product for every positive
one in the nth bit of the multiplier and placing $\bar{a}_0$ and
$\bar{a}_1$ in the mth and $m+1^{st}$ bitplane positions respectively
for each negative one in the mth position of the
multiplier. The value of the constant is given by:

$$C = \sum 2^{m+1} + 2^m$$

for the bit value m corresponding to the locations in
the multiplier where there is a negative 1.

In performing the cross correlation function between a
reference image and an image consisting of two
bitplanes of the input image, a summation is performed
where each term of the sum is the product of the
reference image element $q(i,j)$ and the two bitplane
image translated i elements in the x direction, and j
elements in the y direction. For this cross
correlation task, the sumation is performed for i and j
varying between 0 and 19 corresponding to the 20x20
reference image. Using the above multiplication
technique, each product corresponding to $q(i,j)$, is
too large by the constant $C(i,j)$ when considering only

the bit plane values. Thus in the summation of the bitplanes we have the sum S' given by:

$$S' = \sum_{i=0}^{19} \sum_{j=0}^{19} P(i,j) + C(i,j)$$

where $P(i,j)$ is the actual product of $q(i,j)$ times the two bitplanes image. To obtain the correct sum, S, one subtracts the summation of the $C(i,j)$'s from S' as seen below:

$$S = S' - \sum_{i=0}^{19} \sum_{j=0}^{19} C(i,j)$$

Since the reference image is known ahead of time, the summation of the $C(i,j)$'s can be calculated in advance. Thus only one subtraction need be done at the end of the whole cross correlation computation.

The summing of the products of $q(i,j)$ and the shifted two bit image is performed in the MPP using the shift register stack to accumulate the partial sum. The use of the shift register stack saves a memory storage step which can reduce addition time by up to 33%. The summing process is illustrated in the data flow diagram in Figure 6. This process requires 12 clock cycles in the Array Unit to add the eight bits of the product to the partial sum stored in the shift register and to translate the two bit planes corresponding to the values of i and j. To use this time efficiently, the shift register is set to a length of 12 bitplanes. The translation operations are performed during the last four cycles (indicated by the zero valued bitplanes of the product input in the flow diagram). At the same time, the carries are rippled through the last 4 bitplanes of the partial sum and the summation bitplanes are moved up until the least significant bit is at the top of the stack ready for the next summation.

With six bit data in the reference image, it is possible for the partial sum to exceed 12 bits when the 20th product is added. Thus, after every 20th product is added, the six most significant bitplanes of the partial sum are removed from the shift register and stored in the memory. It is more efficient to store these bitplanes and sum them later than to extend the length of the shift register to accommodate a larger sum. During each phase, 19 sets of six most significant bitplanes are stored in memory.

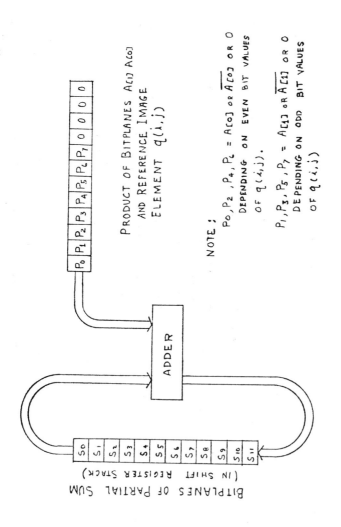

Fig. 6. Data flow for the addition of the partial sum stored in the shift register stack and the product obtained by multiplying the two bitplane image given by $A_0 \times 2^0 + A_1 \times 2^1$ by element $q(i,j)$ in the reference image.

After the last step in each phase, there is no further addition possible to the two least significant bitplanes of the cross correlation function since the next phase operates on the next two least significant bitplanes of the image. Thus, after this step, the two least significant bitplanes of the partial sum are stored in memory along with the 4 most significant bit planes (making a total of six).

After phase three is complete, all of the most significant portions of the partial sum , stored in memory are added to the values in the shift register. Finally, the summation of the constants corresponding to all the elements in the reference image is subtracted from this sum giving the cross correlation function. (In practice, this step may not be necessary if only the location of the maximum is required.)

3.4   Image Resampling Algorithm

The image resampling algorithm creates an output image in which each picture element is a bilinear interpolation of four neighboring picture elements in the input image. The four neighboring picture elements lie within plus or minus 8 picture elements in the X and Y directions from the output image element. The resampling is based on two distortion functions, $D(x)$ for the X direction and $D(y)$ for the Y direction. For each element in the output image the distortion functions indicate the x and y position of that element in the input image.

In the resampling algorithm, the distance that each input element must be translated to its location in the output image is calculated. Two translation distance images $T(x)$ and $T(y)$ are generated by subtracting the distortion functions $D(x)$ and $D(y)$ from the X and Y coordinate position of each output image element. That is, $T(x) = X-D(x)$ and $Ty=Y-D(y)$ where image X and Y are the X-coordinate and Y-coordinate images described previously in Section 3.2.

In order to compute the interpolated value of each output image element, the values of the four neighboring elements in the input image must be translated to that output element. The four neighboring input picture elements are illustrated in Figure 6 where they surround the projection of an output point back into the input image. The brightness of each of the four elements is labeled P_{00}, P_{01}, P_{10}, and P_{11}

according to their relative position with respect to
the projected output element. The relative distance
from the output element to its projected position in
the input image is given by T(x) and T(y). In Figure
7, let T(x) = A+a and T(y) = B+b where A and B are the
integer part and a and b are the fractional part. The
amount of translation required to cause each of the
four neighboring points arrive at the output point is
given by the table in Figure 7. For instance, if A+a
were 7.5 and B+b were -6.3 then a translation of 8 in
the positive X-direction and 6 in the negative Y
direction would be required to translate the
neighboring element with brightness P_{00} to the output
element.

Once the translations have been completed, the
brightness B of the output element is computed as:

$$B = P_{00} + a(P_{10}-P_{00}) + b((P_{01}-P_{00}) + a(P_{11}-P_{01}-P_{10}+P_{00}))$$

which is the bilinear interpolation equation. In the
MPP, four images labeled P_{00}, P_{01}, P_{10} and P_{11} are
generated which supply the required values to compute
the brightness of the output image. The bulk of the

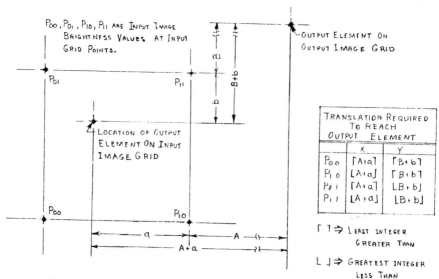

Fig. 7. Demonstration of the amount of translation required
to transfer the four input image values to an output image element
which are necessary to compute the bilinear interpolation function.

algorithm for resampling on the MPP is devoted to performing the necessary translations of the input image and the assignment of input image values to the proper output image elements to generate images P_{00}, P_{01}, P_{10} and P_{11}.

This is done systematically on the MPP by first translating the input image 8 elements in the positive X direction and 8 elements in the negative Y direction and then translating the input image in raster scan fashion over the 17x17 square area corresponding to the + 8 element maximum allowed distortion. Since the scanning is always in the negative X (westward) direction and the positive Y (northward) direction, the first input neighbor to arrive at an output element will have brightness P_{01}. The next will have brightness P_{11}. The third and forth will have brightness values P_{00} and P_{10} respectively.

The first step in performing the resampling algorithm is to generate 17 "masks" which are used to assign the translated input image elements to the corresponding P_{00}, P_{01}, P_{10}, and P_{11} images at each output image element. The 17 masks correspond to the integer portions of the translations in the X direction $\lfloor T(x) \rfloor$. They are generated in the Array Unit shift register
        In generating the masks, the shift register is initially loaded with 2 bitplanes of all 1's on the bottom two levels and 16 bitplanes of all zeros on the sixteen levels above as shown in Figure 9. The values of $\lfloor T(x) \rfloor$ can range from -8 to +8 represented by 4 bitplanes and a sign bitplane. Let x_0, x_1, x_2 and x_3 be the 4 bitplanes from the least significant to most significant forming $\lfloor T(x) \rfloor$. Let S be the sign bitplane. The 17 masks are generated by shifting elements in the shift register stack upward by:

(1 level where x_0=1) + (2 levels where x_1=1) + (4 levels where x_2=1) + (8 levels where x_3=0) + (16 levels where $S=x_0=x_1=x_2=0$ and x_3=1, ie., where $T(x) = 8$)

After this operation, the masks are located in the shift register as shown in Figure 9. The ith mask has ones corresponding to P_{11} and P_{10} for an X translation of i elements and corresponding to P_{01} and P_{00} for an X translation of i - 1.

algorithm for resampling on the MPP is devoted to
performing the necessary translations of the input
image and the assignment of input image values to the
proper output image elements to generate images P_{00},
P_{01}, P_{10} and P_{11}.

This is done systematically on the MPP by first
translating the input image 8 elements in the positive
X direction and 8 elements in the negative Y direction
and then translating the input image in raster scan
fashion over the 17x17 square area corresponding to the
$\pm$ 8 element maximum allowed distortion. Since the
scanning is always in the negative X (westward)
direction and the positive Y (northward) direction, the
first input neighbor to arrive at an output element
will have brightness P_{01}. The next will have
brightness P_{11}. The third and forth will have
brightness values P_{00} and P_{10} respectively.

The first step in performing the resampling algorithm
is to generate 17 "masks" which are used to assign the
translated input image elements to the corresponding
P_{00}, P_{01}, P_{10}, and P_{11} images at each output image
element. The 17 masks correspond to the integer
portions of the translations in the X direction $\lfloor T(x) \rfloor$.
They are generated in the Array Unit shift register
task. In generating the masks, the shift register is
initially loaded with 2 bitplanes of all 1's on the
bottom two levels and 16 bitplanes of all zeros on the
sixteen levels above as shown in Figure 9. The values
of $\lfloor T(x) \rfloor$ can range from -8 to +8 represented by 4
bitplanes and a sign bitplane. Let x_0, x_1, x_2 and x_3
be the 4 bitplanes from the least significant to most
significant forming $\lfloor T(x) \rfloor$. Let S be the sign bitplane.
The 17 masks are generated by shifting elements in the
shift register stack upward by:

(1 level where x_0=1) + (2 levels where x_1=1) +
(4 levels where x_2=1) + (8 levels where x_3=1) +
(16 levels where $S=x_0=x_1=x_2=0$ and x_3=1, ie., where
$T(x)$ =8)

After this operation, the masks are located in the
shift register as shown in Figure 9. The ith mask has
ones corresponding to P_{11} and P_{10} for an X translation
of i elements and corresponding to P_{01} and P_{00} for an
X translation of i - 1.

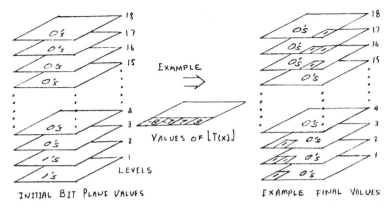

Fig. 8. Initial and final status of the shift register stack when generating translation masks for example values of T(x) =-8, -7,7,and 8.

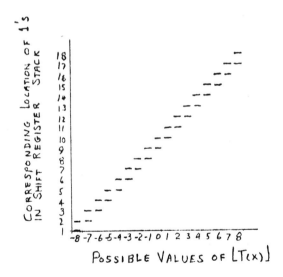

Fig. 9. Location of "1" values in the translation masks in the shift register stack corresponding to values of T(x) between -8 and 8.

The following discusses the translation procedure developed by Goodyear to assign the values P_{00}, P_{01}, P_{10}, and P_{11} to seperate images. The discussion will cover only one bitplane of P_{00}, P_{01}, P_{10}, and P_{11}. For 8 bit brithtness levels, this procedure will be repeated 8 times to obtain images P_{00}, P_{01}, P_{10}, and P_{11}.

Before going into the details of the technique, let us consider first a straight forward technique using the 17 masks corresponding to the translations in the X direction derived above, an additional 17 masks for the Y direction derived in the same way, and the shift register of the Array Unit. The procedure is shown below:

At each translation i in the X direction and j in the Y direction:

1. "AND" the X direction mask (j) with the Y direction mask (i).

2. Where the resulting mask =1, store the value of the translated bitplane into the bottom level of the shift register.

3. Also, where the resulting mask =1, shift the elements in the shift register up one level in the stack.

Remembering the scanning technique and the resulting order of the arrival of the neighbor values, one finds at the end of all the translations that level 4 in the shift register contains P_{01}, level 3 contains P_{11}, level 2 contains P_{00} and level 1 contains P_{10}.
order of the arrival of the neighbor values, one finds at the end of all the translations that level 4 in the shift register contains PO_1, level 3 contains P_{11}, level 2 contains POO and level 1 contains P_1O.

The Goodyear approach speeds this process by removing the "AND" operation in step 1 and using only the X mask as the mask in steps 2 and 3. Because the Y direction mask is no longer "AND"ed with the X direction mask, 34 samples are input at each element in the shift register after Goodyear's procedure. This is because at any element, some mask(i) and mask (i + 1) contain a 1 and the scan is repeated for 17 rows. Figure 10 shows the locations in the shift register stack of P_{01}, P_{11}, P_{00}, and P_{10} corresponding to values of $\lfloor T(y) \rfloor$. Mask (i) inputs values of P_{01} and P_{11} and mask (i+1) values of P_{00} and P_{10} for a given translation of $\lfloor T(y) \rfloor = i$. To align the values in the shift register stack into four

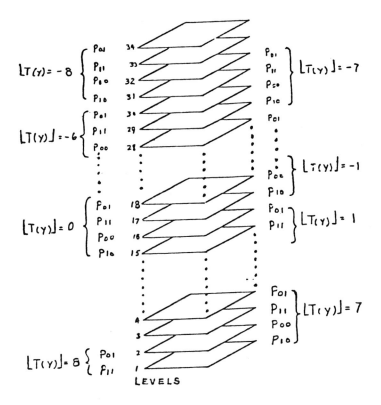

SHIFT REGISTER STACK

Fig. 10. Location in the shift register stack of input image values P_{00}, P_{01}, P_{10}, and P_{11} corresponding to values of $T(y)$ after the scanning process over the 17 × 17 element area assuming a 34 layer shift register stack.

separate bitplanes, one shifts the elements 2×i levels wherever $\lfloor T(y) \rfloor = i$. Where i is positive, the shift is upward. Where i is negative, the shift is downward. When this is accomplished, images P_{10}, P_{00}, P_{11}, and P_{01} will be located at levels 15, 16, 17, and 18 respectively in Figure 10.

This procedure assumes a shift register stack with up to 34 levels and the ability to shift both up and down. The MMp shift register stack has at most 32 levels and con shift only upward. Thus Goodyear has developed a modification to the above technique allowing the use of a shorter shift register and requiring only upward shifts.

This technique performs the same procedure as described above for the first 9 (ie., j = -8 to 0) rows. After the nine rows have been completed, each element in the shift register will have had 18 inputs. In the next 8 rows (j = 1 to 8) the x-direction masks are modified so that 1's continue to exist only where $\lfloor T(y) \rfloor > 0$. In this case, no further inputs will be made into the shift register stack at the elements where the y translation is in the negative direction. In the elements where the Y translation is 0 or in the positive direction, 16 further values will be input to the shift register stack. (The data that was previously in these elements is shifted out above the bottom 16 levels.) After translations in the last row have been completed, the locations in the shift register of PO_1, P_{11}, P00, and P_1O corresponding to all of the Y translations are as shown in Figure 11. One can see that the bottom 18 levels correspond to both positive and negative translations in the Y direction.

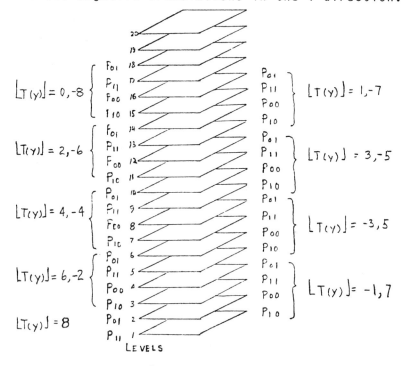

SHIFT REGISTER STACK

Fig. 11. Location in the shift register stack of input image values P_{00}, P_{01}, P_{10}, and P_{11} corresponding to values of $\lfloor T(y) \rfloor$ using the Goodyear technique.

A simple procedure can now be applied to align all the values corresponding to P_{00}, P_{01}, P_{10}, and P_{11} into separate bitplanes shifting only in the upward direction of the shift register stack.

From Figure 11, one can see that alignment can be obtained by shifting the elements corresponding to Y translations of -7 and 1 up two levels, those corresponding to -6 and 2 up 4 levels, and so on where the amount of shifting is given by 2 times the positive y translation. this alignment can be accomplished in a manner similar to the one used to generate the masks. Note that the both the positive and negative values of each pair of Y translations have the same biary representation in the three least significant bits. (For example, 1 = 001 and -7 = 001) Thus except for $\lfloor T(y) \rfloor$ = 8 the three least significant bitplanes of $\lfloor T(y) \rfloor$ y_0, y_1 and y_2 can be used as masks to control the shifting. Using these bitplanes as masks and a special mask for $\lfloor T(y) \rfloor$ = 8, the elements in the shift register stack are shifted upward by:

(2 where y_0=1) + (4 where Y_1=1) +
(8 where y_2=1) + (16 where y_0=y_1=y_2=y_4=0
and y_3=1, ie., where $\lfloor T(y) \rfloor$=8)

When the alignment procedure is completed, the P_{01}, P_{11}, P_{00}, and P_{11} bitplanes will be located in levels 18, 17, 16, and 15 respectively in the shift register stack.

These bitplanes can now be stored and the avove process repeated 8 times corresponding to the 8 bitplanes making up the input image. Once images P_{01}, P_{11}, P_{00} and P_{10} have been generated, the values of the output image can be calculated. This is a straight forward procedure involving additions and multiplications of the fractional parts a and b of the translation images and images P_{01}, P_{11}, P_{00}, and P_{10} according to the bilinear interpolation equation given in the first part of this section.

3.5 Rotation Algorithm

The rotation algorithm rotates a binary 128x128 image about a point located at coordinates p, q (not necessarily integers) in the array by an angle θ

of up to ± 45°. The rotation operation is illustrated
in Figure 12 where the brightness value at output image
element I, J (both I and J being integers) comes from
location m,n (generally not integers) in the input
image. Coordinates m and n are given by the equations
below as a function of output coordinates I and J.

$$m = (I-p) \, Sin\theta \; + (J-q) \, Cos\theta \quad +p \qquad (1)$$

$$n = -(I-p) \, Cos\theta + (J-q) \, Sin\theta \quad +q \qquad (2)$$

The integer values M and N closest to m and n are the
coordinates of the element whose brightness value is
assigned at coordinates I,J in the output image.

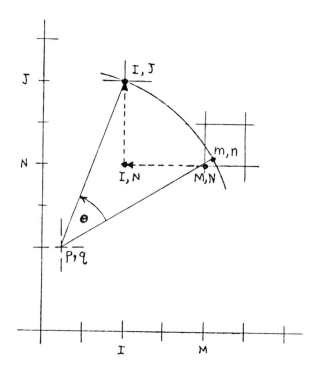

Fig. 12. Location of the nearest element in the input image
which is to be assigned to each output element. Also, showing the
horizontal and vertical paths used by the Goodyear rotation algorithm to transfer the input value to each output image element.

The rotation algorithm developed by Goodyear performs the rotation task in two stages. In the first stage, the brightness value at each element M,N is translated horizontally to element I,N (see figure 12). In the second stage, the brightness value at each element I,N is translated vertically to element I,J. These translations are performed for all elements in the image simulataneously. To perform the first stage operation, the required horizontal transfer distance must first be calculated at each intermediary element, I,N. That is, at each element I,N, the value M-I must be calculated. The value of M as a function of I and J is obtained by rounding the results of equation 1. The value for J as a function of I and N can be computed by solving for J in equation 2. In the MPP, J is computed by inserting the coordinate pairs I,N of each intermediary element into the equation:

$$J = \text{Round}((N-q)\text{Sec } \theta + (I-p)\text{Tan } \theta - q$$
$$- (\text{Sec } \theta)/2 + 1) \qquad (3)$$

Note that the solution for J from equation 2 does not include the term $-\text{Sec}\theta/2 + 1$. The addition of this term will be discussed later. The resulting value of J at each intermediary element is then inserted into equation 1 along with that element's I coordinate value to compute M. The horizontal distance M-I is then calculated by subtracting the element's I coordinate value from the computed value of M. To perform the horizontal translations, the input image is translated horizontally in increments of one element. At each increment i, the image value is assigned at the elements where M-I=i.

To perform the second stage of the algorithm transfering the image values from intermediary elements I,N to elements I,J, the distance J-N must be computed at each element in the output image. To do this, the coordinates I,J of each element in the output image are used to compute the value N using equation 2 (rounding the result). This result is subtracted from the row coordinate J of each element to obtain the required vertical translation distance. Finally the intermediate image obtained in the first stage is translated verticlaly in one element increments. At each increment i the intermediate image values are assigned to the output image elements where J-N=i.

In this discussion, the fact that intermediary elements
I,N might be required to receive more than one input
brightness value was not considered. The following
discussion demonstrates that indeed some intermediary
elements will be required to receive two brightness
values but not more than two.

Consider Figure 13 where the center of Row N, and the
lines defined by N + 1/2 and N - 1/2 in the input image
are shown rotated in the output image. At column
coordinate I of the output image, the center of row N
intersects at value j. The upper line N + 1/2
intersects at [j + (SecΘ)/2] and the lower line N - 1/2
intersects at [j - (SecΘ)/2]. In this illustration,
only one integer value J lies between [j + (SecΘ)/2]
and [j - (SecΘ)/2] and will be the only output
coordinate in column I value having N as its closest
coordinate in the input image. (elements I,J+1 and
I,J-1 being assigned row coordinates N + 1 and N - 1
respectively.) For angles up to 45°, the limits [j +
(SecΘ)/2] and [j - (SecΘ)/2] define an interval up to
1.414 units long. Therefore, two integer values of J
can lie within these limits (but no more than two) as
illustrated by Figure 13. In this case, both elements
I,J and I, J+1 will have row N as an input coordinate
and therefore element I,N must store two values. The
lower coordinate J within the interval is computed with
equation 3. The two values which must be stored are
shown in figures 14a and 14b. In Figure 14a, the
coordinate value m is shown computed from equation 1
(repeated below) for the lower element I,J in the
interval.

$$m = (I-p) \, Sin\Theta \; + \; (J-q) \, Cos\Theta \; + \; p \qquad\qquad (1)$$

Coordinate m' corresponding to element I, J+1 within
the interval is computed as:

$$m' = (I-p) \, Sin\Theta \; + \; (J+1-q) \, Cos\Theta \; + \; p \qquad (4)$$

In figure 14a, the closest integer value to m is M.
Integer M' is the closest to m' and in this case, M' =
M + 1. In Figure 14b, M' = M. Since m' - m is equal
to CosΘ,m' - m will always be less than or equal to
1. Thus, M' will always be either equal to M or equal
to M + 1.

In order to make both brightness values available at
element I, N, the Goodyear algorithm generates two
intermediate images T and U. In image T, each element

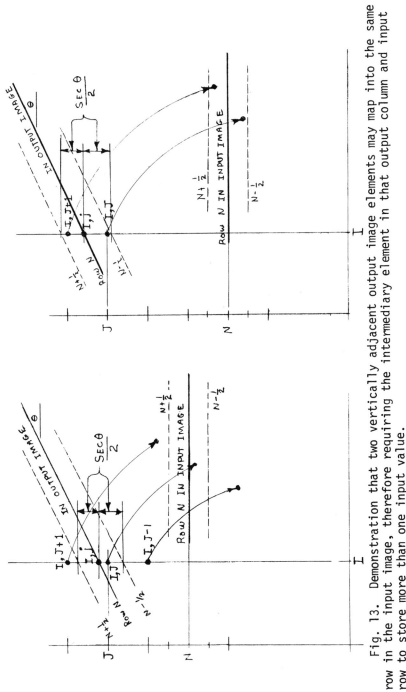

Fig. 13.   Demonstration that two vertically adjacent output image elements may map into the same row in the input image, therefore requiring the intermediary element in that output column and input row to store more than one input value.

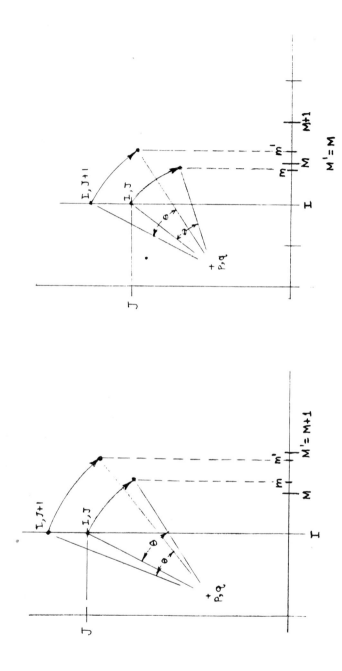

Fig. 14. Demonstration that two vertically adjacent output elements require input values from either the same column in the input image or from immediately adjacent columns. Consequently, if they require input values from the same row, the two values will come from either the same input image element or from immediately adjacent ones in the same row.

I,N contains the brightness value a distance M-I away
in the input image. In image U, each element I,N
contains the input brightness value a distance M' - I
away. In the vertical transfer process, element I,J
selects brightness values from iamge T if element I,J-1
obtains its brightness value from a different row in
the input image than element I,J. (corresponding to
element I,J+1 in figure 13a being outside the interval
[j $\pm$ (Secθ/2]). Element I,J selects brightness values
from image U if element I,J-1 obtains its brightness
value from the same row as element I,J (corresponding
to both elements I,J+1 and I,J in Figure 13b being
within the interval.

We will now discuss the Goodyear algorithm for
performing the rotation operation. For this
discussion, the distances M-I and J-N calculated
previously will exist as "images" labeled M-I and J-N.
Image M-I is used to generate masks assigning input
brightness values to images T and U as the input image
is translated horizontally. Image J-N is used to
generate masks for assigning values to the output image
as U and T are translated vertically. Before
discussing the generation and use of these masks some
assumptions will be stated. First, it will be assumed
that if the value of M-I is positive at element I,N,
the input brightness to that element will come from an
input image element to the right and will be transfered
to element I,N by westward or leftward translations of
the input image. Second, it will be assumed that the
"wrap around" mode of connection in the MPP is used
such that values exiting from the left-most column of
the array, re-enter at the right-most. Third, we
assume that the center of rotation p,q lies within the
image and therefore, the values of M-I lie within the
range -127 to +127. Based on these assumptions, Figure
15 shows the number of western translations in the
"wrap around" mode required to transfer the input image
to distances ranging over values of M-I between -127
and +127. In this figure, we can see for example, that
a translation of 3 does the necessary translation for
M-I=-125 as well as M-I=3. This halves the total
number of masks required. Also, the fact that the
seven bit binary number for -125 is the same as that
for 3 makes it possible to form a translation mask for
both values at the same time in the manner used in the
resampling task.

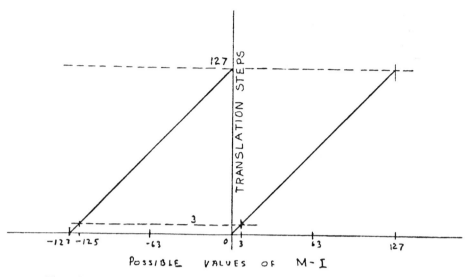

Fig. 15. The amount of westward translation in the wraparound mode on the MPP required to transfer input image values over the range of values of M-I from -127 to 127.

Let us now consider a straight forward method to generate images T and U on the MPP using the shift register stack and 128 masks. We will initially define image U to contain input elements M + 1 - I away at any element I,N corresponding to M' = M + 1. We will later modify image U to equal image T at the elements where M' = M. As the input image is translated westward, the first input value assigned to element I,N will always go to image T. The value at the next translation step is assigned to image U. That is at step i, the input value will go image T at those elements where M-I=i using Mask i. These same elements will receive data for image U at step i + 1 using Mask i + 1. Both Mask i and Mask i + 1 will have 1's at these elements. Since only two data values are input at any element, the bottom level of the shift register will contain image U and the level above will contain image T at the end of 128 steps.

To correct image U where M' = M, image M' is computed by rounding the results from equation 4. A mask is generated which has values of 1 where M' = M. Values in image T are transfered to image U where the mask has the value 1.

The technique just described while straight forward, requires the generation of 128 masks which is time consuming. Goodyear has developed a technique which requires the generation of only 8 masks but requires a realignment procedure similar to that used in the resampling task. Consider the three least significant bits of the seven bit value for M-I. The values defined by these bits are plotted in figure 16 relative to the values defined by all seven bits. One can see that a given value "a" defined by the 3 bits corresponds to one of 16 possible 7 bit values (a, 8 + a, 16 + a,... 120 + a). The algorithm developed by Goodyear generates the 8 masks corresponding to the 3 least significant bits of M-I. In the translation procedure, the masks are used in a cyclic fashion to shift values from the translated input image into the bottom of the shift register stack. For instance, the mask corresponding to the value of 5 is applied at translation amounts 5, 13, 21, 29, etc. After 127 translations each element in the shift register stack will have had input 32 values (two from each cycle of masks). The values in the odd levels belong to image U and those in the even levels belong to image T as shown

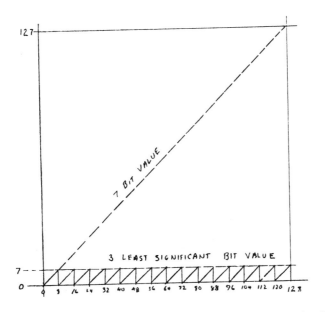

Fig. 16. Values represented by the three least significant bits of M-I (or J-N) over the seven bit range from 0 to 127. As M-I (or J-N) ranges from 0 to 127, each value given by the three least significant bits is repeated 16 times.

in Figure 17.   In order to align the correct values
into one T and U image, the values in the   shift
register stack are shifted upward according to the four
most significant bits of M-I.   Let m_4, m_5, m_6, and m_7 be
the 4th, 5th, 6th, and 7th bitplanes respectively of
image M-I.   Then the shift register stack is shifted
upward using the bitplanes as masks by the amount:

$$(2 \text{ where } m_4 =1) + (4 \text{ where } m_5 =1) +$$
$$(8 \text{ where } m_6 =1) + (16 \text{ where } m_7 =1)$$

after this shifting operation, images T and U will
reside at levels 32 and 31 of the shift register stack
respectively.

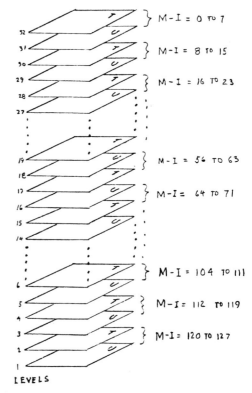

SHIFT   REGISTER   STACK

Fig. 17.   The location in the shift register stack of values
for image T and U corresponding to values of M-I (or J-N) over the
range from 0 to 127.

The 8 masks are generated in the shift register stack.
Since both masks i and i + 1 (Mod 8) have 1's where
M-I=i (Mod 8), the shift register stack is initially
loaded with two bit planes of all 1's in levels 1 and 2
(as in the resampling task) and with bitplanes of
zero's in levels 3 through 8. Then using the 3 least
significant bit planes of M-I (m_0, m_1, and m_2) as
masks, the elements in the shift register are shifted
up by:

$$(1 \text{ where } m_0 = 1) + (2 \text{ where } m_1 = 1) +$$
$$(4 \text{ where } m_2 = 1)$$

The resulting bitplanes are located in levels 1 through
8. It should be noted that the shift register must be
connected as an 8 layer "ring" so that the output from
level 8 is input back to level 1. This is necessary to
assure that masks i and i + 1 (mod 8) have 1's wehre
M-I=i(Mod 8).

After images T and U are generated, the vertical
transfers are performed. These are done in the same
manner as the horizontal ones except that each output
element is assigned only one brightness value from
either image T or image U. To select which output
elements obtain data from image T and which from image
U, image N previously calculated to obtain image J-N is
used. The values of the elements in each row of N are
compared to the values in the row below. At the
elements where they are equal, mask U = 1 indicating
that at these elements the output image obtain its
values from image U. At the elements where the values
are different, mask T = 1 indicating that at these
values the output image obtains its values from image
T. The vertical translation masks are generated
similarly to the horizontal ones using the three least
significant bitplanes of image J-N except that the
shift register stack is initiated with only one
bitplane of 1's at the bottom of the stack.

When image T is translated vertically, the 8 vertical
shift masks are first "AND"ed with mask T to assure the
assignment of values of iamge T to the proper output
elements. When image U is translated, the 8 vertical
masks are first "AND"ed with mask U. At the end of the
vertical shift operation, 16 levels of the shift
register stack receive data. The values in the shift
register stack are realigned using the four most
significant bits planes of J-N (n_4, n_5, n_6, and n_7) as
masks. The amount of shifting at each element is given
by:

$$(1 \text{ where } n_4 = 1) + (2 \text{ where } n_5 = 1) +$$
$$(4 \text{ where } n_6 = 1) + (8 \text{ where } n_7 = 1)$$

The output image will be located at level 16 of the stack.

The final part of the rotation algorithm deals with elements in the output image which come from beyond the edge of the input image. Because of the wraparound feature of the MPP is used in the translation operation, elements from within the input image are placed in elements which should come from beyond the edge. These elements can be "cleared" (assuming zero values beyond the edge) by generating image EM from equation 1 and image EN from equation 2 where J is the row coordinate at each element and I is the column coordinate. The legitimate elements in the output image are those where

$$0 \leq EN \leq 127 \text{ and } 0 \leq EM \leq 127$$

A mask having values of 1 at the legitimate elements is generated and used to clear the non legitimate elements in the output image.

4. CONCLUSION

The algorithms developed by Goodyear to perform the benchmark tasks illustrate how the architecture of the Array Unit can be utilized to the fullest extent to achieve maximum efficiency. The resulting programs are performed several orders of magnitude faster than can be done on general purpose serial computers. This type of algorithm development will be required for NASA tasks requiring a maximum image data throughput.

For developing new image processing algorithms, this type of programmingwill be very difficult and expensive. To aleviate this problem for the general user interestd in developing new image processing algorithms, a library of very efficient standard image processing programs is being generated at the University of Illinois. The programs in this library can be called upon from user programs.

In addition, as indicated previously, a high level language is also being developed. Even with the inefficiencies generally accompanying a high level language, the inherent speed of the MPP will allow the generation and execution of new very complex image processing algorithms which because of long computation times on serial machines is today not possible.

ACKNOWLEDGEMENTS

The author is indebted to Mr. David H. Schaefer for his aid in removing many "bugs" from this paper.

REFERENCES

1. D. H. Schaefer, J. R. Fischer, K. R. Wallgren, "The Massively Parallel Processor", AIAA Sensor Systems for the 80's Confrence, Dec. 1980, pp.187-190.

2. K. E. Batcher, "Design of a Massively Parallel Processor", IEEE Trans. Comput., Sept. 1980, pp. 836-840.

3. D. H. Schaefer, J. R. Fischer, "Spatially Parallel Computers-Ensembles of Two-Dimensional Components", To appear.

4. Goodyear Aerospace Corp., "The Massively Parallel Processor", Phase I Final Report, July 1979.

IMAGE CORRELATION: PROCESSING REQUIREMENTS
AND IMPLEMENTATION STRUCTURES
ON A FLEXIBLE IMAGE PROCESSING SYSTEM (FLIP)

Peter Gemmar

Research Institute for Information Processing
and Pattern Recognition
Karlsruhe, FRG

A flexible image processing system and favourable processing structures for image correlation are described. Typical system requirements are analyzed for object tracking and stereo mapping based on
image correlation. Processing configurations were designed considering data storage, data transmission and data processing. An
image correlation system consisting of the multiprocessor system
FLIP and a host computer is described in terms of memory organization, data flow and process organization. Efficient processing
structures for the multiprocessor system were investigated for
computing the correlation function. Significant characteristics of
different processing structures are discussed. Simulation results
were used to estimate overall processing times for image correlation.

I. INTRODUCTION

Image correlation ranks among the most important image processing tasks. There are a lot of approved applications of image
correlation known for different areas of image evaluation like map
data processing, processing of aerial photographs, and processing
of image series, etc. In most cases image correlation is used to
detect and/or locate objects (patterns) within a picture. Patterns
to be detected vary in size and shape from simple ones like line
elements in binary images to complex ones like textured objects
possibly combined with foreground and background distortions in
grey value images. Therefore, the complexity of correlation algorithms is subject to the application requirements.

MULTICOMPUTERS AND IMAGE PROCESSING
ALGORITHMS AND PROGRAMS

87

In the following, two important applications of image correlation, object tracking and stereo correlation, and their realization on a flexible multiprocessor system (FLIP) will be described. The processing requirements given by each of these tasks would use unreasonable time when implemented on a general purpose computer with e.g. 1 MIPS computing power. Besides a fast processing there have been different goals on the implementation of both tasks on FLIP, e.g., simulation of a special correlation system already being designed (Gemmar, 1979).

II. FLEXIBLE IMAGE PROCESSING SYSTEM FLIP

The FLIP system operates as a peripheral device in conjunction with a host computer. FLIP consists of the central processing unit FIP (Flexible Individual Processors) and the data exchange processor PEP (Peripheral data Exchange Processor). The FIP is built with 16 IPs (Individual Processor). The PEP, which is mainly a fast buffer device, comprises three internal processors to provide the high data rate required by FIP. Additionally, FLIP is directly connected to a MOS image memory (768 k Byte) via a high speed data path. Beside several mass storage devices (disk, tape) and image input/output devices the host provides two additional links to another minicomputer and a VAX 11/780 general purpose computer, respectively (Fig. 1).

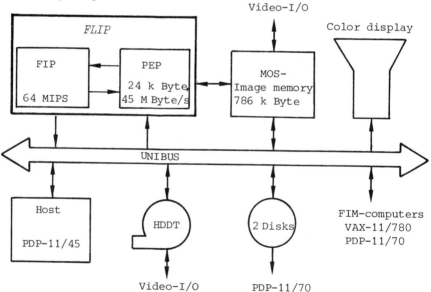

FIGURE 1. FLIP image processing system

All individual processors (IP's) of FIP are identical. They are able to perform all basic arithmetical and logical instructions. Internal data representation is by 8 bit data words. The FIP system provides a flexible and programmable structure. This gives the possibility to arrange the ensemble of 16 IP's according to the topology of a task to be performed. The system is data-flow controlled. An internal bus system connects each IP to all the others. Each processor has two independent input ports and one output port. The data flow between FLIP and the host computer is accomplished by the PEP. The input data stream to FLIP is controlled by the PEP-input/output controller and transferred to the PEP-data memory for intermediate storage (e.g., by DMA). These data then are transmitted by the three PEP-processors to the FIP in a prefetch-like manner. To achieve a suitable data rate PEP and FIP are connected together by three high speed buses feeding 16 medium speed buses providing a maximum data rate of 45 Mbyte/s.

Nearly any desired processing structure of FIP-processors can be established from a pipeline over a cascade up to other parallel structures. To establish a processing structure on FLIP, all the FIP-processors as well as all the PEP-processors which are required by this structure, must be programmed (MIMD multiprocessor system, Luetjen et al., 1980).

The FLIP is programmed at an assembler language level providing convenient macro extensions. Typical FLIP applications are homogeneous operations which are very often used for picture preprocessing, feature extraction, image correlation, etc.

III. IMAGE CORRELATION

In general image correlation represents a measure of similarity between two patterns. Roughly spoken, one image (e.g., actual object) is compared with a second one (e.g., searched object or reference). Although different mathematical computations for image correlation have been specified in the literature, in the following only one of them will be regarded. It is the normalized version of the correlation function (1), which has shown best results for the chosen applications.

Local digital correlation is performed by comparing a mask M (reference) with an image section S (search area) step by step. To calculate the correlation function the mask is shifted systematically over the search area (e.g., raster image) and the correlation value (1) is computed at each displacement position. The main advantage of the normalized version is its limited range of values $1 \geqq C \geqq -1$, greatest similarity is given by the maximum of

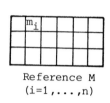

Reference M
(i=1,...,n)

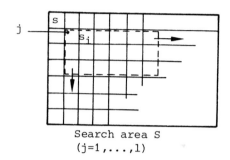

Search area S
(j=1,...,l)

$$C(j) = \{1/n \cdot \Sigma \ m_i s_i - 1/n \cdot \Sigma \ m_i \cdot 1/n \ \Sigma \ s_i \}$$
$$/\{ (1/n \cdot \Sigma \ m_i^2 - (1/n \cdot \Sigma \ m_i)^2) \cdot (1/n \cdot \Sigma \ s_i^2 - (1/n \cdot \Sigma \ s_i)^2) \} \quad (1)$$

FIGURE 2. Local digital correlation

the correlation function, and its independence from a linear transformation of the intensity levels (e.g., caused by changes in the
illumination of the object) (Fig. 2). However, computing   C is
the most time consuming part of the correlation process. Let l be
the number of displacement positions of M in S and n the number of
image elements in M, then the total number of arithmetic operations to calculate C for all l positions results in more than 5nl
operations. The number of operands that must be fetched and/or
stored during the execution of these operations amounts to about
2nl picture elemente (pixels). Moreover, sequential general purpose computers require additional processing time for bookkeeping
actions. In most cases, such computers are inadequate to carry out
image correlation, especially for the applications described in
the following sections.

A. Object Tracking

Tracking of moving objects requires object detection in image
sequences. In existing systems the actual scene is continually monitored by a sensor (e.g., TV camera) and the sensor signals are
processed on-line by an image processing unit. An operator can influence the processing by additional information about the object
and the environment. The result of the processing is the position
of the selected object in each image of the sequence. This position can be used to control the sensor and its platform.

In applications with moving objects on ground with small contrast to background only tracking systems based on the correlation

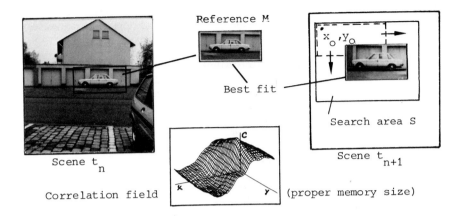

Reference M

Best fit

Search area S

Scene t_n

Scene t_{n+1}

Correlation field

(proper memory size)

FIGURE 3. Correlation tracker

function have shown reliable results. Fig. 3 shows the principle mode of operation of this system. An operator detects an interesting object (car) in a scene (t_n), this object is stored in a memory as a reference and in the next image the correlation function is computed between the actual scene (t_{n+1}) and the reference. The maximum of the correlation values is used to detect the object by a decision criterion and also to steer the update of the reference memory. These updates are due to changes of the object, the background and/or the foreground, respectively. For experimental verification a complex correlation system was implemented which consists of a correlator (for computing the correlation values) combined with a memory for the actual scene and a reference memory. This reference is modified during correlation by a correlation template and an update template. The correlator is connected to an evaluation module which assembles the update criterion, the decision criterion and motion analysis (Bers et al., 1980).

Monitoring of a real scene normally lasts 16 s which is actually represented by an image sequence of N = 400 images. The image sizes applied can be 256x256 or 512x512 pixels with 6 or 8 bit grey values each. From these images references are taken with a size in the order of n = 800. A template T of same size is used for modification of the reference during correlation. The search area for correlation is described by a maximum number of search positions of l = 961. Applying these parameters the simulation of the above discussed tracking system on a sequential general purpose computer (with 1 MIPS) requires about 1.8 hours. During this run the system consumes 3% of the total time to get the image data from a mass storage device (e.g., disc) and to prepare them for correlation, 91% to compute the correlation values, and finally 6% to evaluate the correlation function and for updating the system parameters. Due to these facts it was reasonable to implement only the correlation part on the FLIP, whereas the other processes remained on the host computer. By this, the correlator (FLIP) takes only 25% of the total processing time. Figure 4 depicts the memory organization and data flow for object correlation on FLIP. Image data of image lines, which cover a predicted search area in a scene, are taken by the host computer and transmitted to the PEP. The reference M and the correlation template T to modify the reference are also stored in the PEP memory. Additionally the PEP processors are supported with actualized system parameters (n, l) and the coordinates x_o, y_o of the search area in the image section. Concurrent performance of data processing and delivery is achieved in multiple ways: First by double buffering of the image data within the PEP memory, second by prefetching of the operands to be processed by the correlator and also by transmission of the computed correlation values back to the host computer. The processing structures of FLIP for computing the correlation values (or terms of it) will be discussed in the following chapter.

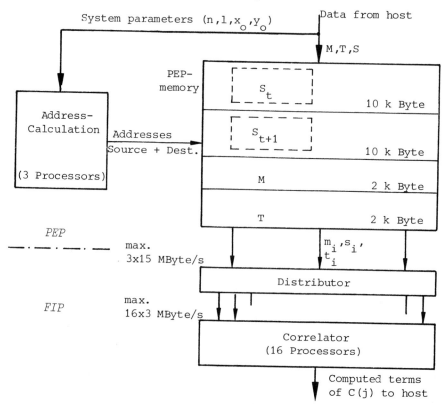

FIGURE 4. FLIP organization for object tracking

B. *Stereo Image Correlation*

Stereo image correlation is a fundamental problem in digital mapping because of its expenditure of computing time. Till now, production times for maps or map substitutes (e.g., DTM, orthophoto, etc.) range from 3 to 8 hours when semiautomatic or analogous methods are applied. On the other hand, new fully automatic and digital methods are providing higher precision and efficiency, and are reducing production times to the order of minutes (Hobrough, 1978). Main problems that arise in this field are the processing power required for image rectification and image correlation, as well as storage requirements. Memory requirements amount to 2x64 Mbyte (1 byte = 8 bit) for a stereo image pair of 23 cm each scanned at a resolution of 30 μm or 12 line-pairs/mm.

The detection of corresponding picture elements in overlapping images of a stereo pair is a fundamental problem in any stereo correlation process. Because there is a one-to-one correspondence between points on corresponding epipolar lines, points on the left and the right stereo image are searched on epipolar lines (Fig. 5). Epipolar lines are straight lines which are generated by the intersection of the two focal planes and any epipolar plane defined by the straight line connecting the two exposure points. Furthermore, correlation requirements are drastically reduced by a prediction of match points in accordance with the terrain slope already evaluated. The following operational parameters are typical for the implementation of the stereo image correlation system: Images U, V represent a stereo pair with 1024x1024 pixels each. Mask size and search area for correlation are variable with n = 24, 45, 95 and l = 15, 27, 45, respectively. The number of epipolar li-

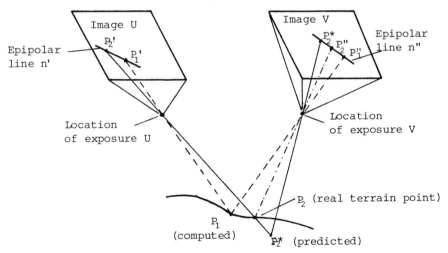

FIGURE 5. *Correlation on epipolar lines*

nes (1024 pixels) lies between 50 and 100 and the distance between search points (foot points) on each line ranges from 10 to 20 points.

The implementation of stereo image correlation on FLIP exploits the inherent parallelism of the correlation task in a suitable manner. Different procedures are performed on parallel hardware processors, additionally the time consuming computation of the correlation values is performed on a parallel processing structure (see also Chap. IV), and data storage and transmission is performed simultaneously with processing (Fig. 6). The PEP memory is divided into 4 sections (PS1-4). The first section holds the actual epipolar lines of image U just being correlated. The second section serves for double buffering during processing. The third section stores the corresponding epipolar lines of image V. A fourth section is provided for storage of the correlation values. Three PEP processors are available for evaluation of the correlation function, for address calculation (prediction), and for controlling the transmission of pixels m_i, s_i to the FIP, which itself computes concurrently the correlation values. Since FIP requires only 20% of the data rate available from PEP, there is sufficient processing power of PEP left for match point prediction and evaluation of the correlation function.

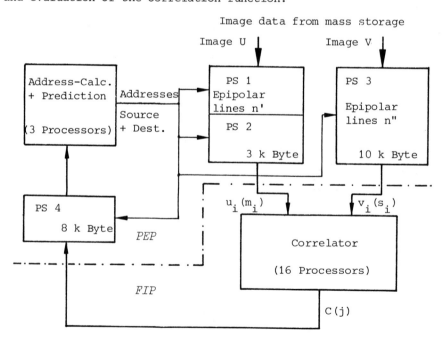

FIGURE 6. Stereo correlation: Principle organization on FLIP

IV. FLIP PROCESSING STRUCTURES FOR CORRELATORS

The computation of the correlation values (1) for one search area can be subdivided into two parts. The first part consists of the solitary computation of the terms denoted in (2) which are only dependent from pixels of the reference and the correlation mask T if specified for object tracking. The second part comprises all terms of (3) which are dependent from both the pixels of the reference, the correlation mask and the search area, and which must be computed repeatedly for all search positions.

$$A = 1/n \cdot \Sigma\, m_i^2 - (1/n \cdot \Sigma\, m_i)^2 \tag{2}$$

$$B = \Sigma\, s_i, \quad D = \Sigma\, m_i s_i, \quad G = \Sigma\, s_i^2, \quad E = 1/n\,(D - 1/nBG)$$

$$F = A(1/nG - (1/nB)^2), \quad C = E/\sqrt{F} \tag{3}$$

A main design goal is the distribution of the overall processing load among the processors in such a way, that the greatest performance, e.g., processing efficiency and processing rate, is achieved with the FLIP. Such a processing structure with 15 identical processors working in parallel is depicted in figure 7. In this configuration processor IP∅ computes A within an interval of $(10.1 \cdot n + 24) \cdot t_c$, where $t_c = 300$ ns represents the maximum cycle time for one complete instruction. At the same time processors IP1-IP15 compute terms B, D and G for 15 subsequent search positions. This calculation takes $(22.17 + 11)t_c$. Processor IP∅ collects all quantities B, D, G for computation of E and F. IP∅ requires $(151.2 + 15) \cdot t_c$ for computation of E, F and for transmission of these terms to the host computer. The final computation of C (square root, division) is performed by the host computer. This is reasonable especially for object correlation, because FIP-processors would use too much time for computing division and square root of digital numbers represented by up to four bytes. With this structure excellent performance will be achieved for correlation applications with a re-

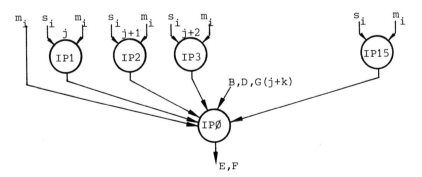

FIGURE 7.   Parallel processing structure for correlator

ference size n greater than 102. On the contrary, stereo correlation uses references of smaller size and doesn't use correlation
templates. Now we achieve our goal, if processors IP1 - IP15 compute also E and F, whereas processor IPØ only collects the results
and transmits them directly to the host computer.

A totally different structure is a cascade for computing the
correlation coefficients as shown in figure 8. In contrast to the
parallel processing structure, computation of C is decomposed
in its basic operations. The computation of B, D, and G is distributed among the processors IP1 - IP14 of the processing cascade.
Additionally calculation of E and F is broken down. This results
in a good system usage and short processing times for small values
of n. The different behaviour of FLIP processing structures for
correlators is represented in Table 1. Obviously, for small values
of n up to 45, the processing cascade is more advantageous. With n
greater than 45 the parallel structure should be preferred.

The input data rates required by FIP-processing structures for
computing the correlation values (correlator) are far below the
maximum data rates available from PEP. Therefore, the total processing times for correlation are determined by the FIP and the
possibilities of the host for data transmission and storage. Since
processing is performed by FLIP on the flow of the data, processing time will be saved by maintaining a continuous data stream
through the system. The processing times of a VAX-FLIP combination
are described in Table 2 for object tracking and stereo image correlation with typical application parameters.

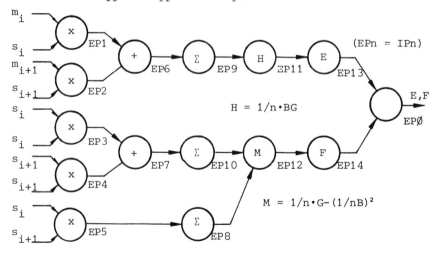

FIGURE 8. Processing cascade for correlator (n < 100)

TABLE 1. *FLIP processing rates for correlator*

Processing structure	Reference size n	IP's used	Input data rate	Output data rate	Computing time correlator
Parallel	>108	16	$1.4/t_c$	$0.7/nt_c$	$(1.5nl+0.7l+151)t_c$
	34	16	$1.1/t_c$	$0.02/t_c$	$(45.5l+334)t_c$
	45	16	$1.2/t_c$	$0.01/t_c$	$(76.6l+548)t_c$
	95	16	$1.3/t_c$	$0.01/t_c$	$(150.5l+1056)t_c$
Cascade	> 57	11	$1.9/t_c$	$0.4nt_c$	$(2.7nl+4.2l+151)t_c$
	34	15	$3.0/t_c$	$0.03/t_c$	$(39.6(l+1))t_c$
	45	15	$3.0/t_c$	$0.01/t_c$	$(74.2(l+1))t_c$
	95	15	$3.0/t_c$	$0.01/t_c$	$(156.7(l+1))t_c$

$t_c = 300$ ns

TABLE 2. *Processing times for image correlation*

Application	Data I/O	VAX (CPU)	FLIP (FIP+PEP)	Total processing time for correlation (VAX+FLIP)
Object tracking	0.7 s	1 s	0.4 s	1.6 s
Stereo correlation	4 s	1 s	4 s	5 s

Object tracking: $n = 800$, $l = 961$, one scene
Stereo correlation: $n = 45$, $l = 30$, 5000 foot points

V. CONCLUSIONS

The processing requirements on a digital image processing system have been discussed for image correlation. A suitable processing system consisting of the multiprocessor system FLIP and a host computer was shown for stereo image correlation and object tracking. Emphasis was laid in the design of efficient processing structures of the multiprocessor system to speed up the computation of the correlation function. It was found, that parallel processing structures as well as processing cascades exhibit nearly same performance. However, the parallel processing structure is more flexible and easier to implement (no structural considerations) for complex tasks. The implementation of correlation on FLIP reduced

correlation times (including I/O) by more than a factor of 10 if
compared to correlation on a one MIPS general purpose computer.

REFERENCES

Bers, K.H., Bohner, M., Gerlach, H. (1980). In "Proceedings Pat-
    tern Recognition", Vol. 2, p. 1320, IEEE Catalog No.
    80 CH 1499-3
Gemmar, P. (1979). In "Angewandte Szenenanalyse", p. 315, Infor-
    matik Fachberichte, Springer Verlag, Berlin
Hobrough, G. (1978). In "Bildmessung und Luftbildwesen", Heft 3,
    p. 79, Zeitschrift für Photogrammetrie und Fernerkundung,
    Karlsruhe
Luetjen, K., Gemmar, P., Ischen, H. (1980). In "Proceedings Pat-
    tern Recognition", Vol. 2, p. 326, IEEE Catalog No.
    80 CH 1499-3

A TYPICAL PROPAGATION ALGORITHM ON THE LINE-PROCESSOR SY.MP.A.T.I. : THE REGION LABELLING

Jean-Luc Basille
Serge Castan
Bernard Delres
Jean-Yves Latil

Laboratoire C.E.R.F.I.A.
Université Paul Sabatier
Toulouse, France

I. INTRODUCTION

SY.MP.A.T.I. is built around one or several memories on which some treatments can be processed in an associative way (Basille 80). The S.I.M.D. structure of each of these memories is particularly well fit to run propagation algorithms where local operators are working line by line. The number of blocks used is at least 16 in order to satisfy the T.V. rate but it could be extended to 512 that is one block for each column. It would then grow up the execution speed of the algorithms with a number of elementary processors rather small in relation to the array processors for instance.

In order to explain the working of a memory we will first give a rather fast but however detailled description of the different parts of an image memory. Then we will take an algorithm, the region labelling, which is characteristic of the propagation algorithms such a line-processor is convenient for.

II. THE DOUBLE STRUCTURE

In fact the line-processor we are talking about is the only memory part of SYMPATI, which is of S.I.M.D. type. The higher level is of M.I.M.D. type and this double level architecture has

MULTICOMPUTERS AND IMAGE PROCESSING
ALGORITHMS AND PROGRAMS

Jean-Luc Basille *et al.*

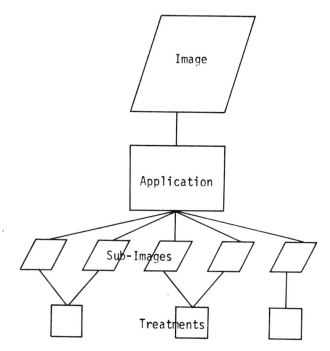

<u>Figure 1-a</u>

The region level treatments in image processing take
advantage of a M.I.M.D. architecture

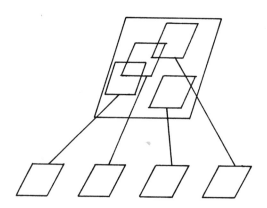

<u>Figure 1-b</u>

A S.I.M.D. structured image memory gives facilities for
memory access and offers some possibilities for some treatments

been suggested to us by the following reasons :

    - First (fig. 1a) one can notice that, in Image Processing,
a lot of applications need not the whole image but only sub-images
to work on. The treatments considered may be either :

    . similar for different sub-images
    . or specific for each sub-image.

    Any way it seems that the best structure to work at such a re-
gion level is of M.I.M.D. type.

    These se procedures must be consequent enough in order not to
make bottle-neck with a lot of exchanges of information.

    - On the second hand (fig. 1b), it is very usefull to have a
memory whose structure gives facilities to get easily information,
that is, makes it possible to get any sub-image from an image in
an easy and fast way.

    It is the first aim for the image memory, SYMPATI is built
around.

    We adopted a S.I.M.D. structure for this memory and it makes
it possible to run procedures at pixel level, especially algorithms proceeding row by row.

III. THE S.I.M.D. MEMORY STRUCTURE

III.1. The column structure

    The memory is structured in an S.I.M.D. way. Each column of
the image is contained in a memory-block and to each block is attached a processing unit. All the processing units are communicating with each other through a shifting loop. (fig. 2)

    So the neighbours of the same column are in the same block
and the close-line neighbours are in the neighbour blocks.

    It seems it would be satisfying, but it is not sure, to have
only one column in each memory-block, that is to have 512 blocks
as we are working on 512 x 512 images. But for economical reasons
and for a simplification purpose we took only 16 blocks. It is the
minimum number in order to be compatible with TV rate considering
the dynamic RAM we chose to realize a memory (1 µs cycle).

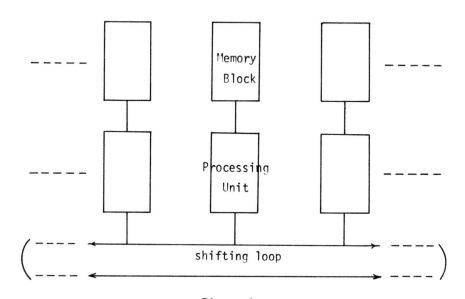

Figure 2

The column structure of one S.I.M.D. image memory of
SY.MP.A.T.I.

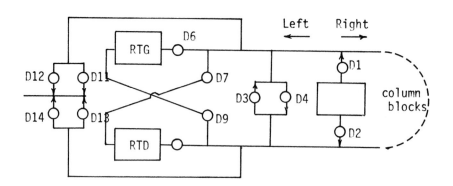

Figure 3

The extremity of the shifting-loop. RTG and RTD are used to
memorize the extremity of the considered segment. R.I.D. is
used to initialize the blocks. D11,12,13,14 make the memory
communicating with the fast bus of the system.

III.2. The shifting loop

So we have to work modulo 16 which is made possible with the shifting loop. (fig. 3) The extremity of the loop is composed of :

- two registers to memorize the left or the right point of the considered 16 points-segment
- one register to initialize some or all of the 16 registers of the shifting loop
- a set of gates which make it possible to put on communication with the fast bus, shifting left or right, input or output and to shift circularly.

III.3. The processing - Units structure

Each processing unit consists of the following components : (fig. 4)

- an arithmetical and logical unit in order to process some plain local expressions.
- an 8 register scratch pad to memorize some temporary results or to avoid memory accesses.
- an indicator set where the ALU indicators may be memorized. This is necessary to process conditional calculus as the 16 ALUs work in an associative way.

This structure shows that memory access and ALU processing cannot be run simultaneously. But memory access and shifting loop or ALU processing and shifting loop may be run so.

III.4. The Command Unit

It manages the 16 blocks (memory and processing unit) and the shifting loop. Procedures running on are microprogrammed with twolevel micro-instructions :

- the "short" micro-instructions that give the sequence of the micro-program
- the "long" micro-instructions that command the different parts of the image-memory.

a) The "short" Micro Instruction (μ.c.)

A program counter gives the address of the μ.c. to be executed. The μ.c. gives the address (bits 0-9) and the length (bits 10-12) of the "long" micro-instructions segment to be processed. Such a segment is composed at the most of one "long" micro-instruction of each type (i.e. 5)

Jean-Luc Basille *et al.*

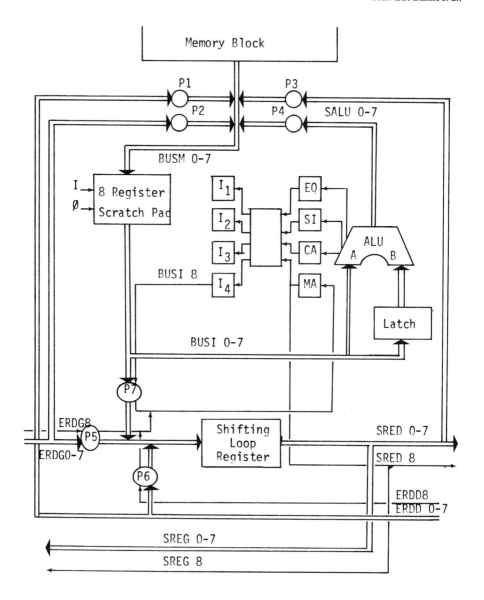

Figure 4

The Processing Units structure

The μ.c. also gives the sequence (bits 13-23) for the "long" micro-instructions of the segment to be processed. It can be noticed that some "long" micro-instructions may be executed simultaneously.

b) *The "Long" Micro Instructions* (μ.ℓ.)

Each μ.l. is coded on 3 bytes. The first 3 bits give the signature of the μ.l.

1) *The access-memory* μ.ℓ.

It specifies the register (bits 4-7) into or from which the data is input or output (bit 3). It also gives the relative address memory (bits 11-20) and, if it is conditional, the indicator considered (bits 8-10).

2) *The shifting* μ.ℓ.

It gives the number of steps (bits 4-8) to the left or to the right (bit 3) and specifies the data path.

3) *The processing unit* μ.ℓ.

It specifies the registers used (bits 15-22) the function applied (bits 3-8) and the indicators to set (bits 12-14) or to consider if it is conditional (bits 9-11).

4) *The command unit* μ.ℓ.

It gives the fonction (bits 3-8) to apply to the operand (address or value : bits 13-20) and the data path (bit 9-12) to lead the result to the considered automaton (number of shifting steps, memory address...)

5) *The* μ.c. *sequencing* μ.ℓ.

It is used to manage the break points in the micro-program and so specifies the absolute or relative branching address (bits 7-21), the indicator (bits 3-6) of the command unit used if it is conditional. One level subprograms can be performed (bits 22-23).

4. THE REGION LABELLING

Many algorithms in image processing are propagation algorithms proceeding one row after the other. The S.I.M.D. structure of the image memory of SY.MP.A.T.I. is rather adapted to perform such algorithms. The region labelling is a complete example for these

a-

```
Sup. line    0    0    0    0    0    2    2    0    0    3    0    1

Inf. line    0   -1   -1    0   -1   -1    0   -1   -1   -1   -1    0
         Top│Down                                      Joining point

Inf. line    0   -1   -1    0    2    2    0    2    3    3    1    0
        Left│Right                           Joining point

Inf. line    0   -1   -1    0    2    2    0    2    2    2    1    0
       Right│Left                                      Joining point

Inf. line    0   -1   -1    0    2    2    0    1    1    1    1    0
         New│labels

Inf. line    0    L    L    0    2    2    0    1    1    1    1    0
```

A Joining point found ⇒ going upwards.

b-

Figure 5

a- The line above has been labelled

b- The different stages to label the considered line

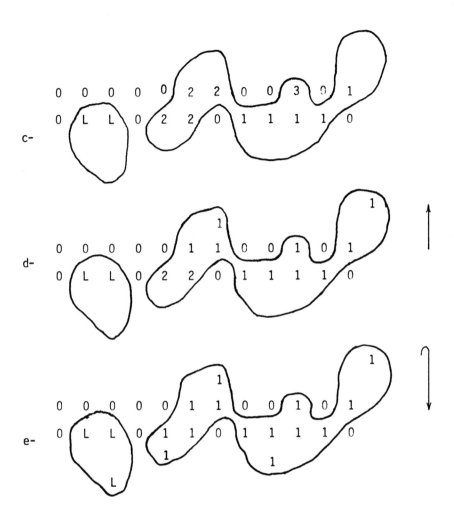

Figure 5

c- The considered line is labelled

d- As a joining point has been found the direction is changed
   upwards

e- When no propagation occurs the direction is changed
   downwards

procedures, and shows how the propagation can run, either downwards or upwards.

This algorithm proceeds on a binary image and gives a different label to each connected component. As we have 8 bits for a pixel we are limited to 255 regions but with use of the memory extension we could label 65 535 different regions.

It proceeds row after row by comparing two adjacent rows according to the following general principles :

(1) A label has to be produced for each pixel and propagated to its connected neighbours :

   a) *From top to bottom* :

       - Projection of the line above onto the considered line. For each pixel, the three neighbours of the line above are taken into account. If two different labels are found, the minimum is kept.
       - Propagation to the right and to the left. The same minimum rule is applied.
       - Production of new labels for the remaining unlabelled pixels.

   b) *From bottom to top* :

       - Projection of the line below onto the considered line as previously.
       - Propagation to the right and to the left.

(2) The direction of the propagation is initialized from top to bottom. It is changed to the opposite in case of the following occurences :

   a) *When going down*, it is changed upwards if joining points are encountered.

   b) *When going up*, it is changed downwards if no propagation occurs from the line below onto the considered line.

The figure 5 shows how the algorithm works from an example.

V. SIMULATION

Before completing the memory, a simulation has been undertaken. The first results confirm the execution times evaluated for simple algorithms. The region labelling algorithm has not yet been programmed and run on the simulation. It should be noted that the

```
**********************************************************************

TEMPS TOTAL:        321514880 NANOSEG

************

TEMPS D'INITIALISATION DES MC:          98582400 NANOSEG  =  30 % DU TEMPS TOTAL

TEMPS D'INITIALISATION DES ML:         126412800 NANOSEG  =  39 % DU TEMPS TOTAL

TEMPS D'EXECUTION DES:

 - ML DE TRAITEMENT:           34298880 NANOSEG  =  10 % DU TEMPS TOTAL

 - ML DE DECALAGE :             4281600 NANOSEG  =   1 % DU TEMPS TOTAL

 - ML DE D'ACC MEM:            42960000 NANOSEG  =  13 % DU TEMPS TOTAL

 - ML DE SEGU DE MC:           17833600 NANOSEG  =   5 % DU TEMPS TOTAL

TEMPS GAGNE PAR LE // ENTRE ML:        2854400 NANOSEG

**********************************************************************
```

Figure 6

Example of the simulation results (Gradient)

execution time will depend on the image considered.

The results for the gradient are shown in figure 6. It must be noticed that the access time for the micro-instructions memory has been fixed on 150 ns. A faster memory of 50 ns could be used. It would carry out an improvement of 147 ms (321 x 0.69 x 2/3) and so give an execution time of 174 ms. Furthermore the use of 32 or 64 blocks instead of 16 would divide the total execution time by 2 or 4 and give a final time of 87 ms or 43.5 ms.

VI. CONCLUSION

The temptation is great to try to solve the main problems in Image Processing with an only S.I.M.D. machine. But region level working requires rather a M.I.M.D. architecture. It accounts for our double structure where the S.I.M.D. memory makes easy and fast the access to windows of any shape and furthermore makes it possible to perform simple treatments such as a gradient or even more elaborate procedures such as a region labelling.

VII. BIBLIOGRAPHIE

Barnes, G.H. et al, (1968). "The ILLIAC IV computer". Trans. IEEE
        on Comp. C-17, 746.
Basille, J.L. and Castan, S.and Latil, J.Y. (1979). "Structure
        logique et physique de l'information dans un multiproces-
        seur adapté au traitement d'images". GRETSI, 7ème colloque
        sur le traitement du signal et ses applications. Nice.
Basille, J.L., Castan, S. and Latil, J.Y. (1980). "Système Multi-
        processeur adapté au Traitement d'Images". Workshop on New
        Computer Architecture and Image Processing, Ischia, Italy.
Bernstein, A.J. (1966). "Analysis of programs for parallel pro-
        cessing". Trans. IEEE on Elect. Comp. EC-15, 757-763.
Comte, D. and Durrieu, G. (1975). "Techniques et exploitations de
        l'assignation unique, 5-7-75. Contrat SESORI 74.167.
Duff, M.J.B., Watson, D.M. and Deutsch, E.S. (1974). "A parallel
        computer for array processing". Proc. IFIP Congress 1974,
        Stockholm, Sweden, pp. 94-97.
Levialdi, S., Maggiolo-Schettini, A., Napoli, M. and Uccella, G.,
        (1979). "Considerations on parallel machines and their
        Languages." Workshop on High Level Languages for Image
        Processing 4-8 June 1979, Windsor U.K.

GEODESIC SEGMENTATION

Christian Lantuéjoul

Centre de Géostatistique
et de Morphologie Mathématique
Ecole des Mines de Paris
Fontainebleau, France

I. INTRODUCTION

Over the past few years, there has been a surge of interest
in the geodesic methods of image analysis. This can be accounted
for by at least three reasons : firstly, geodesy provides a natural conceptual framework to the study of non-planar images. Secondly, it is a convenient and simplifying tool to deal with topological and metric properties of discrete images (Kak and Rosenfeld,
1976; Maisonneuve, 1981). Thirdly, the introduction of a geodesic
distance makes it possible to give a rigourous definition to intu-
itive notions such as the ends of an object, or as the legnth of
an object (Lantuéjoul and Beucher, 1981).

The aim of this paper is to show that geodesy is applicable
to solve segmentation problems. The geodesic segmentation methods
rest on two image transformations, which are the detection of the
regional extrema and the construction of the watersheds of a grey-
tone image. These two image transformations can be carried out
using geodesic dilations, erosions and skeletonizations. Two examples are given as illustration. The first one is the separation
of balls at the beginning of a sintering process. The second one
is the detection of gas bubbles on a microradiograph of a nuclear
pellet.

MULTICOMPUTERS AND IMAGE PROCESSING
ALGORITHMS AND PROGRAMS

II. GEODESIC DISTANCE FUNCTION AND OBJECT RECONSTRUCTION

Let us consider a two-phased image. The first phase, say X, is a population of objects. The second phase is the background. Let x and y be two points within X. If x and y belong to the same object, then there are several paths in X linking them. The shortest one is called "geodesic arc", and its length is denoted as $d_X(x,y)$. If the two points x and y belong to two distinct objects, there is no path in X linking them, and we can write $d_X(x,y) = + \infty$.

FIGURE 1.   *(1) Paths within an object*

(2) Geodesic arc

It can easily be shown that the function d_X satisfies all the properties of a distance function :

i) $d_X(x,y) \geqslant 0$ and $d_X(x,y) = 0$ if and only if x = y.

ii) $d_X(x,y) = d_X(y,x)$.

iii) $d_X(x,z) \leqslant d_X(x,y) + d_X(y,z)$.

In what follows, the function d_X is termed "geodesic distance function". The reader can compare on Figures 2-1 and 2-2 the disks $B_X(x,\lambda)$ and $B(x,\lambda)$ with centre x and radius λ, with the geodesic metric d_X and with the natural Euclidian metric d of the space $\mathbb{R}^2$ in which X is embedded. Obviously $d \leqslant d_X$.

In order to have a better understanding of Figure 2-1, imagine that the objects are a group of islands, and that a fire is set in one of them. Little by little, the fire is spreading, and we observe its propagation at successive moments.

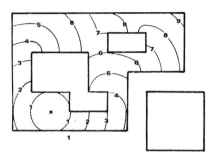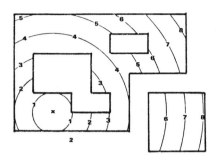

FIGURE 2.   (1) Disks with geodesic metric
(2) Disk with Euclidian metric

From the metric d_X, we can define the geodesic distance between a point x of X , and a part Y of X. $d_X(x,Y)$ is the smallest
geodesic distance between x and any point y of Y :

$$d_X(x,Y) = \inf_{y \in Y} \; d_X(x,y)$$

One of the main interests of the geodesic distance function
is that it is perfectly suited to deal with connectivity problems.
An illustration of this is provided by the following example. Consider two biological images X and Y. X is a population of cells
with parts of broken cells, artefacts, etc. Y is the population
of the nuclei. X and Y are obtained using a double staining technique (see Figure 3). The only cells that must be studied are the
complete cells containing a nucleus. The other ones are just artefacts and must be disregarded. How can we detect cells of X having a nucleus ?

X=Cells   **Y=Nuclei**

FIGURE 3.   Detection of cells from a population of cells
and artefacts, cells being marked by their
nuclei.

Let x be a point of a cell that contains a nucleus. There
exists a path in the cell linking x and a point y of the nucleus.
Hence the geodesic distance between x and the nuclei is finite.
Conversely, if x is a point of a cell without a nucleus, there
exists no path in the cells between x and any point of the nuclei,
so the geodesic distance between x and the nuclei is infinite.
To sum up, the population of cells with a nucleus can be formally
written :

$$\{x \in X \mid d_X(x,Y) < +\infty\}$$

III. MORPHOLOGICAL TRANSFORMATIONS AND GEODESIC DISTANCE FUNCTION

In the phase X with the geodesic metric d_X , it is possible
to generalize the transformations commonly made in mathematical
morphology (Serra, 1981).

i) if $Y \subset X$, points at a geodesic distance less than λ from
Y constitute a set called " λ-dilated set from Y on X" and denoted $D_\lambda(Y;X)$ (see Figure 4).

$$D_\lambda(Y;X) = \{ x \in X \mid d_X(x,Y) \leqslant \lambda\}$$

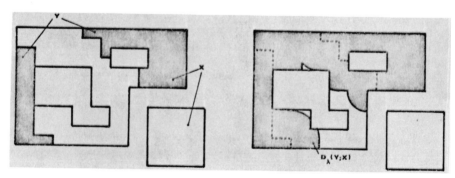

FIGURE 4.   Geodesic dilation. $Y = initial$ set. $D_\lambda(Y;X)$
$= \lambda$-dilated set from Y in X.

In the case where X is two-dimensional Euclidian space $\mathbb{R}^2$,
$d_X = d$ and $D_\lambda(Y;\mathbb{R}^2)$ is simply the classical dilated set from Y by
a disk with radius λ. The trivial but important semi-group equa-
lity

$$D_{\lambda+\mu} (Y;X) = D_\lambda \left[D_\mu (Y;X);X \right]$$

and the fact that for very small λ-values geodesic dilation and Euclidian dilation are approximately the same

$$D_\lambda(Y;X) \sim D_\lambda(Y;\mathbb{R}^2) \cap X \qquad \text{for } \lambda \sim 0$$

make the implementation of geodesic dilations possible.

ii) the points x of X such that $B_X(x,\lambda)$ is totally included within Y, constitute a set called λ-eroded set from Y in X and denoted $E_\lambda(Y;X)$ (see Figure 5).

$$E_\lambda(Y;X) = \{ x \in X \mid B_X(x,\lambda) \subset Y \}$$

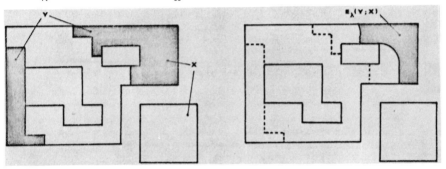

FIGURE 5.   Geodesic erosion. $Y = initial\ set.\ E_\lambda(Y;X)$
          $= \lambda\text{-}eroded\ set\ from\ Y\ in\ X.$

It should be noted that X is invariant under dilations and erosions in X :

$$D_\lambda(X;X) = E_\lambda(X;X) = X$$

On the other hand, dilation and erosion are dual transformations : dilating Y in X is equivalent to eroding in X the complementary set of Y in X (noted X/Y).

$$E_\lambda(X/Y ; Y) = X/D_\lambda(Y;X)$$

iv) Suppose now that Y is made up of n distinct parts :

$$\begin{cases} Y = \bigcup_{p=1}^{n} K_p \\ p \neq q \Rightarrow K_p \cap K_q \neq \emptyset \end{cases}$$

A point x of X is said to belong to the skeleton by zone of influence of Y with respect to X if and only if its geodesic distance to Y is reached for at least two different parts of Y :

$$x \in S(Y;X) \iff \exists \; p,q \leqslant n, \; p \neq q \; : \; d_X(x,Y) = d_X(x,K_p)$$
$$= d_X(x,K_q) < + \infty$$

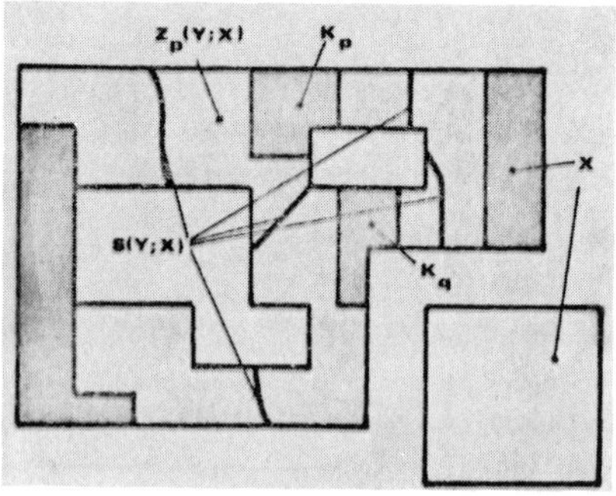

FIGURE 6.   Geodesic skeletonization by zone of influence.
Y = initial set. $Y = \bigcup_p K_p$. $Z_p(X;Y)$ = zone of
influence of K_p in X. S(Y;X) = skeleton of Y in X.

The skeleton S(Y;X) bounds the zones of influence. By definition, the zone of influence of K_p is a set denoted $Z_p(Y;X)$, and made up of points of X at a finite geodesic distance from Y and geodesically closer to K_p than to any other K_q :

$$x \in Z_p(Y;X) \iff \begin{cases} d_X(x,Y) < + \infty \\ \\ \forall \, q \leqslant n, \; p \neq q \Rightarrow d_X(x,K_p) < d_X(x,K_q) \end{cases}$$

It should be noted that the zone of influence and the skeleton do not necessarily partition X. The reason is that X can have points at an infinite distance from Y. It is also easy to show that the skeleton by zone of influence is not necessarily connected. The extensive use of the skeletonization by zone of influence does not stem from its connectivity properties, but from its separation properties (Lantuéjoul, 1980).

The geodesic extension of the concept of medial axis is given in the Appendix.

IV. REGIONAL EXTREMA AND WATERSHEDS

From now on, we are concerned with grey-tone images. For the
sake of clarity, we are going to use a geographic vocabulary.

The surface of a relief f is approximated and represented by
a family of sets X_1, X_2,...,X_n at various levels $h_1 < h_2 <..< h_n$.
X_i is the set of points under h_i :

$$X_i = \{ x \in \mathbb{R}^2 \mid f(x) \leqslant h_i \}$$

Of course we have $X_1 \subset X_2 \subset ... \subset X_n$. For simplicity, we
also write $X_o = \emptyset$ and $X_{n+1} = \mathbb{R}^2$.

A. *Regional Extrema*

The concepts of regional maxima and regional minima correspond to the geographic notions of peaks and chasms. Regional extrema are local extrema, but the converse is not true. A plateau
is a local extremum, but not necessarily a regional one (see
Figure 7).

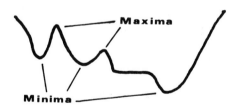

FIGURE 7. Regional maxima and minima.

By definition, the point x is a regional minimum of the relief f if for any other point y such that $f(y) < f(x)$, there
exists no path on the surface of f linking x and y, and that always descend(in the large sense)

Let $x \in X_i$. x is not a regional minimum at height h_i if and
only if there exist $y \in X_{i-1}$, and a path in X_i linking x and y.
As a consequence, the set of the regional minima at height h_i is

$$M_i = \{ x \in \mathbb{R}^2 \mid d_{X_i} (x, X_{i-1}) = + \infty \}$$

Regional maxima can be defined in a similar way.

B.   *Watersheds and Catchment Basins*

The descriptive presentation of watersheds and catchment basins is due to Beucher (1981). Let us imagine that there is a pond at each regional minima of the relief, and that for some cataclysmic reason (tremendous rainfalls, ice breaking up, etc..) the water level is rising. All of the ponds are spreading. To avoid having two different ponds merging into a single one, we build dams between them. They are located so as to prevent water from one pond from running into another.

As soon as the whole relief is under water, all of the ponds are surrounded by the dams, and they can spread no more. The domains occupied by the ponds and their boundaries are respectively called "catchment basins" and "watersheds" (see Figure 8).

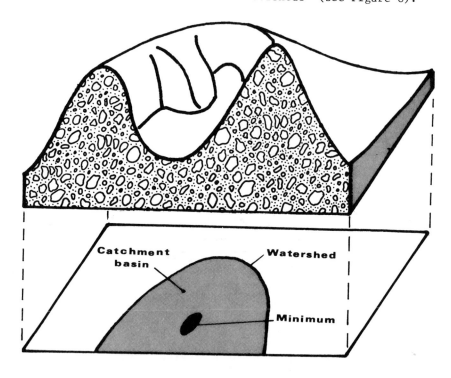

FIGURE 8.   Watersheds and catchment basins.

Let Z be the watersheds, and Z_p the part of them which is detected at a height less than or equal to h_p . Obviously we have

$$\emptyset = Z_0 \subset Z_1 \subset Z_2 \subset \ldots \subset Z_n = Z$$

Suppose that we know Z_p. We want to determine Z_{p+1}.

X_p/Z_p is the set of points whose weight is less than or equal to h_p, and who belong to one and only one catchment basin. Let x be a point at height h_{p+1}. If $d_{X_{p+1}}(x, X_p/Z_p) < +\infty$, and if this distance is the same for two different connected components of X_p/Z_p, x appears to be equidistant from two different catchment basins, consequently x must be considered as a point of Z (see Figure 9). In other words, we have

$$Z_{p+1} = S(X_p/Z_p \; ; \; X_{p+1})$$

The effective construction of watersheds and catchment basins is based upon this iterative formula.

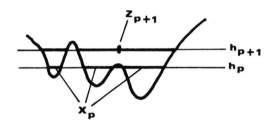

FIGURE 9.   Iterative construction of watersheds.

V.   TWO EXAMPLES OF APPLICATION

A.   Automatic Separation of Disks

In this section, we are concerned with a structure which is a population of touching disks. This may be a polished section of metallic or ceramic balls at the beginning of a sintering process, or latex balls seen by transmission, etc.. In order to perform certain types of measurements, it is essential to separate the disks by drawing contact lines between them (Chermant et Al., 1980).

Let X be the population of disks (see Figure 10-1). The surface of the relief generated by the function f such that

$$f(x) = -d(x, X^c)$$

is a flat surface, perforated by a collection of cones (see Figure 10-2).

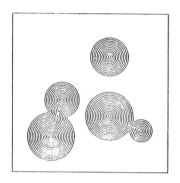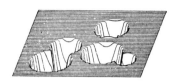

FIGURE 10.   (1) Population of disks

(2) Surface generated from the disks.

We define the contact lines between disks as the watersheds of the relief f. Such a definition is quite easy to implement, since the successive thresholds of f are obtained by simple Euclidian erosions :

$$\{ x \mid f(x) \leqslant -\lambda \} = E_\lambda(X; \mathbb{R}^2)$$

It is also noteworthy that the centers of the disks are the regional minima of f.

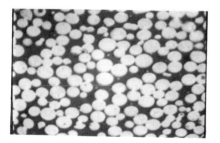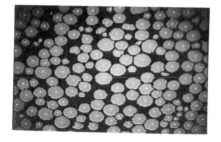

FIGURE 11.   A population of disks before and after
their identification.

B. *Detection of Blobs in a Radiograph*

Consider on Figure 12 a microradiograph of a nuclear pellet. The dark blobs are precipitated gas bubbles. Their detection is not obvious for at least two reasons : firstly, all of the bubbles have not the same grey level, and therefore a threshold technique is completely ineffective. Secondly, all of the bubbles are surrounded by diffraction haloes which give them fuzzy contours.

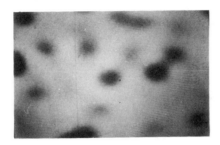

FIGURE 12. The initial image.

In the present paper, it is not our aim to detect the bubbles by taking into account the physical process which has led to the formation of the image. We are just content with proposing a detection method to obtain results which are both satisfactory for the user, and robust.

In order to detect the bubbles, we are going to build their contours. Since a contour is characterized by a sharp contrast, we are first concerned, not by the original image f but by the image g which is the magnitude of the gradient of f. We define the contours of the watersheds as g (see Figure 13). Such a definition is non-parametric, and provides (locally) connected contours.

Unfortunately, such an image is oversegmented. Why ? The catchment basins are associated with the regional minima of the gradient g. But to say that $g(x) = 0$ means that x either belongs to a local extremum or x is a saddle point. On the other hand, x can belong to a regional minimum of g even if $g(x) \neq 0$.

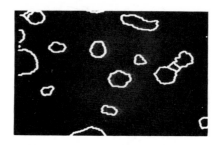

FIGURE 13.   *The contours of the bubbles are defined as the*
watersheds of the magnitude of the gradient of
the initial image.

How can we eliminate the unnecessary contours ? In the initial image, a bubble is perceived as a dark blob. So, the regional
minima of f can be used as markers for the bubbles (see Figure
14).

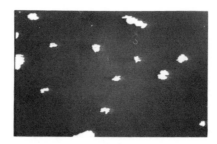

FIGURE 14.   *The regional minima of the initial image are*
used as markers for the bubbles.

Finally, a bubble is defined as a catchment basin of the gradient image, which contains a regional minimum of the initial
image. Figure 15 shows the image of the bubbles, using such a
definition.

The concept of watersheds has many other applications that
we have not presented here. For reference, we suggest Beucher
and Hersant (1979) in metallurgy, and Meyer and Van Driel (1980)
in biology.

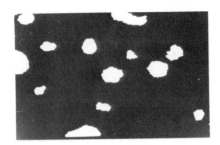

FIGURE 15.   The bubbles are the catchment basins of the
gradient image containing a regional minimum
of the initial image.

REFERENCES

Beucher, S. (1981). Ligne de Partage des Eaux. Comment l'expli-
citer en termes de transformation fonctionnelle. Internal
Report, CGMM, Fontainebleau.
Beucher, S., and Hersant, T. (1979). Analyse quantitative de sur-
faces non planes. Application à la description de faciès de
rupture fragile par clivage. Final report, D.G.R.S.T. n°
76.7.1209.1210.
Blum, H. (1973) J. Theor. Biol. 38. 205.
Chermant, J.L., Coster, M., Jernot, J.P. and Dupain, J.L. (1981).
J. of Microscopy, 121, Pt 1, 89.
Kak, A.C. and Rosenfeld, A. (1981). "Digital Picture Processing",
Academic Press, London.
Lantuéjoul, C. (1980). In "Issues in Digital Image Processing"
(R.M. Haralick and J.C. Simon, ed.) NATO A.S.I. Series, p.
107. Sijthoff and Noordhoff, Netherlands.
Lantuéjoul, C. and Beucher, S. (1981). J. of Microscopy, 121, Pt
1, 39.
Maisonneuve, F. (1981). Convexité et pseudo-convexité en trame
hexagonale. Internal report, CGMM, Fontainebleau.
Meyer, F. and Van Driel, A. (1980). Microscopica Acta, Suppl. 4.
Serra, J. (1981). "Image Analysis and Mathematical Morphology".
Academic Press, London.

APPENDIX : GEODESIC MEDIAL AXIS

The concept of medial axis introduced by Blum (1973) in the Euclidian space can be extended to a geodesic framework. Let X and Y be two subsets of $\mathbb{R}^2$ with $Y \subset X$, and let $x \in Y$. The point x is said to belong to the "medial axis of Y in X" if there exist two disjoint geodesic balls $B_X(y, \mu)$ and $B_X(z, \mu)$ such that

$$d_X(x, X/Y) = d_X\left[x, B_X(y, \mu)/Y\right] = d_X\left[x, B_X(z, \mu)/Y\right]$$

In other words, the medial axis of Y in X is made of points whose geodesic distance to the outside of Y (i.e. X/Y) is reached for at least two distinct parts of the outside of Y in X (see Figure 16).

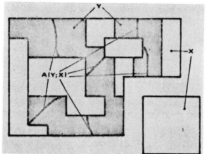

FIGURE 16.   Y = initial set. A(Y;X) = medial axis of Y in X.

The geodesic medial axis has certain properties of connectivity which make it quite useful in image analysis. From the inclusion $Y \subset X$, we have three inclusions : $X/Y \subset X$, $X^c \subset Y^c$ and $Y^c/X^c = X/Y \subset Y^c$. The comparison of the 4 medial axes $A(Y;X)$, $A(X/Y;X)$, $A(X^c;Y^c)$ and $A(X/Y;Y^c)$ is quite informative (see Figure 17).

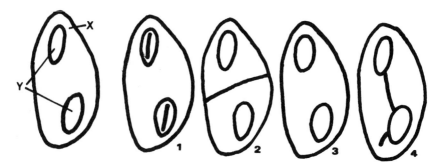

FIGURE 17 : Comparison of the 4 medial axes.

A FAST ALGORITHM FOR PROCESSING
SYNTHETIC APERTURE RADAR SIGNALS
WITHOUT DATA TRANSPOSITION

Morio Onoe
Ichiro Kubota

Institute of Industrial Science
University of Tokyo
Roppongi, Tokyo, Japan

I. INTRODUCTION

A synthetic aperture radar (SAR) can achieve a very high resolution without excessive peak power and antenna size. The high
resolution is achieved by a chirp pulse compression along the
range and an azimuth compression along the track. The processing
of both compressions heavily depend on Fourier transform, which
has been provided by lenses in an optical processor (Cutrona,
1970).

The recent introduction of digital processing by computer
have yielded far better results in resolution and accuracy, but
in much slower processing speed (Kirk, 1975; Wu, 1976;
van de Lindt, 1977; Wu, 1980).

A notable example is the SAR of the SEASAT sattelite. The
volume of data is very large: 13,680 along the range swath of 100
(the acutal time-bandwidth product is the order of 4,000) km and
20,000 along every 100 km of azimuthal direction. At the present
processing speed, it would take more than a decade to process all
the data taken by SEASAT during her short life of only three
months. So there is a clear need for efficient algorithms and
fast parallel processors (ESA, 1980).

This paper presents an efficient algorithm for processing SAR
data, which is based on non-transposing scheme used in the previously reported algorithm for two-dimensional transform (Onoe,
1975a, 1975b).

The paper will first review the method for computing large-
scale two-dimensional Fourier transform without transposing data
matrix and then discuss its application to SAR data processing.

MULTICOMPUTERS AND IMAGE PROCESSING
ALGORITHMS AND PROGRAMS

125

II.   TWO-DIMENSIONAL TRANSFORM WITHOUT DATA TRANSPOSITION

It is well-known that two-dimensional Fourier transform can be decomposed into two one-dimensional transforms: namely, rowwise transforms of data matrix followed by columnwise transforms. When the whole data is stored within a main random access memory, there is a trivial difference between rowwise and columnwise transforms. A large data, however, has to be stored, say row by row, in auxiliary sequential access memories, such as disks.

The rowwise transforms are easily done. The columnwise transforms, however, are incompatible with sequential row access memories. A conventional approach to solve this difficulty is first transposing the intermediate data matrix and then performing another rowwise transforms. Drawbacks of this approach are as follows:

(1)   Transposition requires extra processing and time.
(2)   The result is given in a transposed form.

The transposition can be eliminated, if characteristics of FFT are carefully examined. An essence of FFT is in the decomposition of original transform of size $N = R^m$ into m steps of N/R transforms of size R. At each step, intermediate results are corrected by twiddle factors and regrouped into N/R sets of R points, which become inputs for transforms in the next step.

Based on these facts, the columnwise transform of two-dimensional data, stored row by row, can be performed in m steps. At each step, a set of R rows is read into the main memory and a corresponding step of columnwise transform is performed in parallel over the entire row. The resultant R rows are written back in the auxiliary memories. Then the next set of R rows is read into the main storage. During one step this process is repeated N/R times.

The following simple example shows how this algorithm works. Fig. 1 is a signal flow graph for a N-16 one-dimensional Fourier transform, which is decomposed into m = 2 steps of N/R = 4 transforms of size R = 4. According to this pattern, the two-dimensional transform is performed as follows:

(1)   The first step (m=1)

    (a)   the rows, 0,4,8 and 12, are read into the main memory and a 16-point one-dimensional transform is performed for each row;

    (b)   a 4-point transform is performed on the first element of each row; next, the transform is repeated on the second element, then on the third, and so on; the result is returned to the auxiliary memories

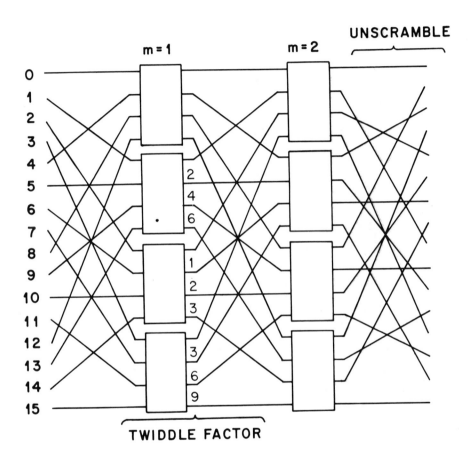

FIGURE 1

(c)   the two preceding procedures are applied on the
      rows 1,5,9 and 13, then on the rows 2,6,10, and
      14, and finally on the rows 3,7,11 and 15; the
      multiplication of twiddle factors is done either
      separately or jointly with transforms.

(2)   The second step (m=2):

      the same procedures as 1-b) and 1-c) are applied on the
      row 0-3, 4-7, 8-11, and finally on 12-15.

(3)   The unscrambling step:

      This step can be skipped in the SAR processing.

   Various modifications of this basic algorithm are possible.
Instead of performing the rowwise transforms in the first step,
it can be broken down into m steps.   This may be advantageous in
designing a dedicated hardware, because all the operation will
be almost identical for both rows and columsn.
   It should be noted that this algorithm is also effective in
a virtual memory environment, because the paging of virtual
memory is controlled according to the nearness of data addresses
involved (Onoe and Kaneko, 1980).

III.   CONVENTIONAL SAR DATA PROCESSING

   The processing of SAR data consists of two steps: the pulse
compression along the range followed by the azimuth compression
along the track.   Both steps can be performed by either the convolution in temporal and spacial domains or the filtering in the
frequency domains through Fourier transforms.   When the data
volume is large, as in the present case, the processing in frequency domains is more computationally efficient.   Fig. 2 (a) is
the flow chart of the conventional method based on frequency
domain filterings (Wu, 1976).
   The SAR data is stored in rowwise along the range and columnwise along the track.   The pulse compression is performed by a
straightforward filtering: the multiplication of range transfer
function sandwiched by rowwise Fourier transform (FFT) and its
inverse (IFFT).   The resultant data matrix is transposed in order
to make the azimuth compression, which is originally to be performed columnwise, compatible with the rowwise access of auxiliary memories.   This transposition is often called corner
turning in SAR data processing.   Then the azimuth compression is
performed also by the multiplication of azimuth transfer function
sandwiched by rowwise (originally columnwise) FFT and IFFT.

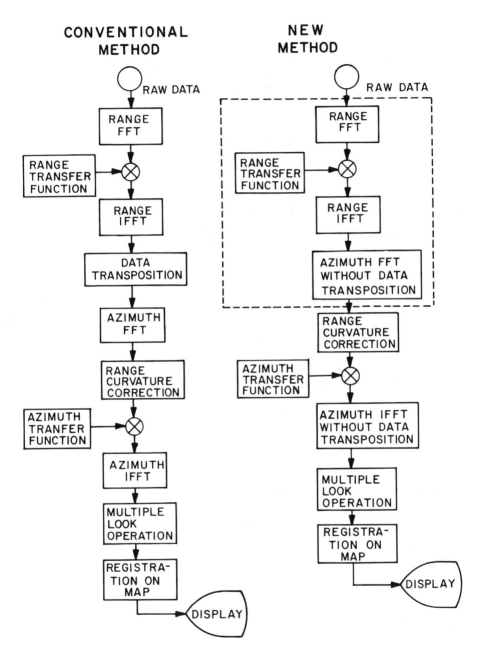

FIGURE 2 (a)

There are two peculiarities associated with the azimuth compression: namely the range curvature correction and the multiple look operation. The former is the correction of the locus of echoes from a point target from a parabola to a straight line. In the azimuth frequency domain, this correction is performed by shifts of data along the range within each frequency band. A number of rows corresponding to the amount of dispersion along the range (now columnwise) have to be handled at the same time (WU, 1976).

The multiple look operation is a process to reduce speckle noise due to coherent illumination. The azimuth aperture is divided into sub-apertures called looks, which are compressed and detected independently, and incoherently summed up. Although the flow chart shows only the last summing up step, parts of the processes migrated in the filtering before IFFT. In the azimuth frequency domain, the operation corresponds to bandpass filterings, which are essentially selections of particular frequency bands. Hence it would be more easily handled, if data were arranged in the original format: rowwise along the range and columnwise along the track (azimuth frequency).

IV. SAR DATA PROCESSING WITHOUT TRANSPOSITION

The algorithm shown in Section II allows columnwise transforms only with rowwise access. The application of this algorithm to FFT and IFFT in the azimuth compression eliminates the need of data transposition.

Advantages of this approach are as follows:

(1) Time for transposition is saved.

(2) No separate program or processing for transposition is required. This makes a design of dedicated processor easier.

(3) Data remains always in the original format: rowwise along the range and columnwise along the track. This makes the range correction and the multiple look operation simpler.

(4) The above mentioned format is compatible with the continuous flow of SAR data along the track. This makes a flexible selection of area to be processed and a making of mosaic of several areas easier.

	CONVENTIONAL METHOD	NEW METHOD
TIME FOR DISK TRANSFER	$T_1 \times 2^{N_{az}} \times N_{az} \times$ $(1+C_{az} \times 2^{N_{az}}) \times (1+C_{ra} \times 2^{N_{ra}+1})$	$T_1 \times 2^{N_{az}} \times N_{az} \times$ $(1+C_{az} \times 2^{N_{az}}) \times (1+C_{ra} \times 2^{N_{ra}+1})$
TIME FOR FFT	$T_2 \times 2^{N_{az}+N_{ra}+1} \times (N_{az}+N_{ra})$	$T_2 \times 2^{N_{az}+N_{ra}+1} \times (N_{az}+N_{ra})$
TIME FOR TRANSPOSITION	$T_3 \times 2^{N_{az}+N_{ra}-1} \times N_{az}$	0

T_1 Average Transfer One Record Time $2^{N_{az}}$ Pixels in Azimuth $(N_{az} \leq N_{ra})$

T_2 Average Multiplication Time $2^{N_{ra}}$ Pixels in Range

T_3 Average Load and Store Time C_{az} Coefficient Azimuth Length

                                                  C_{ra} Coefficient Range Length

Table 1

$T_1 = 57\,ms$

$T_2 = 0.23\,ms$ { Case 1 : Without Array Processor (ANALOGIC, AP400)

       $0.024\,ms$ Case 2 : With Array Processor

$T = 0.72\,ms$

$C_{az} = 2.4 \times 10^{-4}$

$C_{ra} = 4.3 \times 10^{-4}$

Example :

     IF: $N_{az} = 12$, $N_{ra} = 12$

       THEN:

               $(T_c - T_n)/T_c = 0.26$ { Case 1

               $(T_c - T_n)/T_c = 0.62$ Case 2

Table 2

A comparison of processing time by the conventional method and the new method is shown in Table 1. The conventional method uses a fast algorithm for transposition (Eklundh, 1972). The time is estimated by a product of the number of typical operations and the average time required. There is little difference in the time for data transfer and FFT in both methods. Hence the time saving is mostly due to transposition.

A numerical example based on empirical data obtained in a mini-computer system (Hewlett-Packard 2112) is shown in Table 2. An attachment of an array processor (Analogic AP400) improves one order the time for FFT. In the case of SAR data of 4096 × 4096, the time saving is 26 percent without the array processor and 62 percent with the processor. This is understandable because the array processor improves the speed for such complicated operations as FFT, but not much for simple operations as transposition, which is essentially a load and store type operation. Hence it is expected that the more powerful the processor, the more time saving.

V.   CONCLUSION

It is shown that non-transposing algorithm previously used in two-dimensional transforms is applicable to the processing of SAR data. The elimination of data transposition saves both time and memory for a separate processing. The new algorithm maintains the original data format, rowwise along the range and columnwise along the track, throughout its execution. This makes the range curvature correction, the multiple look operation, the selection of area, and the continuous processing of data flow easier. The algorithm has been implemented in several systems. Its time saving is proved by an experiment in a minicomputer system. The algorithm is suitable for parallel processing. As the more powerful dedicated processor will appear, the more time saving will be obtained.

ACKNOWLEDGMENTS

The authors wish to thank Dr. S. Hanaki of NEC Co. for his courtesy in providing SAR data.

REFERENCES

Cutrona, L. J. (1970). "Synthetic Aperture Radar" in "Radar Handbook" (M.I. Skolnik, ed.), Chapter 23. McGraw Hill, New York.

Eklundh, J. O. (1972). "A fast computer method for matrix transposing," IEEE Trans. Comput., C-21, p. 801-803.

ESA (1980). "Instrumentation for Preprocessing of SAR Data to Image Form," SP-1031.

Kirk, J. C. (1975). "A discussion of digital processing for SAR," IEEE Trans. Aerospace, AES-11, p. 326-327.

Onoe, M. (1975a). "A method for computing large-scale two-dimensional transform without transposing data matrix," Proc. IEEE, 63, p. 196-197.

Onoe, M. (1975b). "A fast algorithm for two-dimensional transform and its application to digital image processing," Proc. 2nd. U.S.-Japan Comput. Conf., p. 97-101.

Onoe, M. and Kaneko, M. (1980). "Effectiveness of non-transposing algorithm for two-dimensional transform in virtual memory environment," Nat. Conv. Inst. Elec. Comm. Eng. Japan, 5-112 (in Japanese).

van de Lindt, W. J. (1977). "Digital technique for generating SAR images," IBM Jour. Res. Dev., 1, 5, p. 415-432.

Wu, C. (1976). "A digital system to produce imagery from SAR data," Proc. AIAA Systems Design Driven by Sensors Conf., 76-968, Pasadena, California.

Wu, C. (1980). "A software-based system to produce SEASAT SAR imagery," ESA, SP-1031, p. 7-13.

CELLULAR LOGIC ALGORITHMS FOR GRAYLEVEL IMAGE PROCESSING

Kendall Preston, Jr.

Department of Electrical Engineering
Carnegie-Mellon University

Department of Radiation Health
University of Pittsburgh

Pittsburgh, Pennsylvania

ABSTRACT

Modern cellular automata, such as CLIP4 (Cellular Logic Image Processor model 4) at University College London, are capable of executing one picture point operation per nanosecond – (Fountain, 1981). This is equivalent to 50 billion instructions per second in a general purpose computer. It is important that the image processing community begin to knowledgeably exploit the power of these machines.

Often these machines are unjustifiably criticized as having limited capabilities since they process images using logical neighborhood operators limited to small neighborhoods. This chapter demonstrates that such machines can readily be programmed to perform graylevel image processing tasks ordinarily requiring large neighborhoods through the technique of multilevel thresholding and multilevel iterative processing. This makes possible digital filtering of large (96x96) graylevel images in from 0.1 to 1.0 ms using the cellular logic equivalent of maxmin propagation filtering and in less than 10 ms for more general filtering operations.

The cellular logic filter has certain unique properties, namely, it is a constant-phase filter which passes absolutely no signal beyond a cutoff frequency which decreases inversely with increasing iterations. These unusual properties are illustrated in several examples.

MULTICOMPUTERS AND IMAGE PROCESSING
ALGORITHMS AND PROGRAMS

1. INTRODUCTION

Ordinarily images are digitally filtered either by convolution in the spatial domain or by multiplication in the spatial frequency domain. The latter method employs the Fourier transform. Convolution by a square (NxN) kernel leads to a sinc(x)-sinc(y) transfer function which produces a 180 degree phase shift beyond its cutoff frequency returning to 0 phase at a frequency twice its cutoff frequency and oscillating thereafter between 180 and 0 . There is a finite response at all frequencies except for those frequencies equal exactly to the cutoff frequency and its multiples. Another common filter is produced using a pyramidal kernel which, when made 2Nx2N, provides a $\text{sinc}^2(x)\text{sinc}^2(y)$ transfer function with exactly the same cutoff frequency as the NxN square kernel. This filter generates no phase shifts and, although a finite response occurs at all frequencies except for the cutoff and its multiples, the response is considerably lower.

All of these filters operate upon arrays of integers which represent the digitized and spatially sampled image. In order to process a similar array by means of cellular logic, thresholding is used to produce the required binary images. Usually thresholds are selected at evenly spaced intervals across the range of the probability distribution function of the integer array input. In the most general case a different algorithm is applied to the binary image produced at each treshold. In the situation to be described here, however, the same algorithm is applied to each of the binary images and then the binary images are re-combined in order to create an output array of integers.

2. METHODS

In this chapter binary functions are called "logical functions" because their element values are either true or false (0 or 1). If the image $s(x,y)$ is the function to be thresholded, then $s_L(x,y)$ is a logical function derived from $s(x,y)$ produced according to the following equation,

$$s_L^{\cdot}(x,y) = \begin{cases} 1 & \text{if } s(x,y) \geq T \\ 0 & \text{otherwise} \end{cases} \tag{1}$$

where T is the threshold utilized. Frequently a family of logical functions are derived from $s(x,y)$ by using multiple thresholds. As mentioned above, most of these methods depend upon an analysis of the probability density function (PDF) of the values of $s(x,y)$. In many cases the range of the PDF is divided into a multiplicity of equal increments and thresholds are utilized at each of these increments. In other more specified cases, e.g., when the PDF is multimodal, thresholds may be selected at the modes and at the minima which exists between the modes.

Thresholding is considered a point of operation which, when the number of thresholds is small, may be calculated by table lookup. For example, if $s(x,y)$ is initially digitized at 8 bits per pixel, then a table of 256 addresses is constructed. In this table all addresses corresponding to the values of $s(x,y)$ above threshold have the contents 1; 0, when below threshold. Using this method of thresholding it is possible to compute the value of a single point in the logical image in the time required to address the table and assemble the resulting 1-bit results over the entire array. Typically such an operation may be carried out in less than 1 second for a 512x512 array. The total number of logical images derived from each $s(x,y)$ vary considerably depending on the task to be performed. In most cases cellular logic filtering is carried out using either 16 or 32 logical images which is entirely satisfactory to represent the resultant output function.

2.1 Convolution

The simplest cellular logic filter of the function $s_L(x,y)$ may be expressed as a convolution as given below

$$r(x_i,y_j) = \sum\sum_{\ell m} s_L(x_i,y_j)k_L(x_i-\ell\delta x, y_j-m\delta y) \qquad (2)$$

where δx and δy are the sampling intervals along the x,y axes. The function $k_L(x,y)$ is the kernel function or the "neighborhood" function. There are numerous kernel functions which cannot be discussed in this chapter due to lack of space. As is well known from computational geometry, only the triangle, the hexagon, and the square can be used to uniformly tile the plane. The kernels which result from these tessellations are the most common. They are the 3x3 square and the 6-element hexagon. These two kernels are symmetric and their properties are reasonably well understood. (The 3-element triangle is not considered because its use as a kernel is limited.) Let us assume that the kernel is the 9-element square. In order to construct the logical result function from the logical transform (equation 2), the intermediate integer result $r(x_j,y_j)$ is tresholded using the parameter Ξ as follows

$$r_L(x_i, y_j) = \begin{cases} 1 & \text{when } r(x_i, y_j) \geq \Xi \\ 0 & \text{otherwise} \end{cases} \tag{3}$$

If the value of Ξ is equal to 9, then all elements of $r_L(x,y)$ are identically 0 in that no value of this function can be greater than 9. Thus all elements of this function are reduced to the logical value of 0. When Ξ lies in the range of 5 through 8, some elements of $r(x,y)$ are reduced to 0 but others retain the value of 1. Typically, interior points retain this value and edge points which are at a concavity. This is known as "reduction." On the other hand, if the value Ξ lies in the range 0 through 3, then the number of logical 1's in $r_L(x,y)$ is larger than the number of logical 1's in $s_L(x,y)$. This is called "augmentation."

Augmentation and reduction cause the inward or outward propagation of boundaries which surround contiguous clusters of 1's in the logical image. The rate of boundary propagation depends upon the value of Ξ and the angle which the boundary makes with the x,y axes. Boundary displacement curves are shown in the Figure 1.

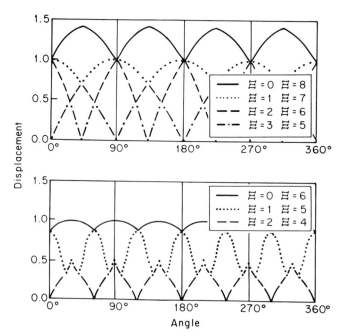

Figure 1 - Boundary displacement curves for the two-dimensional cellular logic transform in both the square (above) and hexagonal (below) tessellations. (See Preston, submitted.)

As can be seen, for the 3x3 square kernel the boundary displacement for $\Xi = 8$ (reduction) or $\Xi = 0$ (augmentation) is larger
by factor of $\sqrt{2}$ at $\pm 45°$ than at $0°$, $90°$ and $180°$. For
augmentation and reduction with the values of $\Xi = 1$ and $\Xi = 7$, respectively, then we find that the boundary displacement is greatest at $0°$, $90°$, and $180°$ while it is a minimum at $\pm 45°$. This
fact aides in obtaining uniform boundary displacement by iterating the logical transform with values of $\Xi = 7$ and $\Xi = 8$ for
reduction and $\Xi = 0$ and $\Xi = 1$ for augmentation.

2.2 Case-In-Point

For illustrative purposes consider the case where $s(x_i, y_j)$
is the zoneplate (Figure 2).

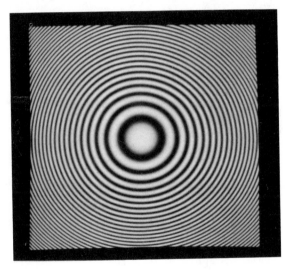

Figure 2 - Graylevel test pattern (512x512) generated using a
sinusoid whose spatial frequency linearly increases with radial
distance from the origin.

This signal is angularly symmetric and consists of a function whose spatial frequency increases linearly with radius. In
the particular case shown, the frequency is 0 at the origin and
increases linearly to 32 cycles across the aperture at the
horizontal and vertical margins and 45.25 cycles across the aperture in the corners.

Two cases are presented here. In both cases the number of
thresholds utilized is 16. In the first case the cellular logic
transform is performed at each threshold with $\Xi = 8$ (reduction)
for three iterations. This annihilates regions of binary 1's
where boundaries are close together, i.e., those regions where
high frequencies are present. Due to the fact that using $\Xi = 8$

produces larger displacements at $\pm 45^{\circ}$, the frequency cutoff in these directions is lower than at 0°, 90°, and 180°. After 3 iterations with $\Xi = 8$, 3 additional iterations are conducted with $\Xi = 0$ (augmentation). This restores the boundaries of those groups of binary 1's which were not annihilated. At this point the 16 logical images are recombined to form a graylevel image by simply counting the number of binary 1's at each x,y in order to generate an integer in the range of 0 to 16. The result is shown in Figure 3. The anisotropic performance is clearly evident with the cutoff frequency at 0°, 90° and 180° extending $\sqrt{2}$ beyond that at $\pm 45^{\circ}$.

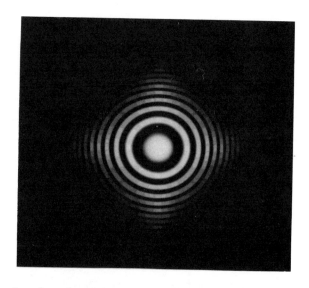

Figure 3 – Result of applying the cellular logic transform using the sequence $\Xi = 8,8,8,0,0,0$ to the test pattern given in Figure 2.

In order to generate isotropic performance one may combine logical transforms using the value of $\Xi = 8$ and $\Xi = 7$ for reduction followed by a similar combination of $\Xi = 0$ and $\Xi = 1$ for augmentation. In the example whose result is shown in Figure 4, the sequence for Ξ was 7,8,7,0,1,0. This provides almost complete isotropy with the cutoff frequency being angularly symmetric. In order to more clearly understand the performance of these filters, it is instructional to view the zoneplate as a a 3-dimensional solid as shown in Figure 5. This figure displays the function s(x,y) in such a way that high values are represented by elevated points in the surface and low values are represented by depressions. The sinusoidal structure of the zoneplate is clearly evident in Figure 5 with its rate of fluctuation increasing linearly with radial distance.

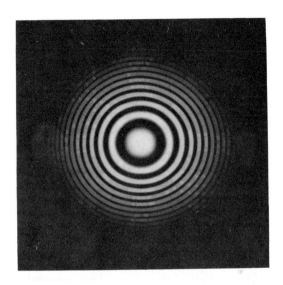

Figure 4 - Results of applying the cellular logic transform using the sequence Ξ = 7,8,7,0,1,0 to the test pattern shown in Figure 2.

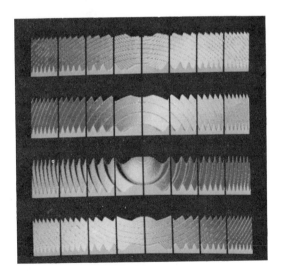

Figure 5 - Three-dimensional surface representing the test pattern shown in Figure 2.

Figure 6 represents the anisotropic result (Figure 2) in a similar manner and it's interesting to note that the attenuation produced by the logical transform filter is carried out by a form of "clipping." In the particular case illustrated clipping is frequency dependent and occurs at the crest of each sine wave. An analysis of this phenomenon is given with reference to Figure 7.

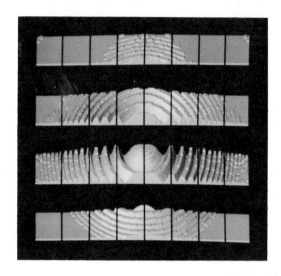

Figure 6 – Three-dimensional surface representing the result of applying the cellular logic transform $\Xi = 8,8,8,0,0,0$ to the function represented three-dimensionally in Figure 5.

Figure 7a shows a cross-sectional view of a small portion of the zoneplate using only 6 thresholds. During the reduction phase of the filtering the six line segments which represent regions of binary 1's are each shortened by a fixed amount which depends on the number of iterations k of this portion of the transform (Figure 7b). In the case shown, the line segment which represents the binary 1's at the sixth threshold is annihilated. In the augmentation part of the cycle, since the same number of iterations are taken as were used in reduction, all those line segments not annihilated are restored to their original length. This has the result of clipping the sinewave at its crest (Figure 7c). In the case shown the reduction in amplitude is approximately 18%.

If P is the period of the sinewave and $N_p = P/\delta x$ is the number of samples over one period, the sinewave will be cut off by this transformation when $k = N_p/2$ yielding a cutoff $k = N_p/2$ x. In general it can be written that the peak-to-peak amplitude (A_{p-p}) is given by

$$A_{p-p} = \tfrac{1}{2}(1 + \cos\pi(k/k_{co})) \tag{4}$$

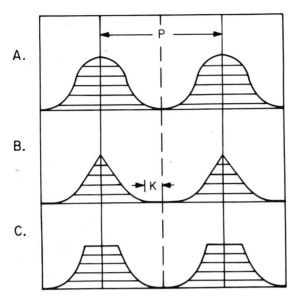

Figure 7 - The two-dimensional cellular logic transform which uses the sequence $\Xi = 8,8,\ldots,0,0\ldots$ causes the peaks of a sinusoidal signal to be clipped.

where $k_{co} = P/2\delta x = N_p/2\delta x$. Thus the transfer function of this cellular logic digital filter is sinusoidal. It is plotted in Figure 8 as a function of 3 values of the number of iterations.

The primary features of this filter is that it passes no signal whatsoever beyond its frequency cutoff, the output is constant in phase, and the cutoff frequency may be adjusted by simply adjusting the number of iterations of the transforms. None of these qualities are obtainable using ordinarily digital filtering. For a reference to prior work in this general area see Goetcherian (1980).

3.   RESULTS

In order to judge the performance of a filter in image processing, it is instructional to apply it to actual images. In this chapter we take two frequently used test cases supplied by the Image Processing Institute of the University of Southern California. One shows the well-known "USC girl" as the $s(x_i,y_i)$ both before and after filtering using both the anisotropic and the isotropic filters described above. In addition a third filter was applied which has the characteristics that instead of driving

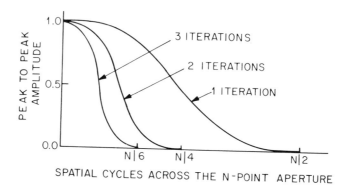

Figure 8. When the peak-to-peak amplitude of the clipped sinusoid (Figure 7) is plotted as a function of the number of iterations in the sequence Ξ = 8,8,...,0,0,... the resulting transfer
function is as shown above.

the signal to 0 beyond frequency cutoff it drives the signal to
its average value. This is accomplished by performing the sequence Ξ = 7,8,7,0,1,0 when the signal is above its average value
and, when the signal is below its average value using, instead,
the sequence Ξ = 0,1,0,7,8,7. The result is that signals above
the average value are clipped at their maxima while while signals
which are below the average value are clipped at their minima.
See Figure 9.

      Figure 10 has a higher spatial frequency content than does
the scene shown in Figure 9. Thus the output of the cellular
logic filter is not as interpretable. Figure 10 shows results of
applying the same cellular logic filters whose outputs are shown
in Figure 9 to the aerial scene show in the upper left frame.
When the anisotropic filter is utilized, there is a distinct
emphasis of horizontal and vertical edges. A similar "blockiness"
occurs in the result for the isotropic filter but there appears
to be a better retention of relative brightness in the scene and
there is clearly a more rounded quality to the boundaries generated by clipping. In the result produced by the filter which
drives the signal to its average value, we find the only output
where there is still some small vestige of the bridges shown
across the river in the original aerial scene.

Figure 9 - Results of applying various types of cellular logic low-pass filters to the portrait of the "USC girl" (see text for details).

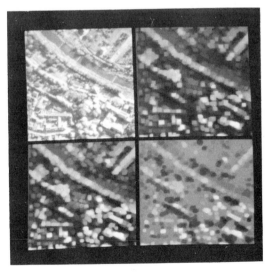

Figure 10 - Results of applying the same cellular logic filters whose outputs are shown in Figure 9 to the aerial scene shown in the upper left frame.

4.   IMPLEMENTATION

All of the results shown in Figures 3 through 10 were obtained using the PHP ("Preston-Herron Processor") produced in a
joint effort between Carnegie-Mellon University and the University of Pittsburgh (Herron et al., submitted). The PHP is a cellular logic machine which interfaces to the Perkin-Elmer 3200--
series computers and receives picture information 16 pixels at a
time via the interface bus. Both the image size and neighborhood
size are under program control allowing a large amount of flexibility. The Perkin-Elmer host computer is used to load the lookup
table required for conducting the particular logical transform.
The host computer also carries out thresholding operations and
loads and unloads binary images from the PHP 32K bits at a time.
Image data is read from local memory and processed by table lookup as an 18-pixel by 3-pixel window. The result of the table
lookup operation is then transferred back to the main memory of
the host computer.

Overall image dimensions are determined not by the PHP but
by the software driving it. For example a square image as small
as 16x16 up to an image slice as large as 3x10912 can be
processed. Any intermediate size can be handled as long as the
number of pixels per line in the image is selected modulo 16. The
PHP processes 16 pixels simultaneously in response to each DR
(Data Request) from the host computer using the 16 identical
1x512 lookup tables fed from the image registers.

The general architecture philosophy of the PHP is to make it
a flexible and fast but "dumb" processor heavily under control of
the computer-resident image processing software. In order to
process an image (or image slice), commands are issued to (1)
load the lookup tables, (2) set the memory address offsets, (3)
load image data memory and (4) commence a processing sequence in
which the contents of the memories are transferred through the
memory registers 48 pixels at a time. At each step in the
processsing cycle 16 pixels and their 16 neighborhoods (54 pixels
total) are available as 16 9-bit individual table addresses (144
total wires) to the lookup tables. The outputs of the lookup
tables provide the 16-pixel result to the host computer.

The PHP is capable of loading and processing 5 megabytes per
second (40 megapixels per second). Thus an image slice consisting
of 32K 1-bit pixels could be processed in 1.6 milliseconds (one
input cycle plus one output cycle). To this must be added the
overhead time required to enter the tables (equivalent to 1024
bytes), load the subfield mask (when required), and load the

memory address offset counters. Depending on the time-sharing load of the computer and other variable factors it now requires from 100 to 200 milliseconds for the PHP to process a 512x512 image (1 iteration). This implies that the logical filtering conducted in preparing the illustrations shown in Figures 3-10 (each being 256x256) image took less than 1 second. Additional time, for course, was required for thresholding and then for recombining the results of the logical transforms themselves.

5. ACKNOWLEDGMENTS

The work on this project was conducted under a grant from the United States Science Foundation (MCS-80-00666) to the Department of Electrical Engineering, Carnegie-Mellon University. The facilities utilized were those of the Biomedical Image Processing Unit of the Department of Radiation Health, School of Public Health, University of Pittsburgh where the PHP is installed. The PHP itself was fabricated using funds furnished by the Optical Technology Division of the Perkin-Elmer Corp. The author would like to acknowledge the assistance of Dr. A. Sanderson, Carnegie-Mellon University who critiqued the results, of Dr. J.M. Herron, University of Pittsburgh, director of the Biomedical Image Processing Unit, and to Mr. J. Karohl of Perkin-Elmer. Thanks are also due to Ms. Cathy Brown (Carnegie-Mellon, Pittsburgh) and Ms. Rose Kabler (Executive Suite, Tucson) for preparing the manuscript.

6. REFERENCES

Fountain, T.J. "CLIP4: A Progress Report", in _Languages_ _and_ _Architectures_ _for_ _Image_ _Processing_ (Duff, M.J.B. and Levialdi, S., eds.) Academic Press, London (1981).

Goetcherian, V., "From Binary to Gray Tone Image Processing Using Fuzzy Logic Concepts," Pattern Recog. _12_, 7-15 (1980).

Herron, J.M., Farley, J., Preston, K., Jr., and Sellner, H., "A General-Purpose, High-Speed Logical Transform Processor," (submitted to IEEE Trans. on Comput.)

Preston, K., Jr., "Multi-Dimensional Logical Transforms," (submitted to IEEE Proc.)

AN EXPERIMENTAL RULE-BASED SYSTEM FOR TESTING LOW LEVEL SEGMENTATION STRATEGIES[1]

Martin D. Levine
Ahmed Nazif

Department of Electrical Engineering
McGill University
Montréal, Québec

I. INTRODUCTION

The objective of computer vision systems is to be able to
understand the contents of pictures originating from three-
dimensional scenes. It is now generally accepted that in order
to do this successfully, it is necessary to employ models of the
domain under consideration. A system which partially accomplishes this task using two-dimensional models to analyse two-
dimensional pictures has been recently reported in (Levine and
Shaheen, in press; Shaheen and Levine, in press). The approach
taken in these references depends strongly on the specific initial segmentation used. That is, an initializer is invoked to hy-
pothesize a set of regions which can subsequently only be merged
or labelled. Obviously in this case, the final solution can only
be as good as the initial partition. This paper presents a rule-
based system for improving the partition by providing an experimental facility for examining different low level segmentation
strategies. To include as much flexibility as possible, the image is described in terms of regions which represent the uniform
neighborhoods in the picture, lines which represent the sharp
changes of intensity, and areas which are either large regions or
groups of regions and lines. The latter are usually indicative
of textured areas, and have in the past proven to be very difficult to analyse. By using rules which only come into play under
specific data conditions in a particular neighborhood in the

[1]*This research was partially supported by the Natural Sciences
and Engineering Research Council under Grant A4156 and the
Department of Education, Province of Québec under grant EQ-633.*

picture, it is possible to selectively segment it. In this way
it is expected that the system will achieve a segmentation which
might approximate what an artist would produce if he were asked
to perform the same task.

The segmentation rules embody general knowledge about regions,
lines, and groups of these and have been referred to as general
purpose models (Zucker et.al., 1975). The objective of these low
level rules is to partition the two-dimensional image into regions. The latter should either correspond to complete objects
or surfaces of objects. However, it is generally accepted that
such a so-called complete segmentation is not possible with low
level techniques. Therefore, what is actually being sought is a
partial segmentation which allows for parts of objects or surfaces to be reported from the computation. But even this less
severe criterion is by no means a simple one to satisfy.

Segmentation is computed on the basis of two properties: the
similarity and disparity between pixels. One of the difficulties
of not using specific models related to the particular image is
that the segmentation is not unique. In fact, the property of
uniqueness of a partial segmentation for an image has not been
defined. Nevertheless, we would like both the partial and ultimately the complete segmentation to correspond to what a human
functioning in the same task would provide as a result. However,
it is unlikely that even humans analysing a specific picture would
produce unique partial segmentations. Does it really matter?
The answer is most likely in the negative if the ultimate goal of
the exercise is a hierarchical analysis of the picture, and it is
intended that top-down feedback will be invoked. Under these
circumstances, errors that have been made at the lower level will
later be undone by a re-analysis of a specific local situation.
Notwithstanding the possiblility of designing such a complex system, it is nevertheless desirable to minimize this type of inter-
vention. It therefore appears to be advantageous to work with as
good a partial segmentation as possible.

The segmentation that we seek contains three kinds of data.
First, there are the regions which are assumed to be connected,
complete, and nonoverlapping. The second are lines which are also assumed to be connected and nonoverlapping, but in general are
incomplete. Finally the third category, which has been neglected
in the literature, includes the focus of attention areas. These
are interesting parts of the image which might be either large
smooth areas or textured ones.

The approach that we have taken to achieve this specific segmentation is to define a set of condition-action rules which embody general knowledge about edges, regions, and areas. We begin
with an initial segmentation and use the rules to iterate to a
final result. The system that we have developed allows us to experiment with different rules and sets of rules in order to find
the best strategy for two-dimensional image segmentation.

II. SYSTEM STRUCTURE

This research is related to the work reported in (Levine and Shaheen, in press; Shaheen and Levine, in press). The system described in these references is capable of both segmenting and interpreting a natural outdoor scene, given a stored set of constraints describing the two-dimensional world in question. The basic paradigm consists of a set of processors, a short term memory, and a long term memory. Data is input to the system and read from it via the short term memory which also acts as a workspace for the collection of processors. The long term memory stores the so-called world model and therefore can only be read by the processors. The latter are meant to function according to the data conditions existing in the short term memory at any given time. They may read from and write into the short term memory, but can only read from the long term memory. The resulting segmentation is a union of regions, each associated with a single object interpretation. To achieve this objective, the system is initialized by grouping pixels into regions having similar colors. From that point onward, regions can only be merged and therefore it is obvious that the final result depends very heavily on this initial segmentation. In this paper, we have maintained the same basic computational structure as in (Levine and Shaheen, in press) but are only concerned with the low level segmentation system. What was represented as a single process in (Levine and Shaheen, in press), is shown in Figure 1 as consisting of many.

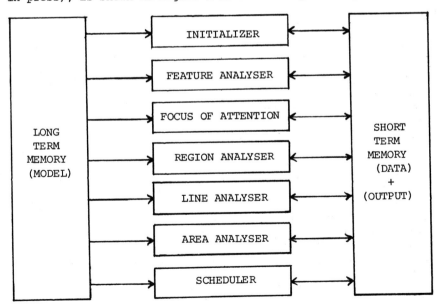

FIGURE 1. System Structure

The short term memory in Figure 1 contains both static and dynamic data. The former category of course, includes the original
three color planes which are assumed to be input to the system.
A gradient image involving both the magnitude and the direction,
is then computed and stored.

Most of the data in the short term memory are dynamic. These
primarily pertain to the three data types of regions, lines, and
areas. The region map is a set of current regions which embodies
the image segmentation at any given time. Also stored are the
average and variance of the color components, the position, area,
and perimeter, plus a detailed listing of the boundary. Spatial
relationships between adjacent regions are also maintained.
Similarly the line map is a set of postulated lines appearing in
the image. Initially, these are obtained by adaptively thresholding the gradient image and performing some minor thinning and
joining operations. The features of the lines which are computed
are the color, the gradient across the line, its position and average direction, its length, and so on. Again the spatial relationships between lines is also recorded, as are the relationships between regions and lines.

At the lowest level, segmentation is obviously concerned with
small regions and lines. However, it quickly becomes apparent
that in order to isolate the surfaces or objects in a picture it
is necessary to group collections of such small regions and lines
into larger areas. We refer to these as focus of attention areas,
and define three types. One is a smooth area which is essentially a large uniform region. Another is a textured area which consists of a significant group of small regions each having similar
feature properties. The third group is not often found and is
defined by a bounding line which has occurred in the line map.

Regions, lines and areas are the data types stored in the
short term memory which directly relate to the image segmentation.
In addition to these dynamic data, the short term memory also includes system performance information and statistics. The complete history of processing is stored; the cycle, the process,
what action was performed, the current region or line, the number
of actions so far, the number of lines or regions in the short
term memory, and so on. We shall discuss in the next section how
this type of information is used by the scheduler.

Returning to Figure 1, it should now be quite evident what
the function is of each of the subprocesses making up the low level segmentation system. The initializer starts the system off by
computing the region and line maps. The feature analyser then
computes the whole collection of features required and of course
is also called into play throughout the analysis in order to update this information. The focus of attention module governs the
order in which the lines, regions and areas are analysed. There
are three analysers, each of which implements the rules associated
with one of the three important data types. Finally, the scheduler is concerned with a set of meta rules which are rules about

rules. These refer to all of the other subprocesses shown in Figure 1 and can be used to implement a particular analysis strategy. As we shall see in the next section, this allows for a very complex definition of the computation and thereby permits a high level of experimental flexibility.

III. RULES

The rules that are used for segmentation are all general purpose in nature. They tend to reflect our own ideas about what we would do given a similar task to perform. We have analysed the literature in detail and have thereby obtained a large set of such rules. These have been stored in the long term memory and are used to perform the analysis.

A set of conditions specified by a rule in the long term memory is matched with the data in the short term memory. When such a match occurs, the rule then prescribes a set of actions which have the effect of modifying the short term memory.

The rule condition is defined by a logical predicate which is optionally preceded by the operator NOT. The predicate may either be a logical function evaluated directly on an object, such as whether an area is smooth or a line is open-ended, or it may be evaluated on an object feature, such as intensity, region histogram, or line orientation. The object feature is modified by a qualifier which permits the system to determine whether the logical predicate is TRUE or FALSE. Examples of such logical evaluation functions are low area, high difference in intensity, and so on. Concatenations of such logical predicates allow for very complex conditions to be specified.

It should be noted that the predicates are defined using symbolic qualifiers rather than numerical ones. All segmentation techniques are extremely sensitive to the selection of certain parameters and it is generally considered to be very difficult to design a segmentation algorithm which is independent of them. Therefore even if a particular system possesses only a single parameter, a whole set of possible segmentations can be obtained by performing the computation over a range of parameter space. Indeed, this property is partially responsible for the fact that a unique segmentation does not exist. In order to circumvent the possibility of "tuning" a particular rule to the image data, we have defined a set of symbolic adjectives which may be used to refer to features and differences in features. These qualifiers are: very low, low, medium, high, and very high. The thresholds defining the scale for these descriptors are selected at the outset and are unchanged throughout the analysis. In addition they are the same for every logical function.

There are three types of rules in the long term memory. The first type embodies the low level general purpose models and

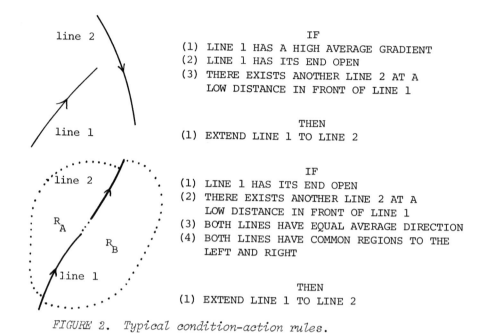

FIGURE 2. *Typical condition-action rules.*

deals with regions, lines and areas. For example, rules about
merging and splitting regions, rules about extending, connecting,
deleting or inserting lines, and rules for creating or deleting
focus of attention areas. Figure 2 illustrates some simple examples of rules which are stored in the system.

A second set of rules in the long term memory are those that
deal with the focus of attention strategy. This concept has been
largely neglected in the literature, although it apparently plays
a significant role in human segmentation strategies (Yarbus,
1967). These rules are responsible for the selection of the next
data item to be considered and therefore have a significant influence on the flow of the analysis. The actions that are carried out by this process are, for example, the switching from one
area of attention to another, the deactivation of a particular
area, and focusing or defocusing. In addition, these rules also
govern the selection of the next adjacent region or line. In this
way it is possible to simulate various parallel and sequential a-
nalysis strategies for segmentation.

The third, and perhaps the most important of the set of long
term memory rules, are those that pertain to the system control,
the so-called meta rules. These are responsible for the running
and stopping of the system, the activation and deactivation of a
particular process, and the scheduling of rules in general. They
make considerable use of the dynamic system information and

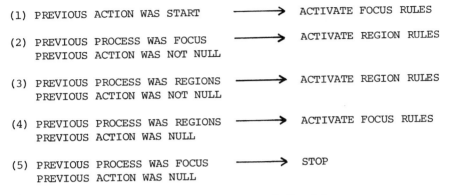

FIGURE 3. A segmentation strategy defined by an ordered circular list of meta rules.

statistics referred to in the previous section. The meta rules are used to design and implement specific strategies for segmentation. New strategies are created by selecting an ordered circular list, as shown in Figure 3, which characterizes a simple merging algorithm.

IV. MEASURE OF SEGMENTATION

As mentioned in Section I, a specific low level segmentation is not unique and its quality can only be judged in comparison to human expectations. However, in order to test and refine the low level model used to produce the segmentations, a performance measure is required that is able to make comparisons and rank the outputs.

A segmentation error measure has been designed and implemented that is essentially a two-dimensional distance measure between two different segmentation outputs. If one of these, taken as a reference, is a segmentation produced manually, then that distance measure would be an indication of how far the other output is from the reference. A higher value of the measure would thus imply a greater amount of error in the tested output relative to the human generated output.

The two-dimensional distance measure referred to above has a component proportional to the amount by which the regions in a tested segmentation overlap those in the reference. The other component signifies the amount by which the tested regions partition the reference regions.

Consider a reference segmentation output containing the set of N regions $\{R_1, R_2, \ldots, R_N\}$ having areas $\{A_1, A_2, \ldots A_N\}$ respectively, and a test segmentation output containing the set of M regions $\{\tau_1, \tau_2, \ldots, \tau_M\}$ having areas $\{\alpha_1, \alpha_2, \ldots, \alpha_M\}$. The under-merging error measure is then given by:

$$UM = \sum_{\substack{\text{all } \tau_j \\ 1 \leqslant j \leqslant M}} \frac{(A_K - \cap [\tau_j, R_K]) \cdot \cap [\tau_j, R_K]}{A_K}$$

where $\cap [\tau_j, R_K]$ is the area of the image shared between τ_j and R_K, i.e., the number of pixels that have label τ_j in the test output and label R_K in the reference output, and R_K is a region in the reference output such that

$$\cap [\tau_j, R_K] = \max_{\substack{\text{all } R_i \\ 1 \leqslant i \leqslant N}} (\cap [\tau_j, R_i])$$

The over-merging error measure is given by:

$$OM = \sum_{\substack{\text{all } \tau_j \\ 1 \leqslant j \leqslant M}} (\alpha_j - \cap [\tau_j, R_K])$$

It can be easily shown that both measures have a lower bound of zero when the test and reference outputs are identical, and a higher bound equal to the area of the image minus one. The two measures can thus be normalized and combined to produce a composite measure:

$$M_1 = [(UM/A_I)^2 + (OM/A_I)^2]^{\frac{1}{2}}$$

or $\quad M_2 = |UM/A_I| + |OM/A_I|$

where A_I is equal to the area of the image.

Given the above measures, a low level model consisting of a set of rules and meta rules can be evaluated as a whole or it can be tuned and refined by testing on a rule by rule basis. Rule performance can be measured by the effect on the measures of running the system with or without that rule.

V. EXPERIMENTS AND DISCUSSION

A close examination of Figure 3 reveals the power of the system in designing and implementing strategies for image segmentation. In this example, the meta rules are used to invoke the region analyser and the focus of attention during processing. As soon as the system is invoked a START action is executed, which in turn will trigger rule 1. As a result, the focus of attention rules will be matched against the short term memory and an action will probably focus on the first region to be tested. If the focus of attention is successful, rule 2 will be triggered and the region rules will be matched to the current region. If a region rule fires, an action will be executed on the current region; alternatively, if the current region does not meet the conditions of any of the rules, the null action is executed. In the latter case, rule 4 will fire on the next cycle and the focus of attention is again brought in to obtain the next region. As we continue to invoke the meta rules in a cyclic fashion, rules 2 and 4 will fire, alternately bringing in regions and testing them. This will occur until the focus of attention fails to provide a new region and rule 5 fires to stop the system.

An interesting role is played by the presence of rule 3. This rule will fire only if in the previous cycle a region analyser rule fired and modified the current region. The action will be to again attempt a match of the region rules to the current region, which has already been modified during the previous cycle. Thus, the strategy defined by this set of meta rules is to keep matching the region rules to the current region until no more matches can occur. The focus of attention is then invoked to obtain the next region. This is seen to be a sequential strategy, as opposed to a parallel one in which a region is acted upon only once before all of the other regions in the short term memory have been tested. This parallel strategy can be easily implemented by modifying the action of rule 3 to be: ACTIVATE FOCUS RULES. Thus the focus of attention will be invoked to specify a new region, irrespective of whether or not the current region has been modified.

The strategies described above have been tested on an image of a typical outdoor scene. Initially, each pixel in the image was taken to be a separate region, and the region analyser had access to only a single rule, namely, one that merges adjacent regions based on a low difference in the three color features. The results are shown in Figure 4 for the parallel and sequential cases. It can be seen that the parallel strategy generally performed better, especially where "leaks" in the boundaries between regions were present, as in the shadow line behind the car. This is because the parallel strategy allows for more global information to propagate through the image during analysis, while the sequential strategy is more susceptible to local errors. The

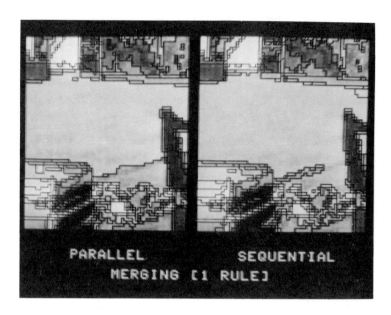

FIGURE 4. A comparison between parallel and sequential
strategies using one merging rule.

FIGURE 5. A comparison between parallel and sequential
strategies using two merging rules.

price, of course, is computation time, since more rule matching would most likely be required for the parallel strategy to achieve the same number of actions.

Figure 5 shows the result of adding a second rule to the region analyser. This rule allows a current region to be merged with an adjacent region if its area is "very low", the adjacency value is "high" and the difference in color features is "not high". The latter is a relaxation of the condition in the first rule which requires the difference to be "low". It is seen that the effect is to eliminate small regions, especially those at boundaries, so that this rule actually represents a noise cleanup operation.

Beginning with the whole image as one region, an experiment was conducted in which the region analyser rules split a region if its histogram showed a "high" degree of bimodality and its area was "high". The region was allowed to split into as many regions as was indicated by its histogram. This results in a good segmentation, but with a large number of scattered small regions. The result is compared in Figure 6 to that of merging regions using a single rule with a parallel strategy. In Figure 7 the output of Figure 6 was used as an input to the merging process, the result being the removal of small regions. This is compared to the output of merging with two rules.

FIGURE 6. A comparison between merging and splitting.

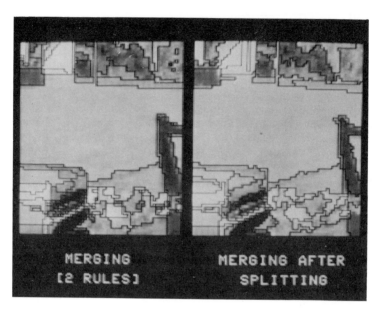

FIGURE 7. Merging with two rules compared with splitting and then merging.

The above experiments demonstrate the flexibility of the system in dealing with a wide variety of possible data configurations. It is observed that the addition or removal of a rule does not require any change in any of the other rules at any level. This is because the meta rules refer to the other rules by their class only and not by their name. It is expected that this experimental facility with such a high degree of modularity will permit us to better understand the image segmentation problem.

REFERENCES

Levine, M.D. and Shaheen, S.I. (in press). A Modular Computer Vision System for Picture Segmentation and Interpretation, IEEE Trans. on Pattern Analysis and Machine Intelligence.
Shaheen, S.I. and Levine, M.D. (in press). Experiments With a Modular Computer Vision System, Pattern Recognition.
Yarbus, A. (1967). "Eye Movements and Vision". Plenum, New York.
Zucker, S.W., Rosenfeld, A., Davis, L.S. (1975). General Purpose Models: Expectations About the Unexpected, Proc. Fourth International Joint Conference on Artificial Intelligence, Tbilisi, Georgia, USSR, Sept. 3-8, 1973, pp. 716-721.

THE BALANCE OF SPECIAL AND CONVENTIONAL COMPUTER ARCHITECTURE REQUIREMENTS IN AN IMAGE PROCESSING APPLICATION

Denis Rutovitz
James Piper

Medical Research Council
Western General Hospital
Edinburgh, Scotland

1. INTRODUCTION

Computers with an architecture oriented towards picture processing can out-perform conventional machinery applied to the same task by many orders of magnitude. But in an image processing application there is always a requirement for conventional computation as well as the processing of the picture proper. The effectiveness of the contribution of specialised architecture will obviously depend on the balance of the different processing requirements. Are we after computer power in general - as in a large mainframe or in an array of uncommitted microprocessors - or do we really need a specialised power of picture analysis? There can be no general answer, but it is perhaps of interest to take a well-known image analysis application and try to analyse its requirements from this point of view. To this end we propose to look in detail at the Edinburgh system of chromosome analysis in relation to the specialised architecture of the CLIP 4 system (1).

2. THE PROBLEM

Chromosome analysis has been much discussed in the image processing literature, but it would be as well to briefly restate the main requirements and the approach taken in the Edinburgh system (3) These comprise essentially the following steps:

a. Finding analysable metaphase cells with a computer controlled microscape stage. An objective of 0.6 NA or less is used, with a 1 μ scanning raster. The distribution of suitable cells varies from one or two per cm^2 to several hundreds. In karyotyping studies (in which it is usually assumed that all the cells will have the same chromosome complement) about 20 good quality cells are required (8).

b. Relocation of cells found in step (a), with an oil-immersion objective for high resolution scanning, raster spacing about 1/8 μ. The relocation operation requires the cell of interest to be brought to the centre of the field and optimally focussed. In addition, its general outlines must be determined so that subsequent processing can be restricted to the chromosomes of that cell, with outlying debris, interphase nuclei and chromosomes of other cells excluded.

c. Digitisation, "pre-processing" and preliminary region definition. It may be objected that these operations could and should be separated, but in the current Edinburgh system they are inter-related. A low spatial resolution scan of the field is thresholded and the above threshold regions determined. Each is then expanded so as to include a certain amount of the surrounding background, then rescanned at high resolution. The purpose of this manoeuvre is to reduce the amount of data which has to be dealt with at one time in any subsequent processing stage. As will appear later, the relatively small size of the CLIP array will make an equivalent procedure necessary in the CLIP 4 context as well (1).

d. Further pre-processing (filtering) (6) and final region definition. The individual regions obtained from step (c) are filtered, thresholded and resegmented. The correctness of the segmentation is assessed by various statistically optimised heuristic criteria, and the segmentation modified by appropriate split or merge algorithms.

e. Feature extraction: determination of the convex hull and the medial axis, calculation of the profile of densities projected onto the axis, location of significant concavities, and identification of centromere region (9).

f. Classification of individual chromosomes from two-parameter Bayesian probability tables. Re-assessment of classification using prior knowledge of expected karyotype (3,4,7,10).

g. Calculation of karyotype of an individual by integration of single cell results (3,11)

The above is a procedure for the analysis of homogeneously stained cells. Much greater general interest attaches currently to chromosome analysis systems which take into account the banding pattern which can be obtained by use of various preparation regimes (8). However, the features of the homogeneous and banded problems are essentially the same in relation to the present purpose, and confining discussion to the homogeneous system will reduce the number of extraneous issues involved.

3. THE PROCESSOR

The CLIP 4 system is in little need of introduction here. However, for completeness, the main features are summarised. Essentially it is a 96 x 96 array of processing elements, each with 32 bits of directly accessible memory, and a prospective further 182 bits addressed in a different mode. The processors are interlinked so as to form either an 8-connected or 6-connected lattice. Boolean operations on selected memory elements and neighbouring processor outputs are loaded to memory, or to neighbour output. Special signals are sent from the missing neighbours of edge cells. All processors obey the same instructions simultaneously, the instructions being derived from internal instruction memory shared between the CLIP array and by a host serial computer - typically a 16-bit machine such as a small PDP 11 computer. There are special facilities for bit-plane arithmetic. If the neighbour output varies with neighbour input, the array will not immediately settle to a steady state. Instead neighbour-to-neighbour signal propagation occurs: but a steady state is always reached eventually, provided the output function does not involve the logical complement of the input function (1). This is one of the most powerful and important features of Clip's architecture, as it enables complex image processing tasks, such as component labelling, to be done by means of very small numbers of propagating instructions.

The version of CLIP 4 which we are considering here will be able to execute a non-propagating instruction in approximately 10 μsecs, and a propagating instruction in 10 +n μsecs, where n is the number of propagation steps required before a steady state is reached. Let us suppose that the serial computer is a 0.5 μsec cycle time machine of conventional design, such that (possibly indexed) memory-to-register addition takes two cycles and register-to-register operations take 1 cycle.

Although it is not possible to make accurate comparative estimates of performance of machines using different algorithms, different compilers etc., it may nevertheless be of some interest to try to set some bounds to the performance ratios which we are likely to encounter. Let us compare the serial machine and the CLIP on two typical neighbour-referencing tasks, namely erosion and 3 x 3 summation.

a. erosion: an efficient assembly code implementation for the serial processor is the following: we employ 9 index registers, and at the beginning of each line of the picture set them so that each addresses one element of a 3 x 3 neighbourhood. We assume that the picture has a "border" sufficient to ensure that the neighbours are meaningful whenever we are within the picture area. Then so long as all the index registers are advanced at each pixel reference, they will continue to address a neighborhood of the current point as the picture is traversed. Erosion amounts to replacing the bit associated with each point by the logical AND of itself and all of its neighbours. Therefore our erosion program comprises, in its inner loop, an initial load register instruction, then a series of 8 sets of instructions of the form:

    i)   load register from memory (indexed)
    ii)   execute register to register "AND"
    iii)   increment index.

Finally there are a store-register-in-memory and increment index instruction to put the calculated value in the output plane. In all we have 35 cycles, with perhaps 5 more to check whether the current line is terminated making 40 cycles, or about 20 μ secs per pixel. CLIP can do this operation on approximately 10 picture elements in 10 μ sec, or about 20000 times faster than the conventional machine (though this is not the best way to do binary operations on a serial machine. The intrinsic word-parallelism of such computers can usefully be employed).

b. 3 x 3 summing: The serial processor time is virtually unchanged. In CLIP's case however we become conscious of the cost of doing grey scale arithmetic in bit-planes. Assuming a 6-bit input picture, 8-bit operations will be sufficient (on average) to add all elements of 3 x 3 neighborhoods to the value of the central point. CLIP will take approximately 10 μ secs to add two bit planes and leave the "carry" plane set for the next operation, and it will take another 10 μ secs to shift the array a single step in any direction. If we add left and right elements to the central point, then the elements above and below, the entire operation is effected in 4 steps, each taking (10 + 10) x 8 μ secs, i.e. a total of about 640 μ secs. Thus the processing power ratio now seems more like 300:1.

For most image processing operations the ratio of the serial m/c to the CLIP pixel processing times will be between these two extremes, i.e. between 20:10 and 20:640 per pixel, with the parallelism and architecture of CLIP giving it an advantage of between 20000:1 and 300:1 when working on the whole image plane. But there is usually no reason why operations need be carried out on the whole image plane. In particular in many biomedical problems involving cells and cell-like objects (or for that matter in the recognition of industrial components), it is possible to distinguish fairly easily between picture, borderline areas, and background proper. The latter can then be discarded, and processing confined to the picture area only, either approximately by use of enclosing rectangles, or exactly, by use of suitable data structures. In the analysis of metaphase cells, we routinely carry out procedure of this type, thereby discarding at least 75% and up to 90% of the nominal picture area. The advantage to be gained by parallelism and special architecture is much reduced because most of what the array processor processes will normally be background. Using an 80% background figure, the effective processing power ratio will be in the range 4000:1 to 60:1. Unfortunately, it is that latter figure which will apply in exactly those applications which take longest to carry out on a serial machine, namely filtering and similar grey scale processing. In fact the processor power ratio for convolution-based filtering operations may be an order of magnitude lower again unless the coefficients are carefully chosen to allow multiplication by means of shift operations. Against this it must be admitted that the serial computer figures ignore the influence of compiler inefficiencies and memory limitations which in practice can be very significant.

4. ANALYSIS

We now review the operations required for the implementation of each stage in the chromosome analysis system and try to assess the proportion of the processing task which one could reasonably expect to be transferable to the CLIP 4 machine. Of the items which could not advantageously be transferred, the most obvious is the "house-keeping" involved in keeping track of measurements, comparing them with decision criteria, and logging object "histories" - traces of their passage through the system essential for the gathering in of statistics for performance optimisation later, and for the execution of complex decision strategies at the time.

a. Finding cells suitable for analysis. This task is at present carried out by a special purpose instrument known as FIP (for "Fast Interval Processor"), which under normally prevailing conditions allows continuous processing of 1mm wide swathes at 1 μ pixel spacing at an effective rate of 8 Megapixels/sec (12). It is of interest to know whether a system built around the CLIP 4 machine for the main image analysis tasks would have need of further specialised architecture in the metaphase search phase, or whether CLIP would cope adequately with all requirements.

It is not difficult to detect candidate metaphase cells. Many systems for doing this threshold the field, detect small objects and then seek regions where small objects cluster together to form an approximately circular mass of around 50 μ diameter (8). The number of contributing small objects in a cluster, and the symmetry and size of the containing region gives a fairly good idea of the "metaphaseness" of the object. If we are scanning with a 1 μ raster, and we wish to delete objects more than, say, 10 μ in diameter, and then to coalesce objects within 10 μ of each other, we shall need to erode and propagate over a span of about 10 picture points. CLIP can do an erode operation in under 10 μsec, and full-field propagation in 35 μsec, but counting the number of small objects contributing to each surviving cluster is a more laborious procedure. Any candidate clusters which pass the first acceptance test, i.e. have sufficient contributory small objects, will have to be dealt with by rather more sophisticated algorithms for measuring convexity and diameter. Nevertheless, it would probably be safe to assume that the whole task could in most cases be accomplished in under 100 CLIP steps, which at present timing would require about 1 msec. If this estimate is valid, the approximately 10^4 picture elements in our 96 x 96 region could be CLIP-processed at a rate of approximately 10 Megapixels/sec, a timing quite commensurate with that of the FIP instrument. Unfortunately however, we are not done there. The problem is the small size of CLIP array. Since metaphase cells extend to about 50 microns, the CLIP field is only twice the size of the objects which we are trying to detect. It will be necessary to detect edge-intersecting clusters and re-input and reprocess corresponding sections of the picture. It follows that we require a frame store, large enough to accommodate a much more substantial picture. If not, for example if we have to do a television scan of each 96 x 96 area, our processing rate will obviously be limited by the camera field rate which on a field this size would make us very slow indeed. Even with the frame store, the processing will be dominated by the requirement for shunting pieces of picture in and out of CLIP-accessible memory unless the arrangements for so doing are very efficient indeed. The Fast Interval Processor employs an efficient data reduction and buffering scheme to smooth out processor demand in areas of the slide dense in cells, or for fields requiring special processing. There does not seem much prospect of doing the same in connection with a field-oriented instrument, so the CLIP timing would have to be based on worst case allowances. Thus there is some doubt as to whether CLIP could equal FIP's performance in carrying out a full search of a slide in about 2 minutes, and in subsequent discussion we have assumed that about twice as long would be required by the CLIP system. On sparse slides, with overall about 100 usable metaphases (on which such instruments are of most use), the CLIP timing is therefore assumed to be about 2 seconds per cell, as against FIP's 1 second.

b.   Cell Relocation, Centering, Focus and Restriction.   To
     avoid frequent objective changes, the usual procedure is to
     find candidate cells at low power, then to switch to an oil
     immersion objective and work through the cell list
     (preferably in order of decreasing quality measure).   If
     the search and relocation phases are taking place on the
     same machine, the slide not having been reloaded in the
     interim, the cells should be repositioned to within one half
     a cell diameter (about 20 μ) or better.   Under other
     circumstances the cells may be somewhat further away from
     the field centre, but experience is that the cell, or most
     of it, will be in a 100 μ diameter region centred about its
     nominal position. If an area of this size be mapped
     directly onto the 96 x 96 CLIP array the effective
     resolution will be rather low for focussing and centering,
     especially if there are extraneous or competing metaphases
     within the scanned area.   With existing equipment this
     operation has been carried out with a 0.5 μ sampling grid,
     and it seems unlikely that this condition could be relaxed.
     Thus once again we will have to use an overlapping scan
     pattern, and the potential efficiency of the CLIP system
     will be reduced.   In this case however, it will not much
     affect our ultimate timing.   There are a number of
     different and easily calculated field parameters which have
     been shown to be effective figures of merit for focus
     purposes.   For example, we could use CLIP to calculate the
     sum of the absolute difference in grey level between each
     point and its neighbours and move in the focal plane so as
     to maximise this parameter.   Many exceptional conditions
     are encountered when attempting to focus at high resolution,
     but providing that the image parameters can be rapidly
     extracted there will be no timing problem here.
          Centering and field restriction: the most obvious way
     to do this is a repeat of the procedure described in the
     section on cell finding, above, though this time at an
     increased resolution.   For dealing with an individual cell
     the CLIP cost of a few milliseconds is quite acceptable,
     even with the processing of overlapped fields as required by
     the CLIP dimensions.

c.   Digitisation and preliminary region definition: with the
     raster spacing of about 0.1 μ used for high resolution
     scanning, chromosome lengths are between 10 and 100 picture
     points, and width between 10 and 30 picture points.   These
     are scattered within an area about 400 picture points in
     diameter. With the CLIP system there is once again a
     choice of treating the picture in overlapping segments, or,
     perhaps better, continuing to employ the coarse
     localisation/fine scan arrangements used in the present
     system.   Unfortunately some of the "Cofins" (coarse object
     finely scanned) will inevitably comprise several touching or

nearby chromosomes and may extend beyond the confines of a
single CLIP field, and special arrangements would have to be
made to deal with such cases.   CLIP's first task in this
connection would therefore be to obtain a good primary
threshold in order to define the coarse regions.   The best
way to do this would be to arrange that a low resolution
sample of the finely scanned field could be presented to the
array.   In other words, although we could not easily change
the magnification or the scan line spacing, it should be
possible to sample the digitised field, sending only every
fourth point (say) into the CLIP buffer for analysis.   The
machine would then have to calculate a threshold, making use
of one or other of the techniques referred to in (5).   In
fact there is little difference between this phase of the
operation and that required in step (a), except that we are
now operating within a mask determined by the initial cell
relocation operation, which could easily be held within a
CLIP bit-plane for low resolution work.   The size of the
CLIP array is about right for coarse region definition
within a cell mask - since a cell will fit within a 50 μ x
50 μ rectangle, it will be possible to map the cell area
onto the CLIP array in such a way that the CLIP pixel
spacing would be equivalent to about 0.5 μ on the slide,
which is the size presently used.   Given suitable frame to
array input arrangements there should be no difficulty in
handling this task in a few msecs, though both calculation
of the threshold and thresholding (i.e., production of a
binary image from the original grey scale image) are fairly
expensive of CLIP operations.

d. Further pre-processing and final region definition: The
    individual components of the binary image generated in the
    preceding step must now be processed one by one.   There are
    existing CLIP procedures for labelling connected regions and
    extracting them in serial fashion.   In each case the first
    step is to define a high resolution binary mask from the low
    resolution component.   This amounts to a dilatation in the
    Euclidian rather than the image processing sense, and might
    require a special parallel algorithm to be developed, though
    this can no doubt be done.   Remembering the 96 x 96 array
    size, we will have to use the host computer to note the
    actual position of the high resolution Cofin, and must rely
    on good windowing facilities to read from a frame store into
    the CLIP buffer.   Next we must apply noise reducing and
    edge enhancement filters to the Cofin.   The filters
    currently employed are a 7 x 7 Laplacian convolution and a
    3 x 3 median filter, which constitute a severe burden to a
    conventional serial processor.   At first sight, CLIP is by
    no means optimal for this type of window operation, as with
    a 6-bit input picture and 6-bit coefficients, the time for
    the 49 6 x 6 bit multiplications and 12 to 15 bit additions

involved could be about 100 msec. However, if we take advantage of the 8-fold symmetry of the convolution matrices used, leave out the corner elements, and scale one of the elements to be 1 and another 100_8, we can see (fig.1) that there are only 9 distinct factors, and only 7 true multiplications.

```
      9  7  4  7  9
   9  8  6  3  6  8  9
   7  6  5  2  5  6  7
   4  3  2  1  2  3  4
   7  6  5  2  5  6  7
   9  8  6  3  6  8  9
      9  7  4  7  9
```

Fig.1

Shifting to bring symmetrically corresponding elements of the array into alignment and adding before multiplying, the number of multiplications is reduced to seven (though these will be 6 x 9 bit multiplies). The time required for the opertion when done in this way will be just over 10 msecs, or under 500 msec for the 40-odd Cofins usually found. If some of these regions can be packed together, as seems likely, to fit into, say, 20 rather than 40 distinct planes, the time may be reduced further for this part of the work.

Timings for thresholding and re-assessment of segmentation will be of the same order of magnitude, or perhaps a bit better as large neighbourhoods are not employed. In this case however processing has to be carried out chromosome by chromosome (probably) so that CLIP times will not be entirely negligible - but still under a second.

To this point the serial processor has mainly been used for controlling the sequence of operations, and occasionally for remembering a variable value, such as the position of the origin of a Cofin. Now a log must be opened in which to record the progress through the system of each of the regions resulting from the segmentation process just carried out and the values of measurements made in association with each object. While it might well be possible to make use of CLIP facilities for such purposes the abuse of memory which would result therefrom, if nothing else, would make it seem unlikely that this could have advantage. At this stage then, we use CLIP to obtain and the host processor to store the following information for each region as it is generated:

    i) area.
    ii) summed density values.
    iii) length, breadth and area of minimum containing rectangle.
    iv) area of convex hull.
    v) size and separation of principal concavities in convex hull.
    vi) centroid of mass, second order principal moments.

Each object is regarded as a candidate chromosome, and a series of tests are applied to the measurements listed

above. Experimentally derived information is used to
determine the probability of each being a chromosome,
composite of several chromosomes, piece, artefact, etc.
Here we seem not likely to be in a region where CLIP can
help. This is part of the inescapable accumulation of
conventional computer housekeeping time which at the end of
the day will not be negligible, however small the individual
contributions. This decision work takes some tens of msec
per chromosome on a 16 bit computer (though this time could
no doubt be reduced with proper attention to program
optimisation). Decisions made, CLIP can handle resulting
actions, such as attempts to split composites, or merge
pieces. The products of these operations must themselves
be measured and assessed, and further action taken as
appropriate.

e. Feature extraction: classification of homogeneously stained
chromosomes is ultimately based on only two features, namely
the size of the individual chromosomes (which depending on
the system may be taken as length, area, or integrated
optical density) and the centromeric index, or ratio of the
short arm size to whole chromosome size. Thus the
recognition procedure consists primarily in recognition of
whether or not one is dealing with a whole single
chromosome, and the location of the centromere region. In
certain applications the recognition of aberrant chromosome
types, such as dicentrics or acentric fragments may be the
object of the exercise. In these cases too, however, one
is concerned to recognise "chromosomeness" and centromere
region(s). The lines of attack on the problem used in the
Edinburgh system (and the same or equivalent procedures are
used in most systems) are on the one hand extraction of a
profile of densities projected and summed onto a medial axis
and on the other the examination of the configuration of
concave and convex parts of the boundary (9).
    In general we find that shape is most useful in
providing a guide to the proper orientation of the medial
axis in cases where this is in doubt, especially for small
acrocentric chromosomes. The medial axis itself is
determined first by choosing as orientation that of the
chord of the convex hull which together with the parallel
support line on the opposite side of the chromosome
minimises the width of the object. The chromosome is then
rotated so as to bring this axis perpendicular to one of the
raster directions, and the medial axis is taken as the
smoothed locus of mid-points of raster line intercepts with
the chromosome in the perpendicular direction. Candidate
centromeres are selected by examination of the minima and
maxima of the profile obtained by adding densities in this
latter direction. A parameter of some importance in
deciding as to the likely correctness of the actions taken

is a measure of the symmetry of the half profiles on either side of the axis. Candidate centromeres are examined and awarded a figure of merit depending on such items as the depth of the centromere defining valley and the sequence of densities measured across the centromere (for example, a double humped profile in this direction would indicate that the candidate line was crossing the chromatid arms rather than being at a centromere). The best candidate is selected and an estimate of the confidence in the entire centromere position is derived from the measurements made in relation to previously acquired statistics. If this is adequate, according to the program's built-in acceptance limits, we proceed to the next stage. If not various actions will ensue depending on the context. For example we may refer the object to an operator for interaction, have a look at a different axis, or try to make inferences from the convexity/concavity analysis. Precise location of the centromere is very important as a difference of a few points either way can affect the centromeric index sufficiently to alter the most likely classification of the chromosome.

For accurate location of dips in the profile, and also for optimal differentiation between good and bad candidate centromeres, it is essential that the projection of the densities onto the axis be accurately done. For this reason rotation is carried out by "reverse interpolation": that is, the position of an an inversely rotated raster point in the original chromosome is calculated, and an appropriate density contribution from each of the four corners of the raster square in which it is located, is taken (2). This avoids the uneven patterning which can result from forward interpolation. It is easily demonstrated that a failure to take such precautions has a noticeably adverse effect on results (2). Now a picture processing array is not a good vehicle on which to carry out accurate grey level interpolated rotation. Indeed it is not particularly easy to do a rotation at all except in 60° increments in a 6-connected lattice or 90° increments in an 8-connected one. It may be possible to carry out the kind of operation I have described, but I suspect that if so it would be by what would essentially amount to serial processing of the picture points. It thus seems that we have come to an image processing operation for which the 10° processors of the CLIP array are unlikely to outperform a single conventional serial processor. If so, we must allow an average 0.5 seconds per chromosome for this operation carried out with a Fortran programmed $1\,\mu sec$ 16 bit computer. Admittedly there may be other approaches to this problem more suited to the facilities of a parallel array, but this paper deals with what can reasonably be forseen rather than with what ingenuity and prolonged experiment might bring in the future. Nevertheless there is perhaps

one road to the employment of the CLIP array for this
purpose for which one could predict success.   If the object
were much magnified, perhaps by a factor of four, then a
non-interpolated picture-point to nearest picture-point
rotation, followed by a smoothing and condensation back to
the original size might well be effective.   However, it
would not be possible to do this for the larger chromosomes
owing to the small size of the array.

Again the host computer must record measurements,
figures of merit and preferably also the history of the
sequence of operations carried out on each chromosome or
other object.

f. Classification of individual chromosomes, and
derivation of karyotype: The first stage in classification
is normalisation, that is, the size of each chromosome must
be expressed as a proportion of the total cell size.   This
simple sounding procedure is bedevilled however by the fact
that we cannot have any knowledge of whether or not the cell
has extra or missing chromosomes, or whether we have
included artefacts amongst the objects.   The Edinburgh
procedure (3,4,7,10) is to make an initial estimate of cell
size, based on median chromosome size: then to normalise
with respect to this size and classify the individual
chromosomes.   Since the mean size of a chromosome in each
of the different classes recognised bears a fixed
relationship to cell size, once a chromosome has been
assigned to a class its size can be used as an estimator of
cell size.   We therefore form a new estimate of cell size
taking account of all the classifications made, and also of
the confidence in them.   This is compared with the cell
size estimate obtained before, and if it differs
significantly the cycle is repeated.   Convergence usually
occurs after about five iterations.   Because of
multiplications and divisions occurring in the cell size
estimate, and because of list-processing and table look-up
operations in the classifier (below), our present Fortran
coded programs take about 10 seconds for this phase, running
on the 16-bit computer mentioned above.   It might be
possible to use the CLIP to speed up the process, because of
the fact that CLIP can carry out arithmetic operations in
parallel.   However the number of objects for which
computation could be carried out in parallel (50 chromosomes
or a smaller number of groups), or the degree of parallelism
available in other ways, may be insufficient to render much
return.

Classification is carried out as follows: the
normalised size and the centromeric index are expressed in
suitable integer ranges (0-50) and an entry acquired from a
look-up table.   This is a packed word giving the first,
second and third most likely classifications of an object at

that position in the table, their relative probabilities and the probability of an object at that position being a normal chromosome at all. Initially the object is assigned to the most likely class. The probabilities of its having been a single complete chromosome, the confidence in the centromeric index and the probability of the group assignment are combined into a single figure of merit which is used as a weighting factor in re-estimating cell size (4,10).

At the end of the primary assignment the karyotype as a whole is re-assessed. Since there is some overlap between the positions of different chromosome classes in the size/centromeric index table, an excess of chromosomes in one group and a deficit in one of its neighbours could well not be due to abnormality in the cell, but simply due to a statistical fluctuation of the measurements. Further, even if an excess in one group is not matched by a deficit in immediately neighbouring groups it may be that chromosomes could plausibly be transferred out of the immediate neighbours into further away groups. Therefore a procedure for revision of the karyotype is entered which examines all plausible pair and cascade transfers of chromosomes between groups to see if a more "normal" karotype can be produced by such action (10). Perhaps this problem too could be solved on CLIP or a similar array, but it has not been done as yet.

g. Calculation of the karyotype of an individual by integration of single cell results: One of the book-keeping tasks of the host processor is to record the karyotypes, and also the probabilities associated with them, for each of a series of cells, and then to build up an overall karyotype by integration of single cell results. Essentially this is done by two means. First, the positions of apparently abnormal chromosomes in the size, centromeric index plane are logged. If at the end of a series of cells we find that there is a cluster of such abnormalities in any one part of the plane, this is regarded as presumptive evidence for the existence of an abnormal chromosome, as against a machine artefact. Second, the histogram of the humber of chromosomes found in each class, with frequencies being the number of cells in which that finding is made, is generated. This histogram is examined and decisions taken as to the likely true number of chromosomes in the class, on the basis of the relation between the first and second modes of the histogram and the total number of cells. This adds a further 2 seconds per cell to the overall processing time. These timings are summarised in table 1.

5.   CONCLUSIONS.

It seems that we might expect the CLIP machine to take less than 2 seconds for the main image processing work of this problem.   This gives a speed 300 times greater than that of the present conventional computer program.

The time taken by our unaided 16-bit conventional processor is about 10 minutes per cell, the heaviest load being filtering, and next heaviest region definition.   Clip can remove the filtering and most of the image processing burden.   It seems however that there will be a residue of about 20 seconds of image processing time, about 10 seconds of classification time and perhaps 10 seconds of book-keeping time, a total of 40 seconds in all, to deal with.   Thus the contribution of a machine with architecture specialised to image processing, even regarded as infinitely fast, is at first sight to increase throughput by only an order of magnitude over that of a single processor.   Further, comparing with our present three-processor transaction processing system, we have a factor of 3 or 4 only.

However the balance of processing requirements could be considerably altered by rethinking the approach to some parts of the problem so as to transfer them to the image processing machine, and by optimisation of the code for the then remaining operations.   It seems unlikely that the host computer requirement will be reduced to less than, say 10 seconds per cell.   While this is a fairly respectable time further gains could only be made by increasing the speed or the number of conventional processors.

Putting the problem in a slightly more general way, if an image processing task requires unit time, of which proportion p can be handed to a specialised processor with capacity R times that of a conventional serial processor executing the same task, the resulting time will be p/R plus $q = 1-p$.   To equalise conventional and special processor times (and we may imagine them as working in parallel) we shall need to enhance the capacity of the serial processor by a factor N, (or use N copies of it) where $\frac{p}{R} = \frac{q}{N}$, i.e. $N = qR/p$.   In our setting we appear to have roughly, $q = 1/15$, $R = 300$, so $N \approx 20$.   If the conventional processing time could be reduced to 10 seconds, giving $q = 1/60$, we would still need five processors to fully match CLIP's performance.

We conclude

a.   If it is hoped to fully exploit the resources of a powerful specialised image processing device, an important consideration is the optimisation of procedures used for all associated tasks such as decision and classification procedures, graph analysis, record keeping and higher level integration of results.

b. An equivalent reduction in time/cell to that obtainable by the use of CLIP and a single conventional processor in the chromosome analysis problem (from 10 minutes to 40 seconds approximately) could be obtained by the use of e.g., a 15 microprocessor transaction processing system. However the throughput of a better optimised single micro system and CLIP could only be equalled by a 60 plus multi-micro system without CLIP.

c. because of the residual requirements for conventional computing, lack of opportunity for background data reduction, IO requirements and mechanical delays, we cannot expect increases in throughput commensurate with CLIP's 10^4 processors. Improvements of the order of 10^2 seem feasible however, and should be obtainable even if the amount of image processing work were much increased. In other words it will be possible without effect on timing to use algorithms of greater variety and complexity than hitherto employed. One can therefore expect a notable improvement in the quality of results as well as the processing rate increases which we have been attempting to estimate.

It may be felt that this analysis is too special to chromosome work to have any general bearing on image analysis problems. But our expectation is that it will be found that in a great many applications requirements of background elimination, precision, house-keeping and classification combine to reduce the effective impact of a specialised picture processor. CLIP also has some particular difficulties, related to the array size. Indeed it should be noted that unless we have excellent selection from frame-store arrangements, CLIP's image processing power will be vitiated by the requirement for repeated rescanning of the field, (or disc access) in order to read in different parts of the picture at appropriate times.

All that is not to say that there are not some applications in which image processing power is essentially all that is required. Indeed we have only to vary the chromosome problem from that of finding the karyotype of an individual to that of scoring cells for aberrant chromosomes (which is probably by far the more important problem anyway) to find that CLIP is just what we need. The reason is that in this case we are mainly concerned with the detection of dicentric chromosomes. We do not need to know the centromeric ratios with accuracy, we do not need to classify the chromosomes in the cell and consequently we do not need to normalise. The role of the host computer diminishes to the point where it is not mismatched to CLIP's capacity.

TABLE 1 : Timings of present serial and projected
CLIP 4 parallel system

Task	Present Timing	Projected Timing with CLIP 4
Finding Cells	Fast Interval Processor < 1 sec.	< 2 sec
Relocation,focus, centering	PDP9: 10 seconds plus interaction, settling time	2 sec, mechanical ?
Digitisation, including coarse segmentation, dilation (finding "Cofins"), conversion to fine raster	35 seconds + interaction + filing time	< 1 msec
Filter, threshold resegment	5 minutes	< 0.5 sec
Assess segmentation,	3 minutes	< 1 sec
Find parameters (axis, profile, convexity, candidate centromeres)	30 secs	? parts of operation possibly not suitable or not economical on CLIP. : 20 secs serial < 0.5 sec, Clip
Classify	10 secs	10 secs serial.
Housekeeping	10 secs	10 secs. serial
Total	≃10 mins	4 secs CLIP 40 secs serial 2 secs mechanical

ACKNOWLEDGEMENT

   The authors would like to acknowledge the large amount of helpful and patient advice which they have received from Michael Duff.

REFERENCES.

1. Duff, M. J. B., Proc. 1978 Nat. Comp. Conf., 1055 (1978).
2. Groen F. C. A., Verbeeck, P. W., and Van Zee, G. A., Proc. 3rd. Int. Joint Conf. Patt. Recog., Coronado, Calif., 547 (1976).
3. Hilditch, C. J., in "Human Population Cytogenetics" (P.A. Jacobs, W.H. Price and P. Law, eds.), p. 297, Edinburgh University Press, Edinburgh (1969).
4. Hilditch, C. J., and Rutovitz, D., Comput. Biol. Med. 2, 107 (1972).
5. Mason, D., Lauder, I., Rutovitz, D., and Spowart, G., Comput. Biol. Med. 5, 179 (1975).
6. Nickolls P., Piper, J., Rutovitz, D., Chisholm, A., Johnstone, I., and Robertson, M., Pattern Recognition, in press.
7. Paton, K., Ann. Hum. Genet. Vol. 33, 177 (1969).
8. Piper J., Granum, E., Rutovitz, D., and Ruttledge H., Signal Processing 2, 203 (1980).
9. Piper J., in "Digital Image Processing and Analysis 1980" (J.C. Simon and R.M. Haralick, eds.), D Reidel, Dordrecht, The Netherlands, in press.
10. Rutovitz, D., in "Machine Intelligence 8" (E.W. Elcock and D. Michie, eds.), p. 455, Ellis Harwood, London, (1977).
11. Rutovitz D., (presented at "Pattern Recognition in Practise", Amsterdam, May 1980).
12. Shippey, G., Bayley, R., and Farrow, S., Anal. Quant. Cytol. 3, 9 (1981).

CONE/PYRAMID PERCEPTION PROGRAMS FOR ARRAYS AND NETWORKS

Leonard Uhr
Larry Schmitt
Pat Hanrahan

Department of Computer Sciences
University of Wisconsin
Madison, Wisconsin

This paper compares converging pyramid/cone algorithms for image-processing and pattern recognition, as executed on:

1) Traditional single-CPU serial computers;

2) Image raster-scanners like the De Anza or Grinnell;

3) Pipelines like Sternberg's 1979 Cytocomputer;

4) Parallel arrays like Duff's 1978 CLIP, Reddaway's 1980 DAP or Batcher's 1980 MPP;

5) Pyramids of computers (e.g., Tanimoto, 1981; Uhr, 1982).

Any conventional 1-CPU serial computer is "general-purpose:" given enough time and memory it can execute any program. But (also general-purpose) highly parallel multi-computers offer far faster potential speeds. Cone and pyramid programs are designed to execute efficiently on parallel array and pyramid networks.

To execute L layer (each with T transforms) perception programs on N-cell images: Serial systems need $O(1.3KNT)$ time (see Table 1 for notation). Raster-scanners need $O(KTL)$ 30 millisecond operations; a P-processor pipe can speed this up by P. Parallel arrays need $O(2TLlogR+TL)$ time. Pyramids need $O(TL)$ time, and can pipeline streams of moving images in $O(T)$ time each.

The "recognition cone" (Uhr, 1972, 1978) is a relatively parallel algorithm, embodying parallelism not only in the basic operations, which might correspond to machine instructions, but also in the control structure of the entire system of executing processes. It is hoped that by analyzing these two different kinds of parallelism one can determine the most appropriate type and number of computers, and how they might best be interconnected. The analysis of a recognition cone provides several advantages over the more traditional analysis of parallel algorithms:

1) It is a complete system that accomplishes meaningful tasks (recognition of objects).

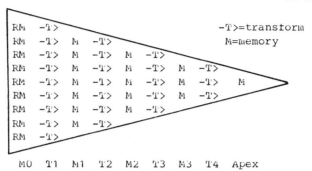

Figure 1. The Overall Structure of a Cone/Pyramid

2) It is a complex system embodying both operational parallelism and control structure parallelism.

3) The complexity of recognition cones can readily be analyzed by breaking them down into their constituent components (which are well defined) and analyzing these independently.

CONE/PYRAMID PROGRAMS, AND THE ALGORITHMS THEY EMPLOY

A cone/pyramid perceptual system (Douglass, 1978; Hanson and Riseman, 1978; Kruse, 1978; Levine, 1978; Tanimoto and Klinger, 1980; Uhr, 1972, 1978) has the following structure:

The raw sensed image (e.g., from a TV camera) is digitized and input to a large "retinal" memory (RM).

One or more "transforms" are applied, each at each cell in this retinal image, as described below. Each transform examines a window or compound of several relatively positioned parts, evaluates the result, and merges one or several implieds into the corresponding cells for those implieds in output arrays.

This process continues, with potentially many transforms applied at each layer; each transform applied at each cell; and layers shrink-ed in size, moving toward the cone's apex.

A transform might look for only 2 or 3, or for many, parts, either in a tightly packed (e.g., 3x3 or 5x5) window, or scattered more or less over the entire array. It might imply only one, or many, possibilities, with any degree of certainty (represented as a numerical weight in the layer). This can be either a discrete truth-value (i.e., "there" vs. "not-there") or probabilistic (i.e., percent certain). A transform can look for mixtures of different kinds of parts (e.g., features, qualities, contexts, sub-objects, objects).

A program is built interactively, using ICON. ICON asks what size and type transform is desired, and prompts the user to specify each part to be looked for, with its weight and relative location. It displays the transform, when completed. The user also

specifies the size of each layer, and the names of the transforms
in each layer. (The user is given a number of options, including
filing transforms under sets of descriptors, using already-filed
transforms, displaying transformed images after each layer, and
collecting statistics on a run.) The constructed pyramid is then
translated to PascalPL (Uhr, 1979), to be compiled for the available hardware (at present a conventional serial computer).

An output array can be as large as the array that stores the
input (this means both belong to the same layer). Or it can be
part of a smaller "shrink-ed" layer, so that, when several
transforms in adjacent cells output the same implied, they are
merged into a single cell in the smaller layer.

A full-blown system will have 5 to 20 layers, each with 1 to
100 transforms. The designer of a particular system chooses the
mix of types of transforms to be used.   The system's specific
knowledge is embodied in the transforms. Typically, transforms
build by compounding several lower-level transforms, thus becoming
successively more global.

Figure 2 shows a few transforms from a much larger set of the
sort needed to recognize "house," "barn," "car," and other related
objects. The layers might be organized to detect: L1) averages;
L2) gradients (like "vertgrade"); L3) short oriented edges,
colors, simple textures; L4) longer and more complex edges, curves
and textures; L5) simple angles (like "Langle"); L6) very simple
objects; L7) more complex objects (like "door"); L8) still more
complex objects (like "house"); L9) groups of objects; L10)...

Far larger sets of more complex, carefully tuned transforms
are needed to handle many different real-world objects, with their
enormous number of unknown non-linear variations. But the early
layers of transforms (which, when serial computers are used, do
most of the processing and take most of the time) contain simple
edge, color and region detectors common to most objects. The
"models" for all the different objects a system can recognize are
stored decomposed into trees of transforms embedded into the cone.
Each transform serves in many such models, giving efficiency in
storing, searching for, and choosing among possible objects.

The strategy for structuring transforms follows:

a) Use the smallest retinal array with enough resolution.

b) Continually converge the successively abstracted internal images (to save space and processing time, and to move contextually
related information successively closer).

c) Decompose several (complex) transforms with local sub-parts
in common (as, e.g., chair, couch and table have legs and edges in
common), and build several cascading tree structures with many
transform nodes in common, but a different root node for each implied higher-level object and concept.

d) Embed these transform trees in several layers, to efficiently
build and compute the complex tree functions.

e) Build transforms as much as possible to look at local regions, and accept alternate possible combinations of parts.

f) Apply transforms as much as possible in parallel (for speed).

TRANSFORM vertgrade THRESHOLD = 25 TYPE = numeric,ratio

10	:10	:20	:10	:30		:10
		:		:		

IMPLIEDS	NAME	WEIGHT	TYPE
	vertgrad	25	attribute

TRANSFORM Langle THRESHOLD = 40 TYPE = symbolic

1	:	2	:	3
edge90deg :10	:		:	
	:		:	
4	:	REF CELL	:	6
edge90deg :10	:edge00deg :5		:	
	:edge90deg :5		:	
	:		:	
7	:	8	:	9
edge90deg :10	:edge00deg :10		:edge00deg :5	
edge00deg :10	:		:	

IMPLIEDS	NAME	WEIGHT	TYPE
	Langle	40	attribute

TRANSFORM door THRESHOLD = 60 TYPE = symbolic

1	:	2	:	3
gamma :10	:edge00deg:7	:revgamma :10
	:	:
4	:	5	:	6
edge90deg:7	:constarea:3	:edge90deg:7
gamma :6	:	:bgamma :6
	:	:
7	:	REF CELL	:	9
Langle :2	:constarea:3	:edge90deg:7
edge90deg:7	:	:bgamma :2
	:	:
10	:	11	:	12
Langle :6	:constarea:3	:edge90deg:7
edge90deg:7	:	:revLangle:6
	:	:
13	:	14	:	15
Langle :10	:edge00deg:7	:revLangle:10
	:	:

IMPLIEDS	NAME	WEIGHT	TYPE
	window	20	attribute
	doorframe	40	attribute

Cone/pyramid systems embody a process structure that is inherently parallel. This structure can be implemented by an MIMD pyramid of successively smaller, converging SIMD arrays. An SIMD array is ideally suited to execute in parallel at every location in the array the sequence of operations that effects a transform.

```
     TRANSFORM house         THRESHOLD = 60   TYPE = symbolic
```

```
|     1    :        2    :        3    :        4    :        5      |
|          :             :             :             :              |
| roof     :4:roof    :8:roof    :8:roof    :8:roof      :4         |
|          :             :             :             :              |
| chimney  :8:          :             :             :chimney   :8   |
|          :             :             :             :              |
|_____6____:_____7____:_____8____:_____9____:_____10_____|
|          :             :             :             :              |
|          :roof      :4:roof    :4:roof    :4:                     |
|          :             :             :             :              |
|____11____:_____12____:___REF_CELL:_____14___:_____15_____|
|          :             :             :             :              |
| vertedge :4:window   :6:window   :6:window   :6:vertedge :4       |
|          :             :             :             :              |
|          :vertedge  :4:wall     :5:vertedge :4:                   |
|          :             :             :             :              |
|          :wall      :5:          :wall      :5:                   |
|          :             :             :             :              |
|____16____:_____17____:_____18____:_____19___:_____20_____|
|          :      •      :             :             :              |
| vertedge :4:horedge  :4:window   :6:horedge  :4:vertedge :4       |
|          :             :             :             :              |
|          :vertedge  :4:door     :8:vertedge :4:                   |
|          :             :             :             :              |
|          :wall      :5:horedge  :4:wall     :5:                   |
|          :             :             :             :              |
|          :             :wall     :5:                              |
|          :             :             :             :              |
|____21____:_____22____:_____23____:_____24___:_____25_____|
|          :             :             :             :              |
| vertedge :4:Langle   :8:door     :8:revLangle:8:vertedge :4       |
|          :             :             :             :              |
|          :horedge   :4:horedge  :4:horedge  :4:                   |
|          :             :             :             :              |
|          :vertedge  :4:wall     :5:vertedge :4:                   |
|          :             :             :             :              |
|          :wall      :5:          :wall      :5:                   |
|          :             :             :             :              |
|_____:_____:_____:_____:_____|
       IMPLIEDS     NAME       WEIGHT       TYPE
                    building      35        attribute
                    house         25        attribute
```

Figure 2. Four simple ICON-built Transforms (Schmitt, 1981). The transform is displayed (preceded by its name, threshold and type) in a 2-dimensional window. Each cell contains (to the left of the colon) objects being looked for at that relative location, and (to the right of the colon) the weight of each object. The window of parts is followed by the implieds.

THE BASIC BUILDING-BLOCKS IN CONE/PYRAMID PROGRAMS

The basic operations executed by transforms, and a cone's control operations, follow. Iterations are executed serially, or all at once when suitable parallel hardware is available.

Basic Transform Operations (Sets of Machine Instructions)

CP) Combine a pair, using one 1-bit logical or 8-bit arithmetic or string-matching instruction. This can combine two numerical weights, or use a logical-AND to test that two parts (e.g., features) in a larger whole are present. (Table 1 summarizes the instructions needed for this and subsequent operations.)

SER (Serial, with random access) will fetch each part, execute the instruction, and store the result.

PAR (Parallel, with near-neighbor links) must fetch and successively shift the parts to the same processor.

Table 1. Instructions Needed for Different Operations

Op/Instr	Fetch	Index	Shift	ALU	Test	Store
CP) SER	2	–	–	1	–	1
CP) PAR	2	–	D	1	–	1
CW) SER	HV	HV+2V	–	2HV	HV+2V	1
CW) PAR	–	–	–	1	–	–
CC) SER	P	2P	–	2P	2P	1
CC) PAR	P	2P	PD	2P	2P	1
E) BOTH	2	–	–	1	1	1
M) BOTH	2I	2	–	I	I	I
S) SER	2	–	–	2	–	2
S) PAR	2	ID	ID	–	1D	–
T) BOTH	PT+2IT	T	–	–	T	–
N) BOTH	N	N+2R	–	–	N+2R	–
L) BOTH	–	L	–	–	L	–
W) BOTH	W	W	–	W	W	–

NOTES: Notation, and useful and plausible variations:
N[R=row,C=col]=array: 64x64; 128x128; 512x512; 1,024x1,024
S=shrink: 1x1[within layer]; 2x2, 3x3[from layer to layer]
W=word size: 1[black,yes], 4-8[grey,symbol], 9-24[color]
T=transforms: 1, 5, 10, 20, 50, 100[per layer]
P=parts: 2x2,3x3,5x5,7x7,3x9,9x3[window]; 2 to 50[compound]
L=layers [in cone/pyramid]   D=average shift distance
I=implieds [in transform]   h=horizontal,V=vertical [window]
X = processors in pipe   t = hN/X [time for one scan]
  k = slowdown from indexing, tests, etc. [2<k<100]
  h = speedup from specialized hardware [2<h<100]

CW) Window operations fetch the value stored in each cell within the H*V window surrounding each cell, and combine these values (and, optionally, their weights).

SER gets the parts in a window with indexes and border tests.

PAR uses special window hardware (e.g., 1 parallel CLIP instruction fetches all 9 cells, operates and stores the result).

CC) Compound operations. Like window operations, except use lists of P X,Y offsets to compute locations of each part of the compound. Offsets can reference any location within the layer, whereas windows are local and tightly packed.

SER can fetch each part.

PAR must shift each part.

E) Evaluate the result of window and compounding operations (typically: a transform succeeds if the combined weight exceeds the threshold).

M) Merge the transform's I implieds (after the transform has been evaluated as succeeding). This entails: fetch each old value from the corresponding cell of the output array, combine it with the new incremental value, store the result.

S) Shrink the results into a smaller output array in the next layer. PAR must shift.

Serial systems with the entire array in random access main memory can efficiently (but very slowly, since serially) compute the relative location of the parts to be combined, and the cell into which results are to be converged and merged. The X and Y offsets are each fetched and added to the X and Y locations of the present cell, then one instruction fetches the part from that cell. This needs roughly 5 instructions, which must be iterated at each cell in the array.

Parallel systems with near-neighbor links must execute a sequence of shifts to bring together the parts to be combined, and to move implied information to the cell into which it must be converged. The sequence of shifts averages (R+C)/2 for shrinking, and up to R+C to combine, window and compound. A cone can (and should) be built to minimize shifting, by applying transforms as much as possible within a single layer (with no shrinking), converging only when necessary, with compounds as local as possible.

Basic Overall Control Structures

T) Iterate (for serial processors) thru all T transforms, executing every transform in each layer at each cell. This entails: one FOR..DO loop, index and test for end.

N) Iterate (when processing serially) thru an N=RxC image array: 2 nested FOR..DO loops, index, test for borders.

L) Iterate (when processing serially) thru all L layers: one FOR..DO loop, index, test for end.

W) Iterate (if the word is smaller than data) thru word-serial loops (bit-serial with 1-bit processors).

Table 2. Instructions/Time Needed for Operations and Programs

Op/Computer:	Serial	Array	Pipe	ImProc	Cone/Pyr
CP)	4N	D+4	tD/hX	4t/hX	logD+4
CW)	5HVN+4VN+N	1	t/nX	tHV/hX	5HV
CC)	7NP+t	7P+PD+1	tPD/hX	tP/hX	7P+PD+1
E)	5N	5	5t/hX	5t/hX	5
M)	5NI+2N	5I+2	5tI/hX	5tI/hX	5I+2
S)	6N	3ID+2	3tID/hX	5t/hX	6

NOTE: See Table 1 for notation.

Table 2 shows the number of instructions needed for particular operations for each kind of computer. Since instructions executed in parallel count as one, this equals time.

ALTERNATIVE ARCHITECTURES

1) Traditional Serial Computers

The conventional computer's single processor executes one instruction at a time, serially. Any information in the entire array can be accessed in one fetch instruction when memory is large enough. Otherwise very slow shifting is needed from and to disk.

2) Commercial Image-Scanning Processors

Raster-scanning computers like the Comtal, De Anza, Grinnell and Stanford have a specialized processor that executes a short pipe and/or window of instructions at each array cell in one TV scan (typically, one 8-bit arithmetic or 1-bit logic operation over a 512x512 array in 30 milliseconds). Some systems execute in parallel in three color channels. Some execute a 3x3 or 5x5 window operation (usually a convolution). Some combine two bytes shifted from any locations.

Thus 250,000 operations are executed in each 30 millisecond scan, or roughly 100 nanoseconds per operation. Each operation consists of two, or more, fetches, adds, multiplies, etc. (e.g., 9 of each in a 3x3 convolution). Since the single scanning processor has random access to all image memories, a single fetch replaces the long sequence of shifts needed with near-neighbor connections. This makes possible cone systems with no shift overhead (see Hanrahan, Schmitt and Uhr, 1981). Indexing, border tests and other scanning overhead are built into hardware.

3) Pipeline Computers of Many Specialized Processors

Long pipelines of computers have been built, at least one with over 100 powerful specialized processors. The Cytocomputer, with 118 processors, executes 3x3 window operations (both arithmetic and logical/structural) over a 1024x1024 array of cells at TV scan speeds. A new LSI system will allow one to lengthen the pipeline indefinitely, and to vary the size of the array processed at each scan. Shifting dominates, since these systems have the basic near-neighbor window capability. Because of their very small intermediate memory it may be hard to program such pipelines with 118 useful instructions.

4) True Hardware-Parallel Arrays

An array of many computers can achieve enormous speedups, by executing the same instruction in parallel at every cell. For example, CLIP4 can compute an "OR", or an omni-directional difference, or an expand, or some other function of all 8 neighbors (in its hardwired 3x3 window) plus the center cell in one 11 microsecond machine instruction for all 9,216 cells in a 96x96 array. A serial computer needs 5 to 100 instructions for each cell!

DAP and MPP fetch only one horizontal or vertical neighbor at a time, and need two instructions for each diagonal neighbor. So they sometimes need up to 15 100-200 nanosecond instructions where CLIP needs only one. But many of the instructions in a perception program test for 2 lower-level features, in a conditional expression that implies a higher-level feature or object. For this DAP and MPP need only 2 to 4 instructions.

If the image array is larger than the computer array, each computer can iterate thru a sub-array stored in its memory. Arrays can be used efficiently only when the array of information is the same size as, or an exact multiple of, the array of processors. So images should be input and shrink-ed to fit the hardware; otherwise much extra indexing and I-O are needed. (This is similar to scaling numbers to fit the word-size.)

Nearest-neighbor shift instructions will quickly dominate, unless algorithms examine regions that are as local as possible. To combine and compound, data must be shifted thru each cell that lies between the cells containing the data to be combined. To shrink, the implied output must similarly be shifted to the converge location in the output array.

Asssume an array is being used to simulate a 10-layer cone with 1,000 transforms, averaging 3 parts each. Each transform needs 1-3 instructions, plus (possibly long) sequences of shifts to bring the parts together. To converge, each part must be shifted on the average $(R+C)/4$ times. E.g., in a 500 by 500 array an average of 250 shift instructions are needed to shrink each of a transform's implieds. This can be minimized by transforming as much as possible within each layer, and shrinking only rarely.

5) Cones of SIMD and MIMD Arrays

A cone of arrays could be designed to combine most of the virtues of these different systems, and minimize their drawbacks. Such a system would form a 2-dimensional converging pipeline appropriate for a continuing stream of images, as input by a TV camera. A 10-layer cone would have its apex processing the results from an image input 10 frames in the past, while its first layer of transforms was processing the most recent image. To minimize links, only one computer in each 2x2 or 3x3 sub-array need be connected to the next layer. Thus each computer now has 1 extra link toward the retina, and 1/4th or 1/9th have 1 extra link toward the apex. (This opens the possibility of links to input, output and mass storage devices for the other 3 or 8.)

A cone needs $RC+RC/2S+RC/(2S)^2+RC/(2S)^3...RC/(2S)^L$, or at most 1.33 times the memory needed for the retina array alone (assuming 2x2 shrinks, every layer with the same total number of bit-planes and no overwriting of memory). E.g., a 1024x1024 retina with 8-bit images and 7 8-bit image-processing transformations with no shrinkage needs 1 MByte of memory for the image, plus 7 MBytes for the transformed images, 2 MBytes for the 2nd layer, and 1/4th as much for each subsequent layer. This gives 8 MBytes for the first layer, and less that 3 MBytes for all subsequent layers of internal images (each with 72 bit-planes to store information).

Each array of computers would have its own controller, and could be specialized as desired (e.g., to process windows, to execute appropriate perceptual processes, to handle 8-bit or 32-bit numbers or symbols, to execute different instructions). Here the cost of making the individual computer's hardware more powerful must be compared with the cost of using more computers. If it were known that T transforms were to be applied at each layer, a system with exactly T computers at each cell layer could effect each layer in one step (one computer would take T steps). More realistically, one 1-bit computer might be assigned to each cell of each layer, with simple near-neighbor links. Convergence thru the cone would eliminate most of the shifting overhead. Toward the apex, more powerful (and expensive) MIMD computers might be used.

SUMMARIZING COMPARISONS

The basic features of the architectures examined include:
a) One vs. several vs. many processors;
b) Word-size of processors (1, 4, 8, 16, 32 or 64 bits);
c) Data fetch from near-neighbors vs. anywhere;
d) Serial data fetch vs. parallel window fetch;
e) Hardwired scanning and indexing functions;
f) Hardwired specialized image-processing operations;
g) One vs. several arrays, identical vs. converging;
h) SIMD vs. MIMD vs. SIMD/MIMD.

1) A serial computer has a powerful processor, but only one. It makes one fetch, anywhere. It has no special scanning or image-processing hardware (but it could). Serial computers are inevitably slow because they must iterate over each cell of each array in each layer. And each transform is a sequence of instructions. They need $O(1.3kNT)$ time ($k>10$, since k includes the transform's sequence of instructions plus indexing and border-testing overhead for iterating). A typical instruction takes 1 to 10 microseconds, and a typical transform will need 10 to 100 instructions. The number of cells in the retinal array dominates.

2) Image-Scanners have built-in scanning of one or a few 8-bit specialized processors with random access. Specialized scanning hardware and image processing instructions can reduce time appreciably. They need $O(30kTL)$ time, since they must execute the entire 30 millisecond scan for each array, no matter how small. A typical instruction takes 100 nanoseconds, and a typical transform needs 1 to 20 instructions. But the 8-bit processor must sometimes handle longer strings and numbers in slow word-serial fashion, compared with a 32-bit serial computer. 30 milliseconds is very slow compared to the time available to process moving images, or to a parallel array that executes comparable operations in 1 microsecond. But random access eliminates shifting. If variable-sized arrays could be processed at 100 nanoseconds per cell, such a system would need $O(33.3kT)$ time, and could apply many transforms on 128x128-base cones in 30 milliseconds.

3) Pipelines have built-in scanning of 100 or more simple image-processors with near-neighbor fetches and windows. A 100-processor pipe can, when it is filled, give up to 100-fold speed-ups over the commercial image processors. The window operation can replace up to 9 instructions. But the processors (because 1-bit and 3-bit) give word-serial slowdowns when larger strings and numbers must be handled. Time is reduced by the number of processors, to $O(30kTL/X)$. Near-neighbor links mean that to compound and shrink shifting dominates, to give $O((30kTL/X)(2logR+1))$.

4) Arrays have many processors, with near-neighbor fetches. Some fetch and execute operations on whole windows of data. But their 1-bit processors must work bit-serially on longer words. They have special logic and bit-serial arithmetic hardware, but no raster-scanning hardware (none is needed when there are as many processors as transforms at each cell in the image). An array as large as the image needs only $O(TL)$ time when all parts of a transform are within a window or can be fetched directly and there is no convergence. But near-neighbor links mean that $O(2TLlogR+TL)$ time is needed and shifting dominates.

5) Cones could be built with all the above features except random memory access by all processors. The converging links from layer to layer can reduce shifting from $O(2TLlogR)$ to zero.

Table 3. Hardware Features of Each Type of Computer

Type:	Proc	Word	Fetch	Window	Scan	ImProc
Serial	1	8,16,32	any	1	none	none
ImScan	1-10	8	any	1,9	yes	yes
Pipe	100	1,3	9	9	yes	yes
Array	10000	1	5,9	1,9	none	yes
Cone	13333	1,8	5	1	none	yes

The cone also serves for efficient message-passing: an array's diameter is R+C, but an array/cone's diameter is logR+logC. A cone with base as large as the input image, and subsequent layers as large as the reduced images, would need only O(TlogR) instructions to process each image, and would perceive sequences of images input to its 2-dimensional pipeline as fast as each layer of transforms could be processed. The fastest speed could be achieved by having a processor for each transform; then a new image could be input at the same rate that one, rather than a whole layer of, transforms could be executed (with specialized window processors, in only 3 to 10 instructions).

DISCUSSION AND CONCLUSIONS

To achieve perception programs fast enough to handle real-time 30 millisecond TV images, large parallel arrays appear to be essential. Built-in parallel operations on a word and/or a window can each give 9, 25 or 32-fold speedups; but the average is usually 2, 3 or 8. Special index and border-test hardware can give 10-fold increases in speed. Specialized image-processors, plus a large random access memory to eliminate shifting, variable array scans (under program control), and a capability to converge and merge output, can greatly speed up a basically serial device.

As image size increases, shifting needs dominate in systems with near-neighbor links. This can be minimized by keeping transforms local, and using non-converging transforms as much as possible, interspersing converging transforms as needed to move information closer.

Tree and cone links between converging layers can eliminate most near-neighbor shifts. Pipelining continuing images adds the parallelism of the several layers.

Ignoring costs, the fastest system would use a cone of true parallel arrays of 32-bit window-parallel computers large enough to assign a separate computer to each transform at each cell of the input image, with all computers sharing a very large common memory. But this would be prohibitively expensive, and it seems unlikely that programs could be written that efficiently used large numbers of such powerful computers. Each computer should almost certainly be made much simpler, probably with little or no parallel window capability (since the full window is rarely used), and with 1-bit words (or 8-bit at most). When 100 simple 1-bit processors can be built for the price of one powerful 32-bit processor, these two alternatives should be carefully compared.

Probably the most effective use of parallelism is to expand the computer array to be as large as the image it processes (thus each processor will always have image cells to work on). Each time array-size quadruples, processing time reduces to 1/4th or less, since each processor must iterate thru a sub-array 1/4th the size, and indexing overhead may be eliminated. In contrast, increasing by N the word-size, window-size or processors at each cell will give an average increase substantially less than N.

Near-neighbor arrays best fit parallel local operations. Pyramids handle shifting and shrinking relatively efficiently. They effect global transformations by applying a converging tree of local operations. They are best suited to algorithms that decompose and spread the load evenly over each array and from layer to layer, using simple, local, ever-more global processes. This is what layered converging cone programs attempt to do.

REFERENCES

Batcher, K. (1980). Proc. 7th ACM Symp. Comp. Arch. 168.
Douglass, R.J. (1978). Comp. Sci. Dept. TR 317. U. Wis.
Duff, M. J. B. (1978). Proc. NCC, 1055.
Hanrahan, P., Schmitt, L. and Uhr, L., (1981). Comp. Sci. Dept. TR. U. Wis.
Hanson, A. and Riseman, E. (1976). COINS TR 76-9. U. Mass.
Kruse, B. (1978). Proc. NCC, 1015.
Levine, M. D. (1978). In "Computer Vision System" (A. Hanson and E. Riseman eds.), p. 335. Academic Press, New York.
Reddaway, S.F. (1980). In "Electronics to Microelectronics" (W.A. Kaiser and W.E. Proebster, Eds.), p. 730. North-Holland, Amsterdam.
Schmitt, L.A. (1981). Comp. Sci. Dept. TR 421. U. Wis.
Sternberg, S.R. (1978). Proc. 8th NBS Pat. Recog. Symp.
Tanimoto, S.L. (1981). Comp. Sci. Dept. TR 81-02-01. U. Wash.
Tanimoto, S.L. and Klinger, A., eds. (1980). "Structured Computer Vision," Academic Press, New York.
Uhr, L. (1972). IEEE Trans. Comput. 21, 758.
Uhr, L. (1978). In "Computer Vision Systems" (A. R. Hanson and E. M. Riseman, eds.), p. 363. Academic Press, New York.
Uhr, L. (1979). Comp. Sci. Dept. TR 365. U. Wis. also: (1981). In "Languages and Architectures for Image Processing" (M.J.B. Duff and S. Levialdi, eds.). Academic Press, London.
Uhr, L. (1982). "Computer Arrays and Networks: Algorithm-Structured Parallel Architectures," Academic Press, New York.

PARALLEL ALGORITHMS FOR INTERPRETING
SPEECH PATTERNS

Renato De Mori
Attilio Giordana

Istituto di Scienze dell'Informazione
Università di Torino
Torino, Italy

Pietro Laface

C.E.N.S.—Istituto di Elettrotecnica Generale
Politecnico di Torino
Torino, Italy

The paper describes a system capable of generating hypotheses
about the phonetic features and the phonemes coded into a speech
waveform. Decoding is performed by a pluralism of interpretation
activities performed by a society of experts.

1. INTRODUCTION

Distributed processing is now practical (Lesser, 1980). This
makes it attractive to consider very complex tasks such as speech
understanding or speech recognition in a framework of distributed
problem solving.

The attention is focused in this paper on a system capable of
generating hypotheses about the phonetic features and the
phonemes coded into a speech waveform.

Decoding is performed by a pluralism of interpretation
activities performed by a society of experts according to the
scientific community metaphor (Kornfeld and Hewitt, 1980).

Following the experience in interpreting speech data gained
by the authors as well as by many other researchers in the world,

the behavior of various experts capable of extracting acoustic cues and phonetic features has been characterized.

The task of hypothesis generation is decomposed into a number of sub-tasks performed by a society of basic experts capable of using knowledge sources and dynamically creating several instantiations of them.

Instantiations are computing agents similar to the Hewitt's actors or sprites, they cooperate by exchanging messages. They are generated by a basic expert, they use its knowledge and communicate with other experts for obtaining all the information they need to achieve their task.

There are two main motivations for using this model.

— The first one is that phonetic and acoustic knowledge acquired by human experts about speech sounds is scattered in a variety of propositions and conjectures. This makes it impossible to avoid generating multiple interpretation hypotheses which grow up concurrently.

— A second motivation is that a distributed model of an expert phonetician is coherent with the models of other components of a speech understanding system. These models must be conceived as distributed problem solvers because the system has to work close to real-time facing ambiguous data and a very large variety of possible solutions.

Many valuable speech recognition systems do not use rule-based experts for phoneme hypothesization. In spite of this, we believe that an attempt to model human knowledge in interpreting speech patterns may contribute to a progress in the speech scientific community for the following reasons:

1. Conceiving models and making experiments with them may help in improving our knowledge about speech perception;

2. automatic speech recognition systems with complex tasks such as voice activated typewriters require highly accurate phoneme hypothesization;

3. rule-based experts for phoneme recognition are promising for speaker-independent, task-independent and perhaps language independent applications provided that suitable rules are given to the system.

Learning can be performed by updating or modifying the rules.

2. SYSTEM STRUCTURE AND KNOWLEDGE MODEL

Fig. 1 shows a portion of the structure of a Speech Under-standing System. The Portion refers to the so-called 'task-independent knowledge' whose organization and use will be dis-cussed in this paper.

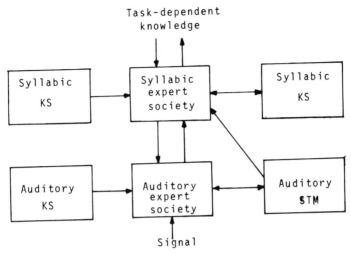

FIGURE 1. Task-independent knowledge of a SUS

Task-independent knowledge is represented by two sets of rules stored into a sort of Long Term Memory (LTM). These two sets are respectively: the Auditory Knowledge Source (AKS) and the Syllabic Knowledge Source (SKS).

Knowledge Sources (KSs) are used by reasoning agents called 'Experts'. An expert receives requests for the generation of hypotheses from other experts.

Hypotheses are data-structures written into a sort of Short Term Memory (STM) associated to each expert.

Experts are grouped into Societies according to their level of expertise.

The task-independent knowledge is structured on two levels corresponding to the Auditory Expert Society (AES) and the Syllabic Expert Society (SES).

The AES has the task of transforming the speech waveform into a three-dimensional pattern from which most of the acoustic cues can be more easily detected than from the waveform.

The most popular type of patterns is the spectrogram which is a succession of short-time spectra built along the dimensions of time, frequency and energy.

Another task of the AES is that of interpreting the requests from the SES. These requests may concern the extraction of acoustic cues for phonetic features like VOCALIC, PLOSIVE-CONSONANT etc. or for phonemes like /p/, /m/ etc.

The behavior of an expert is shown in Fig. 2.

Each expert EXPj is associated a KSj stored into LTMj and a data-base stored into STMj which contains hypotheses.

The STM of an expert society is partitioned into a set of disjoint parts SMTj (j = 1,2,.....J).

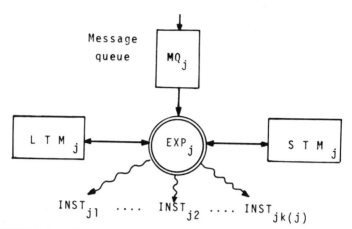

FIGURE 2. Behavior of an expert

Each expert EXPj is also associated a message queue MQj containing the requests made to it from other experts. EXPj reads sequentially these requests. If a request concern some information which has not been requested before, then EXPj creates an instantiation of KSj. Let INSTj1, INSTj2,....,INSTjk,....., INSTjk(j) the instantiations created at a given time tk.

An instantiation is a computing agent that may create other instantiations of KSj or send requests to other experts or send answers to the experts which have made requests to EXPj. In other words, an instantiation INSTjk can send a message MESSjkl to other experts.

Messages for the experts can be stimuli from lower level experts or verification requests from higher level experts or commands from a strategy KS.

When an instantiation has performed its task, it dies and leaves the system.

The Auditory experts use some of the rules of AKS for extracting acoustic cues such as dips or peaks of energies in certain bands or the frequency values of the zones of energy concentrations in a segment for which the phonetic hypothesis VOCALIC has been previously generated. These values are called formant loci.

Rule-driven generation of hypotheses about acoustic cues is a typical example of syntactic (structural) pattern recognition. Furthermore, as the data can be vague and the rules are allowed to be imprecise, the hypotheses are affected by a degree of plausibility making the structural approach a hybrid one.

The syllabic experts first segment the speech pattern into Pseudo-Syllabic-Segments (PSS). In order to perform segmentation, the experts generate hypotheses about broad phonetic features which do not require context-dependent rules for their detection.

Phonetic features are hypothesized by some rules of SK. These rules relate phonetic features and acoustic cues.

As phonetic features are hypothesized with a degree of imprecision, PSS hypotheses may be ambiguous and partially overlapped in a lattice structure. A PSS hypothesis delimits a time interval in which context-dependences due to coarticulation effects are expected.

The details of this rule-based approach are reported in (De Mori, 1981). The rest of this paper will be devoted to the generation and parallel execution of the computing agents.

3. THE SOCIETIES OF THE AUDITORY AND SYLLABIC EXPERTS

Fig. 3 shows the experts of the auditory (below) and syllabic (above) societies.

The speech signal is sampled, quantized and transformed by an expert called 'Auditory Expert for end-Points Detection and Signal Transformation' (AEPDST). AEPDST looks for the starting point of a sentence by using a subset of Rules-11. When this point has been detected, AEPDST starts transforming the signal in order to obtain a frequency-domain representation of it that is stored in the STM of signal transformation.

After a long enough part of the signal has been transformed, a synchronization signal is sent to the Syllabic Vocalic Expert (SVE).

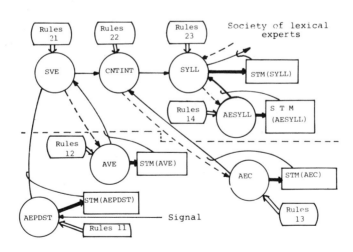

FIGURE 3. Experts of the auditory and syllabic societies

SVE has the task of hypothesizing the presence of vowels in a time interval in which spectral transformations are available. At the same time, AEPDST goes on and transforms another portion of the signal until a sentence end-point is detected by applying another subset of Rules-11.

SVE generates hypotheses about the presence of vowels by applying the following Rule-21:

VOCALIC: = acl1/w1 + acl2/w2 + acl3/w3 + acl1 * acl2/w4 + acl1 * acl3/w5 + acl2 * acl3/w6 + acl1 * acl2 * acl3/w7 (1).

The symbols acli ($i = 1,2,3$) represent acoustic cues for the phonetic feature VOCALIC, while wj ($1 = 1,2....7$) represents the weight of the j-th string in the rule. The weights may vary between 0 and 1 and express the imprecision of the associated string in characterizing the feature VOCALIC.

In order for SVE to be able to generate the vocalic hypotheses in a given time interval, a request (dashed arrow in Fig. 3) is sent to the Acoustic Vocalic Expert (AVE).

AVE applies Rules-12 in the requested time interval. Under the control of these rules, the acoustic cues acli are extracted in each peak of the signal energy. A set of acoutic cues [acli] is sent to SVE for each one of the total energy peaks detected in the requested time interval. An example of acoustic cue is 'high ratio between low and high frequency energies in a peak of total energy'.

As 'high' is a vague concept, the degree of vagueness MU(acli) is defined by a membership function which maps the ratios into the interval [0 - 1].

The cues detected by AVE together with their degrees of vagueness are stored into the short-term-memory of AVE and a reference to them is sent back to SVE.

SVE applies Rules-21 to the received features and computes the plausibility of the hypothesis VOCALIC given the acoustic features. This plausibility MU(VOCALIC) is a value of the interval [0 - 1] computed from the weights of the strings and the memberships of the acoustic cues using the operations of the fuzzy algebra. By collecting the values of the membership MU(VOCALIC) for the vocalic segments and for other peaks of the total energies, the probability density p (VOCALIC/MU(VOCALIC)) can be estimated and a decision criterion based on the value of MU(VOCALIC) can be established in order to ensure that the probability of missing the right hypothesis is kept lower than a desired threshold.

For other phonetic features, the rules are more complex than Rule-21 and can be those of a context-free or a context-sensitive grammar.

Based on the decision criteria, vocalic hypotheses are generated for some of the peaks of the signal energy. As soon as these hypotheses are generated, they are sent to another syllabic expert called CNTINT.

CNTINT considers every possible pair of vowels and assigns some consonantal features to the intervals between pair of vowels. In order to perform these operations, CNTINT sends requests to the Acoustic Expert for Consonantal intervals (AEC) which sends back references to acoustic cues stored into its short-term-memory. CNTINT applies Rules-22 to every set of acoustic cues it receives and generates hypotheses about some phonetic features.

Notice that the Rules-22 relate phonetic features with acoustic cues in a context-independent way; furthermore the rules are affected by degrees of imprecision and CNTINT is capable of combining the degrees of rule imprecision together with the vagueness with which the acoustic cues are detected.

The consonant features detected by CNTINT are reported in Table 1.

Hypotheses about consonantal features are sent as message from CNTINT to the Syllabic classification expert (SYLL).

SYLL uses Rules-23 and the messages from CNTINT for generating hypotheses about the bounds of PSS. Each PSS hypothesis is described in terms of the features of Table 1 and stored into the SYLL's short-term-memory.

Each PSS delimits a time interval in which enough information can be made available in order to apply context-dependent rules. Context-dependent rules describe coarticulation effects.

PSSs are organized in a lattice of syllabic hypotheses. Fig. 4 shows an example of a fragment of such a lattice.

In parallel with the generation of PSS hypotheses, the SYLL expert attempts to generate more detailed hypotheses, about each PSS. For this purpose it sends requests to the Acoustic Expert for Syllabic features (AESYL) which uses Rules-14 for extracting acoustic cues according to the requests it receives. References to the requested cues are sent back to SYLL. SYLL uses a subset of Rules-23 which is invoked by the pattern of the consonantal features representing the PSS for which more detailed hypotheses are looked for.

SYLL can be requested to generate or verify syllabic hypotheses also from the Society of Lexical Experts. It reacts to these requests by establishing dialogs with AESYL.

For each message received by an expert, an instantiation of its knowledge is created. Instantiations are expert activities;

TABLE 1. *Consonantal Features*

SON: *tract with one or more sonorant sounds.*
UT: *tract with more than one nonsonorant sounds, or a nonsonorant sound in a cluster of consonants.*
NC: *single intervocalic nonsonorant continuant consonant.*
NA: *single intervocalic nonsonorant affricate consonant.*
NIL: *a single intervocalic nonsonorant interruped lax consonant.*
NIT: *a nonsonorant interruped tense consonant.*

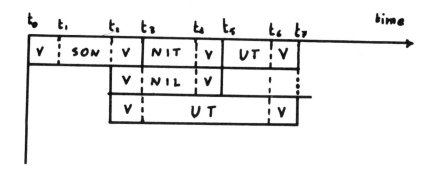

FIGURE 4. Fragment of a syllabic hypotheses lattice

parallel improving the speed of the system. A computational model for parallel execution of KS instantiations will be introduced in Section 5.

4. EXPERT BEHAVIOR

Some of the activities of the syllabic experts can be represented as the solution of a complex problem which can be decomposed into subproblems whose individual solutions are simpler.

Problem Reduction Representation (PRR) is a good formalism (used in Artificial Intelligence) for a better understanding of problem decomposition.

Fig. 5 shows the PRR for assigning the labels which are needed for segmentation to a consonantal interval.

A semantic information is associated to the graph. It establishes that problems are solved 'Top-down', furthermore the leftmost subproblem must be solved first. AND nodes are indicated by circular arrows: dashed arrows indicate the invocation of subproblems of the SYLL expert. Invocation of such subproblems is performed by the labels associated to the dashed arrows. These labels are transmitted together with vocalic hypotheses and degrees of plausibility to SYLL. SYLL uses the messages for segmentation and for invoking subproblems for the generation of detailed phonetic features.

The types of phonetic features detected by the terminal subproblems of CNTINT are reported in Table 2. Each one of these subproblems is solved by the use of context-independent rules relating acoustic cues to the phonetic features which is the label of each subproblem.

Fig. 6 shows the PRR for UT which contains also some of the subproblems invoked by the PRR of CNTINT (NA,NC,NIL,NIT).

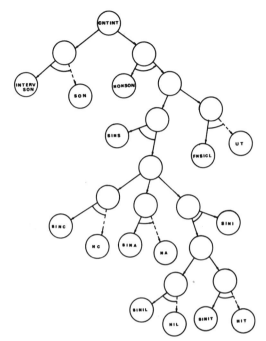

FIGURE 5.   PRR for consonantal interval labelling

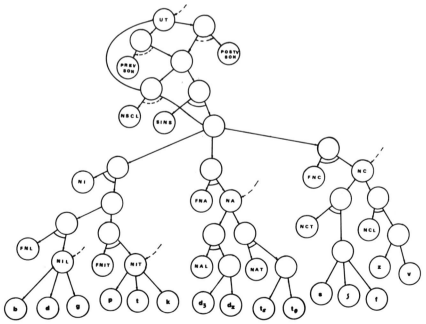

FIGURE 6.   PRR for a UT labelled interval

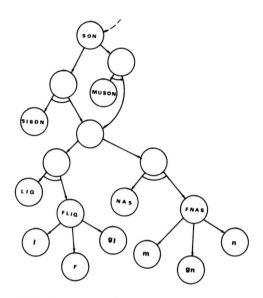

FIGURE 7. PRR for a SON labelled interval

TABLE 2. Types of Phonetic Features

PREVSON: prevocalic sonorant consonant in a cluster.
POSTVSON: postvocalic sonorant consonant in a cluster.
INTERVSON: intervocalic sonorant consonant.
NONS: nonsonorant consonant.
NSCL: nonsonorant consonant in a cluster
SINS: single intervocalic nonsonorant consonant.
SINI: an interrupted SINS.
SINA: an affricate SINS.
SINC: a continuant SINS.
NAT: nonsonorant affricate tense.
NAL: nonsonorant affricate lax.
NCT: nonsonorant continuant tense.
NCL: nonsonorant continuant lax.
SINIT: single nonsonorant interruped tense.
SINIL: single nonsonorant interruped lax.
SISON: single sonorant consonant.
MUSON: multiple sonorant consonant.
LIQ: liquid sonorant consonant.
NAS: nasal sonorant consonant.

Fig. 7 shows the PRR for SON. Again the semantics of the PRR imposes to solve the problem in a Top-down way always starting with the leftmost subproblem.

Table 2 contains also the types of phonetic features detected by the terminal subproblems of UT and SON. Most of these subproblems use context-dependent rules for relating phonetic features with acoustic cues.

A semantic is associated with the PRR which imposes a partial ordering in the solution of the subproblems for ensuring that context are available when context-dependent rules are applied.

In spite of these constraints a fairly good degree of parallelism can be achieved in solving these problems.

Details about the rules can be found in (De Mori, 1981). A learning algorithm for the rules has been provided by (De Mori and Saitta, 1980).

5. THE COMPUTATIONAL MODEL

Fig. 8 shows a scheme of actor creation inside an expert subsystem. A *controller* which is a permanent actor, decodes the messages from MQj. For each message representing a request of information not yet available in STMj, a set of four actors is created from the controller.

Rule-applier is an actor which sends messages to other experts based on the rules it has to apply.

Evaluator is an actor which receives the answers to the questions advanced by the *Rule-applier* and performs evidence composition.

Decision maker receives evidence qualifications from the *Evaluator* and decides whether it is worthwhile to apply again the rules to other data by creating new sets of actors.

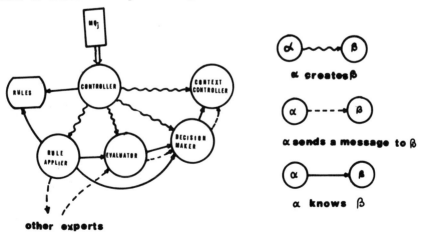

FIGURE 8. *Actor creation inside an expert*

TABLE 3. *Summary of Experimental Results (Four Speakers)*

Feature	Overall error rate
Sonorant/Nonsonorant	5%
Continuant/Interruped Africate	7%
Tense/Lax	6%
Nasals	6%
Plosives	1%
Vowels	3%

Context controller controls the access and the arrival of new hypotheses from the *Decision maker*.

When it is worthwhile for *Decision maker* to continue the problem solving activity, it may create one or more triplets of the first three actor-types introduced above. The *Decision maker* creates more than one triplet when it encounters a disjonction of possible continuations.

CONCLUSIONS

Experiments on several hundreds of syllables uttered in continuous speech by four male and one female speaker gave the right interpretation with the highest evidence value in more than 90% of the cases.

Table 3 summarizes some experimental results obtained in the bottom-up generation of hypotheses about the feature "sonorant" (INTERVSON) and "nonsonorant" (NONS). These results show an average error rate of 5%.

Formant tracking takes much of the computation time and is the bottleneck of the system. This inconvenient is avoided by tracking in parallel the formants for various vocalic intervals.

Simulation results have shown that with an average parallelism degree of 12 the whole process can be done in real-time using standard multi-microprocessor architectures.

REFERENCES

De Mori, R. (1981). "Computer models of speech using fuzzy algorithms" Plenu". Plenum Press, New York.
De Mori, A., and Saitta, L. (1980). "Automatic Learning of Fuzzy Naming Relations over Finite Languages". *Information Sciences*, 93, 139.

Giordana, A., Laface, P., and Saitta, L. (1980). "Modelling Control Strategies for Artificial Intelligence Applications". *Proc. Int. Conf. on Parallel Processing*, Boyne Highlands, 347, 349.

Kornfeld, W. A., and Hewett, C. (1981). "The Scientific Community Metaphor". *IEEE Trans. SMC-11*, 24, 33.

Lesser, V. R., and Corkill, D. D. (1981). "Functionally-Accurate, Cooperative Distributed Systems". *IEEE Trans. SMC-11*, 81, 96.

COMPUTING OCCLUSION WITH LOCALLY
CONNECTED NETWORKS OF PARALLEL PROCESSES

Robert J. Douglass

Department of Applied Mathematics
and Computer Science
University of Virginia
Charlottesville, Virginia

I. INTRODUCTION: PARALLEL ALGORITHMS FOR THREE-DIMENSIONAL COMPUTER VISION

Many applications in image processing and computer vision require the derivation of depth information from an image or sequence of images of a three-dimensional scene. Although several attempts have been made to implement some of the traditional depth cues such as stereopsis, motion parallax, and texture gradients, computing such information involves a combinatorial search. Recent work has focused on the formulation of algorithms for depth perception in terms of parallel processes to speed up the derivation of three-dimensional information by taking advantage of high speed parallel architectures (Ullman, 1979; Marr and Poggio, 1976; Douglass, 1981a).

The parallel occlusion algorithm developed in this paper is typical of a number of image processing and image understanding problems such as relaxation labeling and stereo matching. It can be computed using a class of algorithms known as parallel image graph algorithms (Douglass, 1981b). In an image graph algorithm, one process is assigned to each node of an image described as a graph of image components such as regions or edges.

Briefly stated, image graph algorithms can be characterized as consisting of a closely coupled set of similar or identical processes operating asynchronously on different parts of an image description. Processes typically are assigned to each component of an image description, and the interprocess connections represent the adjacency of image components. For purposes of this discussion an image graph will be considered to consist of a set of region descriptions produced by segmenting an image into connected areas of approximately uniform color and texture. Com-

munication links between processes correspond to adjacencies
between regions in the image. Since regions may be merged or
split apart, and the exact number and configuration of regions
may differ from image to image, a system for processing image
graphs must be dynamically reconfigurable.

Image graph algorithms do not fit cleanly into the SIMD mold
of synchronized arrays of individual operations, one per pixel,
nor can they be accurately characterized as distributed
processing algorithms where distinct processes execute asynchronously and communicate only occasionally. Douglass (1981b) reviews the characteristics of image graph algorithms and gives
examples of parallel programs for processing image graphs.

II. MAC AND THE CREATION OF A NETWORK OF REGION PROCESSES

A. The Programming Language MAC

MAC is an extension of Brinch Hansen's distributed processes
language, and is intended for applications in parallel image
processing (Douglass, 1981b). It uses a PASCAL-like syntax plus
language constructs for defining parallel processes and controlling communication between them. MAC allows a programmer to
define a prototype process, and then replicate instances of that
process on each element of a data structure such as an array or
an image graph. Once activated, processes run asynchronously,
but may communicate via a READSHARED operation which enables them
to examine the state of other processes' data. MAC provides MIMD
control constructs for activating processes and managing the
communication between them.

MAC programs consist of two sections: a PROCESSTYPE section
which defines prototype processes and a PROCESSVAR section which
declares actual instances of processes. A key construct is the
process pointer, which allows the creation of a network of
processes by declaring an array of pointers to processes and
allowing each process to have a process pointer to its neighbors.

The basic element of a MAC program, the process, consists of
a process name followed by definitions of process variables and
subprocesses. Each process includes definitions of procedures
(PROCs) and functions (FUNCs) which operate on the process'
variables and are callable by other processes. In addition,
every process has an initialization section (INIT). When a
program begins execution, processes are activated and perform
their initialization section. Once initialization is finished,
or whenever a process is blocked waiting on a condition to become
true, the process is free to execute any of its procedures or
functions called by other processes.

The programs below illustrate some of the features of MAC.
Additional MAC constructs are explained in program comments as
they are introduced.

B. Creating a Network of Region Processes

The occlusion algorithms described in the next section work with images that have been segmented into regions of approximately uniform color and texture. This section introduces the skeleton of a segmentation routine which produces a new representation of the image in terms of region descriptions. The MAC program below, called 'Create-regions', differs from a typical segmentation routine in that it creates a set of active processes, one per region, rather than producing a data structure which describes the image.

'Create-regions' as described below is a one pass raster scanned algorithm adopted from Yakimovsky (1976). It preprocesses the image into a set of edge values for each pixel and then grows a region by adding pixels to it until a peak of edge values is crossed, whereupon a new region is created. Two points should be noted about 'Create-regions': although 'Create-regions' itself is strictly sequential, it produces a set of independent parallel region processes as its output, and the method used to perform the segmentation is irrelevant to the functioning of the parallel occlusion algorithm developed in the next sections. 'Create-regions' can be defined as follows:

```
PROCESS Create-regions;
   numofregions : 1..maxregions;
   regions : ARRAY[1..maxregions] OF ↑ PROCESS region;
   PROC addpixel(i,j:INTEGER;reg : ↑ PROCESS region);
      {addpixel adds pixel i,j to region reg.                    }
   PROC mergeregions(reg1,reg2 : ↑ PROCESS region);
      {mergeregions merges region reg2 into region reg1          }
   PROC start-occlusion-rules;  {call applyrules in each
                         region process -- described below       }
   PROC start-relaxation;           { initiate the relaxation
                         procedure in the region processes --
                         described below                         }
   INIT     {Create-regions' initialization section goes here}
```

II. OCCLUSION

A. Occlusion as a Depth Cue

Occlusion is a relative depth cue providing information about the position of surfaces in a scene with respect to each other. Relative depth information can be useful for improving the accuracy of information derived from other cues such as linear perspective. It can also provide constraints on the position of surfaces in a scene which can be used to reduce the combinatorial search involved in stereo or motion parallax matching. Douglass and Roberson (1981) reviewed previous attempts at formulating occlusion rules and formulated four new nonsemantic rules based on tests of human subjects' judgments of relative depth relationships. These rules are given below.

B. Four Rules of Occlusion

A formal definition of the four nonsemantic rules of occlusion can be found in Douglass and Roberson (1981). Each of the rules can be applied independently on each region in an image description. The results of the different rules applied to one region are combined linearly into one occlusion confidence value expressing the likelihood that one region occludes a given neighboring region. The four rules are:

(1) "T" junction rule: Where the boundaries of two regions intersect, and one region's boundary is discontinuous in its first derivative and the second region's boundary is continuous in its first derivative at the point of intersection, then the first region appears to be occluded by the second region. Junctions of the above type look like a capital "T".

(2) Continuation of boundary: Given three regions a, b, and c, if the boundaries of region a appear to extend behind region b and intersect the boundaries of region c, and region c intersects b, then region b appears to lie in front of a single region, splitting it into two regions a and c.

(3) Continuity of surface: Given three regions a, b, and c, region b appears to occlude a single region and split it into two regions a and c if a and c are adjacent to b, and a and c have similar characteristics to each other, such as color and texture, but dissimilar to b's surface characteristics.

(4) Concavity: If a region has a convex projection forming a concavity in another region, then the region with the concavity will appear to be occluded by the convex region. Total containment of one region by another is a special case of a concavity in a region.

The four rules are not equal predictors of occlusion, although not enough empirical observations of human judgments of occlusion in a controlled environment have been made to accurately weight the rules with respect to one another. A few qualitative estimates can be made; for example, the "T" junction rule seems to be the best indicator of occlusion. For the purposes of the next section, confidence weights will be attached to each rule, but such weights have little experimental validation at present.

D. Global Consistency

The result of an occlusion analysis of an image is a map of the image indicating the relative depth of each region with respect to its neighbors. Each region can be assigned a depth level which can be adjusted to be nearer or farther from the viewer than the region's neighbors, depending on the evidence from the occlusion rules. One difficulty with this approach is that the relationships between a group of regions as indicated by the local application of the rules may contain global inconsistencies.

Most strategies for detecting occlusion involve a two-step process: the local application of occlusion rules followed by a global process to obtain a consistent relative depth labeling by combining and adjusting local evidence.   The global process can be accomplished by a relaxation algorithm that deletes incompatible relative depth labels until a consistent labeling is obtained, subject to the weighted constraints provided by the occlusion rules (Rosenberg et al., 1978; Douglass, 1981a).   The next section outlines a parallel algorithm for accomplishing this task.

IV.   COMPUTING OCCLUSION IN PARALLEL

A.   The Process 'region'

The outline of a region process is given below. One such process is assigned to each region description generated by the process 'Create-region'. Once Create-regions has completed the scene segmentation, it initiates the procedure 'applyrules' in each region process to compute the likelihood of occlusion between each region and each of the region's neighbors. After the rules have been applied by every region process, Create-regions initiates a relaxation procedure 'adjust' in each region process to label each region with a consistent depth interpretation.

To apply the occlusion rules in parallel, each region process has four subprocesses. The procedure 'applyrules' initiates execution of each of the subprocesses to apply its particular rule to this region.

The result of applying an occlusion rule, r, is a confidence weight or likelihood, $c_{r,i,j}$, which represents the confidence

that region i occludes region j by rule r. The weights given by the different rules are summed into one confidence weight $c_{i,j}$

for each region i with respect to each of its neigbors j.

The region process expressed in MAC is as follows (the relaxation procedure 'adjust' is specified in section IV.B.):
```
PROCESSTYPE
  PROCESS region;
    neigh : ARRAY[1..numneigh] OF ↑ PROCESS region;
      {an array of process pointers which link this region
       process to its neighbors                                    }
```

```
term : PROCESS cd;
  {defined below, this process determines when relaxation
   is complete}
rule1 : ↑ PROCESS trule;   {process to apply the "T"
                                 junction rule            }
rule2 : ↑ PROCESS boundary; {process to detect a
                                 continuation of a boundary }
rule3 : ↑ PROCESS surface; {surface continuity rule        }
rule4 : ↑ PROCESS concavity; {concavity rule               }
newprob, oldprob : ARRAY[1..maxregions] OF REAL;
cij : ARRAY[1..numneigh] OF REAL;
  {cij holds the probability that region i occludes each of
   its neighbors j = 1 to maxneigh.                         }
PROC applyrules;
  { procedure applyrules starts the execution of each of the
    occlusion rule processes                               }
  j : INTEGER;
  BEGIN
    FOR j IN [1..numneigh] :
      BEGIN
        ACTIVATE(rule1);
        ACTIVATE(rule2);
        ACTIVATE(rule3);
        ACTIVATE(rule4);
      END;
      {the following fork statement initiates execution of
       each of the rule processes without waiting for them to
       complete execution.  The rule processes store their
       results in the array cij.                          }
    FORK j IN [1..numneigh] :
      BEGIN
        { To apply a rule a PROC in the rule's associated
          process is executed.  It receives a pointer to this
          region process (self) and a pointer to the process
          on region j and the current confidence weight cij
          that this region occludes its jth neighbor.  Each
          rule up weights or down weights the occlusion con-
          fidence cij based on the results of applying the
          rule.  The '#' separates input parameters from out-
          put parameters (to the right of the '#').       }
        rule1↑.compute(SELF,neigh[j],cij[j]#cij[j]);
        rule2↑.compute(SELF,neigh[j],cij[j]#cij[j]);
        rule3↑.compute(SELF,neigh[j],cij[j]#cij[j]);
        rule4↑.compute(SELF,neigh[j],cij[j]#cij[j]);
      END;
  END;  {PROC applyrules}
PROC adjust;   { procedure adjust is defined below          }
INIT
  {initialization section for process 'region'             }
```

B. Assigning Relative Depths to Regions Using Relaxation
    Labeling

Once each region has been assigned a confidence weight
expressing the likelihood that a region is occluding one of its
neighbors, then can be used to assign a relative depth to each
region. The occlusion information could be combined with absolute cues and used as a constraint on absolute depth estimates,
as done by Douglass (1981a). Alternatively, the occlusion confidence weights could be used by themselves to label each region
with a relative depth class as Rosenberg et al. do (1978). To
illustrate the parallel nature of the process, the procedure for
assigning relative depths will be outlined briefly here following
Rosenberg et al.'s algorithm.

The procedure 'adjust' is given below. Once each region
process has applied its occlusion rules, the process Create-
regions initiates the PROC 'adjust' in each region process.
'Adjust' initially assumes that there are r relative depth labels
L : $\{l_1, l_2, ..., l_r\}$ where r is the number of regions (potentially,
each region could be at a different depth). Each of the labels,
l_i, has a probability associated with it which indicates the
likelihood that it is the correct label for the given region, and
initially all label probabilities are equal. The procedure
'adjust' performs relaxation labeling to increase or decrease the
probabilities associated with each label according to the com-
patibility of each label and the labels of each of the neighboring regions. Two labels are compatible if they agree in terms
of the occlusion relation between their respective regions.

In the MAC algorithm below the procedure 'adjust' in each
region process is called initially by Create-regions. Each time
adjust corrects the probabilities on a region's set of labels, it
calls the adjustment procedures on neighboring regions to correct
their region descriptions. The label probabilities in the array
'oldprob' are used to compute the probabilities for the next
iteration, which are then stored in the array 'newprob'. The
procedure adjust is part of process region and is defined as
follows:

```
PROC adjust;
  k : INTEGER;
  totalchange : REAL;
  BEGIN
    { update the probabilities for each depth class label (new-
      prob[1..number of depth classes] using the current prob-
      abilities (oldprob) plus the probabilities for each depth
      label on each neighboring region subject to the occlusion
      constraint cij between two regions                       }
    FOR k IN [1..maxdepths] :
      BEGIN
        newprob[k] := oldprob[k] + correctionfactor(k);
```

```
            { the correction factor is calculated from the old
              probabilities for every depth label on each
              neighboring region using cij. The newprob[k] must be
              normalized by dividing by the sum of all the
              corrected label probabilities.                      }
        END;
      totalchange := 0.0;
      FOR k IN [1..maxdepths] :
        totalchange := totalchange + ABS(newprob[k] - oldprob[k]);
        { determine if the probabilities have been altered
          significantly from the previous iteration.             }
      IF totalchange > threshold
        THEN
          BEGIN
            { repeat the adjustment procedure on neighboring
              regions                                            }
            FOR k IN [1..maxdepths] : oldprob[k] := newprob[k];
            FORK k IN [1..numneigh] : neigh[k]↑.adjust;
          END;
    END;  {PROC adjust}
```

C. Distributed Termination

One problem with formulating the relaxation algorithm described in the last section as a parallel network of locally
connected processes is that it is difficult to determine when the
network as a whole has converged to a stable state. If a single
process is assigned the task of sequentially polling each region
process to see if it requires adjustment, it is possible that the
relaxation process may be terminated prematurely because the
adjustment process can be reinitiated asychronously by other
processes after the polling process has checked the process for
activity.

The problem of determining when the computation has terminated can be solved by implementing Dijkstra's distributed termination method in the adjustment procedures (Dijkstra, 1968).
This solution requires the addition of two integers c and d to
each region process. c and d are both initially zero and have
the following interpretation: c equals the number of unacknowledged requests to a region process to adjust its label probabilities, and d equals the number of unacknowledged requests a given
region process has made to its neighboring region processes to
adjust their probabilities.

A process is in a neutral state if c = 0 and is in an
engaged state otherwise. Each node has for its parent the node
which moved it from a neutral state to an engaged state. To use
the distributed termination method, the order in which nodes
engage one another must form a tree of activations with a single
root or gate node which was the first process in the network to
be engaged. A network has terminated its computation when the

gate node has passed to a neutral state (i.e. c = 0). In the example of parallel relaxation above, the region processes form a one-level tree of initial activations with the Create-regions process as their parent and the gate node of the network.

Two additional constraints must be imposed on the network for the distributed termination technique to perform properly:

(1) The parent of every engaged node is engaged (which implies that if d > 0, then c > 0).

(2) A node can go to a neutral state only by sending an acknowledgment to its parent.

D. Incorporating Distributed Termination into the Region Process

To use the distributed termination method to signal Create-regions on completion of the relaxation process, a separate sub-process, called 'term', must be defined in each region process. 'Term' is of type cd and is defined as follows:

```
PROCESS cd;
  c,d : INTEGER;
  processname : ARRAY[1..maxcalls] OF ↑ PROCESS region;
  { processname points to all processes which have called the
    region process containing this instance of process cd.
    Process cd must have a pointer to all calling processes to
    acknowledge the calls once they have been serviced        }
  PROC ack;
    BEGIN
      IF c > 1 OR d = 0 THEN
        BEGIN
          c := c - 1;
          processname[c+1]↑.recordack;
        END
      ELSE  { if d is not 0 and c = 1 then the parent node may
              not be acknowledged so wait for d to become 0   }
          WHEN d = 0 : IF c = 1 THEN
                          Create-regions.ackparent;
              { acknowledge the parent node by decreasing the d
                count in the parent process.                   }
    END;
  PROC recordack;
    BEGIN
      d := d - 1;
    END;
  PROC send(thisreg,reg : ↑ PROCESS region);
    BEGIN
      d := d + 1;
      reg↑.term.recordsend(thisreg);
    END;
  PROC recordsend(reg : ↑ PROCESS region);
    BEGIN
      c := c + 1;
```

```
      processname[c] := reg;
   END;
END; { of process cd   }
```
   To incorporate distributed termination into the relaxation
routine, the procedure 'adjust' must be modified as follows:
```
PROC adjust;
   { compute new probabilities as in previous version.  Compute
     the total change in the adjusted probabilities as in the
     previous version                                            }
      IF totalchange > threshold THEN
        BEGIN
          FOR k IN [1..numdepths] :
            oldprob[k] := newprob[k];
          FOR k IN [1..numneigh] :
            BEGIN
              term.send(SELF,neigh[k]↑);
              FORK(neigh[k]↑.adjust);
            END;
        END;
      { acknowledge the calling process for this activation of
        PROC adjust.                                             }
      term.ack;
      {  a parent will be acknowledged only if d = 0             }
   END;
```


V. DISCUSSION

A. Parallelism in the Occlusion Algorithm

   The above algorithm for computing occlusion requires $O(n)$
processes where n is the number of regions in an image. Each
region process has as may connections to other processes as its
corresponding region has adjacent regions. Typically, and image
will contain about 100 to 400 regions with an average
connectivity of 4 to 5. During the application of the four
occlusion rules there will be n region processes active plus 4n
occlusion rule processes for a total of 5n active processes. The
relaxation procedure requires 2n active processes, n region processes and n cd processes.
   Sequential execution of the relaxation procedure requires
$O(I*n)$ execution steps where I is the number of iterations to
reach convergence. The parallel program outlined in the
preceeding sections requires $O(I)$ steps. Applying the occlusion
rules in a sequential program would require $O(n*(r_1+r_2+r_3+r_4))$

where r_i is the program complexity of rule i. The parallel algorithm above requires $O(r_{max})$ where r_{max} is the rule r_1 to r_4, with the maximum number of execution steps.

B. Interprocess Communication and Processor Contention

Theoretical performance gains with a parallel algorithm are tempered in an actual program by the overhead cost inherent in delays for sending and receiving messages between processors and because of contention over shared memory. The characteristics of message frequency and size and the amount of contention have architectural implications.

Each occlusion process must request information about a neighboring region from the region's process. The frequency of interprocess messages and the size of the messages required to apply the occlusion rules depend on the exact way in which the rules are computed, but it is frequently the case that only 6 to 10 integer values need be passed between processes. An occlusion process may obtain data about a neighboring region using the MAC READSHARED operation which allows one process to read the values of another process's variables but not modify them. A READSHARED operation may be implemented with physically shared memory or with message passing by requesting a process to send a copy of the requested data values. In either case, there are $4n$ occlusion processes contending for access to n region descriptions, leading to a large potential delay due to contention. However, since the processes run asynchronously, and since only a small percentage of each occlusion rule is spent in accessing a neighboring region's description, contention over shared variables is reduced.

The relaxation algorithm also involves potential contention between region processes in accessing the label probabilities on neighboring regions. Although the relaxation process involves only $2n$ processes contending for $2n$ sets of process variables, and thus has less potential for contention delays than the occlusion rule processes, the relaxation procedure must communicate with its neighboring regions more frequently and pass relatively larger amounts of information. Each time a node of the relaxation algorithm reads its neighbors' label probabilities, it must examine or transfer about 100 floating point values.

C. Nonconventional Computer Architectures and Parallel Image Graph Algorithms

The parallel occlusion program outlined in the preceding sections typifies a number of image graph algorithms such as the motion parallax algorithm of Ullman (1979). The extent of parallelism for these algorithms is on the order of 100 to 1000 processes. The processes communicate frequently but send

relatively small amounts of information each time, on the order
of 10 to 100 integer or floating point values. Given these
characteristics, a few conclusions on computer architectures for
image graph algorithms can be drawn.

Since image graph algorithms do not conform to an array of
interconnected processes and they operate asynchronously, they
can not be efficiently implemented on SIMD array computers such
as CLIP4 or the MPP (Duff and Levialdi, 1981). However, traditional MIMD multicomputer networks such as the MICRONET of Wittie
et al. (1981) are also unsuitable because they have relatively
long message transmission delays and require long message protocols making interprocess communication inefficient unless
messages are large and infrequent.

An appropriate nonconventional architecture for image graph
algorithms must allow asynchronous execution with very fast message passing or use shared memory, and it must permit dynamic
reconfiguration of process interconnections. Since image graph
algorithms are locally interconnected, the architectures may also
be biased toward local interconnections. A few nonconventional
computers which meet these criteria include the CM*, TRAC at the
University of Texas, and PASM (Siegel, 1981). Other computers
which might be capable of simulating an image graph machine are
the CRYSTAL system proposed by the University of Wisconsin and
the ZMOB system at the University of Maryland (Rieger et al.,
1980). Both of these systems may have overlong interprocessor
communication times and excessive message size overhead to make
them feasible for implementing image graph algorithms.

VI. CONCLUSION

Occlusion can be computed using a network of locally connected asynchronous processes. It is typical of a class of image
processing algorithms, called image graph algorithms, which process images represented in a more abstract form than a pixel
array. These algorithms are locally connected but not regularly
connected like an SIMD array, and they run asynchronously as in a
typical MIMD network, but they are more closely coupled,
frequently passing small messages.

One difficulty with iterative algorithms like relaxation
labeling when computed with a distributed network is that it can
be difficult to determine when the network of processes has
converged on a global equilibrium. Dijkstra's distributed termination algorithm has been modified to enable the region
processes to signal when they have converged as a network.

Image graph algorithms demand a different view of parallelism. They involve replicating similar processes over each element of a large data structure such as an image described in
terms of a set of regions. Such algorithms require new program-

ming languages, new operating systems, and new computer architectures. The nature of the algorithms and their requirements should drive the development of new nonconventional computer systems. The parallel computation of occlusion outlined here provides one example of the interplay between algorithm characteristics and system requirements.

ACKNOWLEDGMENT

The author wishes to thank Janet Holmes Stanford for her careful reading and editing of this chapter and her indefatigable efforts to purge out the occasional infelicitous phrase.

REFERENCES

Dijkstra, E. (1968). "Diffusing Computation with Distributed Termination," ·Technical Report EWD 687a.

Douglass, R. (1981a). "Interpreting Three-Dimensional Scenes: A Model Building Approach," in "Computer Graphics and Image Processing" (in press).

Douglass, R. (1981b). "MAC: A Programming Language for Asynchronous Image Processing," in "Languages and Architectures for Image Processing" (M. Duff and S. Levialdi, eds.). Academic Press, London.

Douglass, R. and M. T. Roberson (1981). "New Rules for Computing Occlusion." DAMACS Technical Report, University of Virginia.

Duff, M. and S. Levialdi, eds. (1981). "Languages and Architectures for Image Processing." Academic Press, London.

Marr, D. and Poggio, T. (1976). "Cooperative Computation of Stereo Disparity." Science, 194, 4262.

Rieger, C., J. Bane, and R. Trigg (1980). "ZMOB: A Highly Parallel Multi-Processor," In "Proceedings of the Workshop on Picture Data Description and Management." Asilomar, CA.

Rosenberg, D., M. Levine, and S. Zucker (1978). "Computing Relative Depth Relationships from Occlusion Cues," in "Proc. of Inter. Joint Conf. on Pattern Recognition." Kyoto.

Siegel, H. J. (1981). "PASM: A Reconfigurable Multi-Microcomputer System for Image Processing," in "Languages and Architectures for Image Processing" (M. Duff and S. Levialdi, eds.). Academic Press, London.

Ullman, S. (1979). "The Interpretation of Visual Motion." MIT Press, Cambridge, Mass.

Wittie, L., R. Curtis, and A. Frank (1981). "MICRONET/MICROS - A Network Computer System for Distributed Applications." In this volume.

Yakimovsky, Y. (1976). "Boundary and Object Detection in Real World Images," JACM, 23.

REFLECTIONS ON LOCAL COMPUTATIONS

Concettina Guerra

Istituto di Automatica
University of Rome

Stefano Levialdi

Istituto Scienze dell'Informazione
University of Bari

I. INTRODUCTION

As seen from previous meetings (IEEE Comp. Soc. 1980a, 1980b,
Onoe,1981) the time for applying SIMD machines to image processing is near. Many reasons point towards this fact: the experience
gained in using existing prototype machines like CLIP, PICAP and
CYTOCOMPUTER; the fast advances of VLSI in the design and construction of complex chips including both processor and memory;
the new high level languages (and concepts) particularly tailored
for expressing parallel algorithms and last but not least, the
improvement in communication between the different communities
of researchers; i.e. modellers, architects and programmers. This
last fact has been accellerated by workshops such as this one and
plays a significant role in the dissemination of the different
experiences that have matured during the design phase of an IP
system and the evaluation of its computational requirements. A
number of papers have appeared which attempt an overall view on
the subject of basic operations to be performed on images,namely,
on the problem of defining primitives. Another approach to the
general problem of computing predicates on images is the one of
establishing a general computational model for recognition (acceptance) of a feature in a pattern. Finally attempts at classifying the different architectures suggested for IP and PR have
been made recently mainly aiming to establish the adequacy of a
machine structure to a computational task, bearing in mind which
parallelism can be exploited for processing a two dimensional
structured data for obtaining the maximum speed-up compared to a

sequential Von Neumann machine.

As it is known from the history of computers, the Turing machine, born long before the first electronic computer was built, gives a computational model of which a sequential computer is a finite state case. The input data is generally imagined on a tape and therefore considered as a string of symbols which can be read out or written in whilst during processing, some of these symbols will evolve in time so that,at the end, a definite final value will be obtained. This computation may be described in terms of a program, i.e. a sequence of instructions which must be unambiguous, exhaustive and self-consistent so that after the last instruction is executed the final result is reached. This description has a one-to-one correspondence with the one making use of the state transition function. If the level of abstraction is reduced so approaching a concrete computing system with real data, these two descriptions diverge, one producing an abstract computational mo- del and the other a program scheme using a formal grammar. It will be seen in the following that abstract models for IP are relevant for expressing time dependency in relatively simple tasks but do not appear natural nor transparent when more complex tasks are described and, more generally, could be better expressed by means of a structured program in a high level language.

It must be remembered that image data have two main proper- ties: a quantitative one and a qualitative one. The first implies that a very large amount of information is present in an image making the problem of time execution a crucial one. The second one is connected to the problem of structure which is present in an image; this structure is sometimes referred to as texture, scene layout, morphological features, color and light patterns, etc. Both facts bear some relevance on the tools which have to be designed for processing this kind of information. Taking these considerations into account, a number of SIMD machines were built, where a correspondence between a pixel (picture element) and a processor was established and, furthermore, each processor could easily communicate with its neighboring ones (4,8 or 6 according to the chosen plane tessellation).

II. SOME MODELS

Many different possibilities arise starting from common pre- mises and, in fact, a number of machines have been suggested and built around the very basic principles previously mentioned,(see Duff and Levialdi, 1981). We will now consider two different ap- proaches aiming towards the modelling of a computation performed on an image data structure. The first one is due to Klette (1980) and has the elegance of simplicity. He assumes that only two basic operations are required in order to express any local computation performable on an array containing the image data: a shift operation and a boolean operation. In his paper three different registers are defined so that their contents may be used in the evaluation of the operation performed on the data. Vector, index and matrix correspond to the result of a processed row, to the loop counter and to the input data respectively. A rather compact and clear way to express a local parallel computation on an image is given by an Algol-like program such as the one that follows:

$$
\begin{aligned}
&begin \quad image \; X,Z_1,Z_2,\ldots,Z_n; \; read \; X; \\
&\quad for \; t:=1 \; step \; 1 \; until \; n \; do \\
&\qquad Z_t:=(X \leftarrow j_t)\uparrow i_t \; od; \\
&\quad X:=f^*(Z_1,Z_2,\ldots,Z_n) \\
&\quad print \; X \\
&end
\end{aligned}
$$

As may be seen, an input image X and successive transforms $Z_1,Z_2,\ldots,Z_n$ may be obtained after a suitable combination of horizontal shifts (given by j_t) and vertical shifts (given by i_t). Their direction is indicated by a sign: leftwards and topwards for positive sign. The number of shifts is given by the n parameter and f^* is a boolean function which has $Z_1,\ldots,Z_n$ as its arguments.

Let us now consider a particular task; is X empty? This predicate will be computed within time $O(\log N)$ and is expressed by

$$Zero(X):=if \; X=0 \; then \; 1 \; else \; 0 \; fi$$

and the program which computes it is

$$
\begin{aligned}
&begin \; vector \; A; \; index \; I; \; image \; X,Z; \\
&\quad int \; I=N/2; \; read \; X; \\
&\quad while \; I\neq 0 \; do \\
&\qquad Z:=X\downarrow I; \; X:=X \lor Z; \; I:=I\rightarrow od; \\
&\quad A:=X; \\
&\quad if \; A=0 \; then \; print \; 1 \; else \; print \; 0 \; fi; \\
&end
\end{aligned}
$$

This program iteratively shifts half the array on itself (N being a power of 2) as long as I > 0   . Each time the boolean OR is executed between the image and its vertically shifted version, and, finally, a test is performed on vector A which contains the last row of X   . For each new iteration I is halved by a one bit right shift.

The paper analyzes typical preprocessing tasks as non linear smoothing, shrinking and expanding which are all shown to be time constant with this model (called PBS for "paralleles Binärbild-verarbeitungs-system"). This result is due to the combination of two machines (the binary processor and the vector processor) into the PBS machine so obtaining a logarithmic or linear time dependency for most tasks. As the author states in the conclusions, for more realistic applications a further extension to grey level images is required since only binary images have been considered.

From a different point of view, many years ago Beyer (1969) in an M.A. thesis developed a model for extracting topological invariants from a binary image by means of a two dimensional bounded automata. Those processors were essentially memory-less and the interactions were vertical and horizontal so that information could only flow along columns and rows. Each state transition was expressed by a function which had, as arguments, the state of each cell and of those belonging to its 4-neighborhood and a number of tasks were considered including connectivity detection, parity evaluation, maze path finding, etc.

The interest of this model lies in its relative simplicity and transparency between the described computation and the evolution of the states of all the concerned cells.

More recently, (Dyer and Rosenfeld 1981) have suggested a memory augmented cellular automata which outperforms Beyer's due to the specific storage increase of each cell. For a number of tasks discussed in their paper the time dependency is shown to be either linear with the region-diameter of the object or with the diameter of the array.

In fact, each cell has a memory proportional to the logarithm of the number of pixels of the image to be processed. More particularly, region labeling techniques are described and the time dependency for fulfilling each task is given in a final table. For instance, the area of a connected region may be computed during the evaluation of the minimum spanning tree (MST) rooted on a given cell of the region. If the intrinsic diameter (defined as the maximum distance between two elements of the region according to a given distance function) is considered for this object, both the MST and the area evaluation may be performed in a time at most equal to the intrinsic diameter. In the example of the paper, 4-connectivity was used and each node of the MST graph could count all its sons so arriving at the total count (area value) at the root node.

In the case of image transforms which are generally used for preprocessing purposes the augmented automata also performs in times proportional to the diameter of the object contained in the image. For instance, in the case of the MAT which is applied both for storage and for perceptual representation of an object, each cell of the object having at least one cell in its neighborhood belonging to the background, will change state and its label will be incremented until no more cells can change state. The values of the labels correspond to those of the distance transform.

In practice, algorithms for skeletonizing images are based on the propagation of the background over the connected objects and are generally applied on elongated shapes. This explains why the time dependency is closely related to the "width" of the object more than to its diameter. An operative definition of "width" may be given as follows. For a given simply connected object, rotate it around its baricenter until both its x and y projections are equal; draw an axis at 45° through its baricenter and compute the longest segment which is perpendicular to this axis and has its end points on the contour of the object. This length may represent the width of the object and can be used as a closer approximation to the time bound for both MAT and some thinning transformations. In fact, in some thinning algorithms, the deletability condition of each element may be tested by means of a local operation so allowing the use of the same computational scheme.

III. SOME ALGORITHMS

Let us consider the skeleton algorithm in (Arcelli *et al.*, 1981) which allows the preservation of connectivity of the original object and is totally based on local computations (on a three by three neighborhood). This transformation can therefore be computed by the cellular augmented automata mentioned above by a time which is proportional to the width (as defined before) of the object to be skeletonized. Turning now to parallel thinning algorithms, we may consider (Arcelli *et al.*, 1980) where eight three by three masks are required for each iteration, the time dependency is again O(width).

In other tasks, typically in smoothing transformations a small number of local operations is applied on the array; this number is independent from the array size. For instance, in computing the mean value within a local window, or the median within the same window (which better preserves the edges of the objects), the number of operations is a function of how many neighbors are contained in the window. For a three by three window, median filtering requires $n \log n$ comparisons, for n equal to the size of the window.

In all these instances, the automata models the computations performed on the image for achieving a specific goal; conversely the computations may also be described in terms of a program scheme. Languages such as Algol or Pascal which have been formally defined and on which program correctness may be proved may turn to be useful tools for this purpose. The need for constructs which naturally reflect the operation of a parallel processing algorithm

plus the availability of high level language compilers has prom-
pted a number of projects in this area. For a comparison of the
different approaches taken in the design of new languages for
image processing, see (Maggiolo-Schettini, 1981). According to
our approach, a language named Pixal, was formally defined and a
number of simulated programs for typical image processing tasks
have been written (Levialdi and Maggiolo-Schettini, 1981; Levialdi
et al., 1980a; Levialdi et al., 1980b).

Let us now discuss the modelling power of the Pixal constructs
when expressing computations on images. Using particular data
structures, special built-in functions and a clear-cut distinction between global parallelism (where one computation is performed on all the elements of the array without interactions between them and their neighbors) and local parallelism (where the
computed predicates operate on a previously defined neighborhood),
Pixal was designed to handle naturally expressions representing
transformations to be performed on images regardless of the fact
that these transformations are sequential ones or parallel ones.

We will now consider in the same order as before, the area
computation, MAT and thinning operations described by Pixal constructs.

A built-in function such as sum could easily represent the
addition of all the elements of an array I and therefore provide a
number equal to the area of the object present on such an array.
This implies that only one object is present on the array, otherwise, an algorithm for object extraction must be firstly executed
(Levialdi et al., 1980b).

```
program  area

comment  array I contains the input image;

binary array

        I [1:100,1:100];

integer  A

  begin

   A:=sum (I)

  end
```

The statement which computes the sum simulates a parallel addition performed on all the elements of I (not realistic in hardware) but when this computation is achieved by means of a sequential machine, its time will be proportional to the total number
of elements in the array, i.e. to the area. In this example the
statement is inadequate to reflect a cellular automata which, as
shown before, will compute the area in a time O(intrinsic diameter). In order to reflect this, a MST construction and node count
should be programmed in Pixal as in (Levialdi et al., 1980a).The
central part of the program will label each 4-neighbor of the

root element with a value depending on the chain coded direction
necessary to reach it so that the time required for these labeling operations is now proportional to the maximum length of the
object (intrinsic diameter).

Referring now to the skeleton transform and considering the
algorithm which was quoted before, we notice that the program is
made of about fifty local operations. This implies that a parallel processor is well matched for accomplishing this task. Let us
now see roughly how a Pixal program may describe some of the basic
operations required for this task. For instance, 4-core extraction and diagonal expansion, are typical local operations which
label each element of the object according to a given neighborhood configuration. The first one extracts all 1-elements which
have no 4-neighbors in the background (0-elements) whilst the
second one will change 0-elements into 1-elements whenever they
have a 1-element as a diagonal neighbor.

> *program* 4-core
>
> *comment* array I contains the input binary image;
>
> *binary array*
>> I [1:100,1:100];
>
> *binary mask*
>> M [1:3,1:3] of (0,1,0,1,1,1,0,0,1,0);
>
> *begin*
>> *par*
>>> *if* sum(overweigh(M,I))=5 *then* I:=1 *else* I:=0
>>
>> *parend*
>
> *end*

and now the program for the diagonal expansion:

program D-expansion

comment this program will expand an object along the diagonals
           of all its 1-elements;

binary array
>   I [1:100,1:100];

mask   M1[1:2,1:2] of (0,0,1,0) on (1,2);

mask   M2[1:2,1:2] of (1,0,0,0) on (2,2);

mask   M3[1:2,1:2] of (0,0,0,1) on (1,1);

mask   M4[1:2,1:2] of (0,1,0,0) on (2,1);

> *begin*

```
par
    if compare (M1,I) then I:=1;
    if compare (M2,I) then I:=1;
    if compare (M3,I) then I:=1;
    if compare (M4,I) then I:=1;
  parend
end
```

In the first program the function overweigh multiplies all the elements contained in the mask area by the mask weights, the sum function will add their values. In the second program the function compare will return a 1 value if the elements in the mask are equal to those enclosed in the same neighborhood around the considered element of I.

The statement(s) inside the *par/parend* control pair is intended to be performed simultaneously for every element of each array appearing within the statement. As the statement may also be a computed statement, an unlimited number of statements can occur inside.

Since all local operations may be described by a construct within the *par/parend* control structure, the skeleton program may be written directly in a Pixal program exhibiting its time independency with respect to the diameter.

In conclusion, the skeleton program written in Pixal is described homogeneously with a cellular automaton that may be used for computing such a transform. In a similar way, a thinning algorithm may be coded in Pixal by means of the usual set of masks which will check the deletability of each single pixel (again with a compare function). These operations require $O(\text{max width})$ time.

IV. CONCLUSIONS

Local operations may be naturally implemented on SIMD machines and modelled by cellular automata. Recently, two computational models based on elementary shifting operations combined with boolean ones and a memory augmented automata have been evaluated on a set of basic image processing transformations. The first model also includes a general purpose Algol-like program that comprises the set of all possible local operations. The second model, when applied to different image transforms, shows a remarkable speed-up factor because of the available memory pointing towards the need of providing the processors of a real array with similar resources. Area computation, skeleton transformation, and thinning are revisited, using specific algorithms, when programmed in a high level language encompassing parallel facilities. These program schemes exhibit a relatively transparent computational description whilst suggesting specific time dependence.

REFERENCES

Arcelli, C., Cordella, L.P., and Levialdi, S. (1981). IEEE Trans. on PAMI, 3, 134.

Arcelli, C., Cordella, L.P., and Levialdi, S. (1980). Electronics Letters, 16, 51.

Beyer, W.T. (1969). MAC TR-66(AD699502), M.I.T., Cambridge.

Danielsson, P.E., and Levialdi, S. (1981). IEEE Computer (to appear).

Duff, M.J.B., and Levialdi, S. (1981). "Languages and Architectures for Image Processing" Academic Press, London.

Dyer, C., and Rosenfeld, A. (1981). IEEE Trans. on PAMI, 3, 29.

IEEE Computer Society (1980a).Proc. Workshop on "Picture Data Description and Management" Asilomar.

IEEE Computer Society (1980b).Proc. of 5th Int. JCPR, Miami Beach.

Klette, R. (1980). CGIP, 14, 145.

Levialdi, S., Isoldi, M., and Uccella, G., (1980a). In Proc. Workshop on "Picture Data Description and Management", p. 74, Asilomar.

Levialdi, S., Isoldi, M., and Uccella, G. (1980b). In Proc. 5th IJCPR, p. 1285, Miami.

Levialdi, S., Maggiolo-Schettini, A., Napoli, M., and Uccella, G. (1981). In "Real-Time/Parallel Computing" (M.Onoe, K.Preston, Jr., and A. Rosenfemd, eds.), p. 131, Plenum Publ. Corp., New York.

Maggiolo-Schettini, A. (1981). In "Languages and Architectures for Image Processing" (M.J.B. Duff and S. Levialdi eds.), Academic Press, London.

Onoe, M., Preston, Jr., K., and Rosenfeld, A. (1981). "Real-Time/ Parallel Computing" Plenum Pub. Corp., New York.

PARALLELISM IN PICAP

Björn Kruse
Björn Gudmundsson

Department of Electrical Engineering
Linköping University
S-581 83 Linköping, Sweden

INTRODUCTION

Image processing is in many ways one of the most resource
demanding activities in modern data- and signal-processing. The
obvious reason for this are

o Large data volumes
o Real-time processing requirements
o Complex data structures
o High speed input devices.

For those who have had some experience in pictorial pattern
recognition in practice these demands do not have to be stressed.
In the following of this paper we will show how some of the
problems have been solved in the PICAP-system. It will be apparent
that to come close to the ideal one has to consider all the
problems simultaneously and design not a processor, nor a frame-
buffer, nor a camera but a system of resources including processors, memory and input devices that can operate efficiently
together as a whole. PICAP is a multi-user image processing system
of modular construction allowing for a variety of processors and
input/output devices controlled by a powerful host computer.

Parallelism is necessary in one way or other to achieve a
substantial speed advantage as compared to general purpose
computers. In PICAP parallelism is used both locally and globally.
On the global level processors operate in parallel, simultaneously
accessing and processing information from memory. On the local
level the processors themselves have an augmented capacity due to
internal parallel processing streams.

MULTICOMPUTERS AND IMAGE PROCESSING
ALGORITHMS AND PROGRAMS

231

The PICAP system is a research vehicle for parallel image processing research developed at the Picture Processing Laboratory, Linköping University, Sweden. The former picture processing system in the laboratory, which has been operational for seven years, will in the following be referred to as PICAP I to distinguish it from the new system that will be described here.

PARALLELISM

There have been numerous approaches to achieve high speed in image processing over the years. In the late sixties and early seventies a number of machines were built that employed neighborhood parallelism (Golay, 1971; Gray, 1972; Kruse, 1973). The local processing of the image elements and their immediate surround was done in parallel in these machines whereas the image as a whole was processed sequentially. Using buffering techniques the operation had the same effect as if the entire image had been processed in one step, disregarding speed of course. In the late seventies and beginning of the eighties large two-dimensional arrays of simple processors were built to perform very large numbers of simple operations in parallel over the image (Reddaway, 1975; Duff, 1976). Another approach uses a large number of relatively simple processors operating in a long pipe-line. In contrast to the processor arrays in which each module performs exactly the same operation at any given instant, the modules of the pipe-line are performing operations according to their position in the line. Thus the picture processing is physically distributed on the basis of operations. The Cyto-computer (Sternberg, 1979) consists of a set of small identical cascaded processors that complete a program in one pass. In the FLIP system (Luetjen, 1980) a network of relatively powerful processors are set up according to the computing structure that best fits the algorithm. Another example of cascaded processor design is the GOP (Granlund, 1980) processor, also developed at the Picture Processing Laboratory in Linköping, Sweden.

Most of the processors mentioned above rely on the multiplicity of fairly simple basic modules and/or parallel operation over a neighborhood. The large number of modules which is required prohibits wordlengths larger than one, that is the individual processors operate on one bit data. These are of course not the only ways to achieve high speed. An interesting calculation to make is the number of one-bit modules that are needed to achieve the speed of a single sixteen by sixteen bit multiplier. The answer is of course that 256 one-bit processors can produce 256 products in 256 time-units which is equivalent to one product per unit. This means that, given equivalent technologies, it is not very hard to design a processor from complex components such as programmable

logic arrays, multipliers and memories that is equivalent to
thousands of one-bit modules. A few such processors and systems
have also been built (Mori, 1978; Gerritzen, 1980). The PICAP
system (Danielsson, Kruse, Gudmundsson et al, 1976-80) that will
be described in the next section belongs to this category. A
systematic investigation and classification of different forms of
parallelism in image processors has been made by Danielsson
(Danielsson, 1980b).

SYSTEM ARCHITECTURE

PICAP is a multiprocessor of type MIMD where the processors
operate on pictures stored in a large random-access picture
memory. An outline of the PICAP system architecture is shown in
Fig. 1.

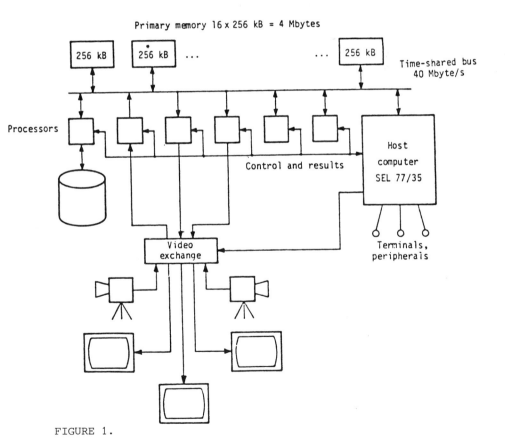

FIGURE 1.

The processors run independently of each other and each of them has a specialized function, e.g. linear and non-linear filtering, segmentation, picture I/O, etc. A processor starts operating when it receives a command from the host, and when the operation is finished the processor signals this by interrupting the host. Since all processors share a common memory it is of vital importance that the datapath between memory and processors is fast. The time-shared bus (Danielsson, 1980b), for which the processors contend on a priority basis, has a bandwidth of 40 Mbyte/s and hardware arbitration among the processors is performed in every bus-cycle.

The picture memory space (4 Mbyte) is linear and interleaved over sixteen separate memory modules.

PICAP is modular and open-ended in the sense that processors can easily be replaced and added. There is room for sixteen processors, one of which is an intelligent interface between the host and the picture memory. This interface has a bandwidth of 3.2 Mbyte/s (determined by the host machine) and it can access picture windows of arbitrary dimensions.

PICAP is a multiuser system and the users are seated at workstations each equipped with a terminal and picture I/O-devices (TV camera, monitor). The users share the processors and picture memory dynamically. If a processor is idle when a user task requests it, the processor is immediately started to perform the users task, otherwise the request is put to wait in a queue. Also, space in picture memory is automatically allocated to the task before the processor is started. In this way each user gets a virtual PICAP system of his/her own.

In order to achieve parallelism on the global level, picture memory as well as the processors must be shared among the users. Since PICAP can handle pictures of arbitrary dimensions it would not be very efficient to partition the picture memory statically. Therefore a scheme for dynamic allocation of picture memory space is utilized. The user is provided with two conceptually different picture storages; one is a temporary workarea and the other is a database for long-term storage.

The picture processors operate on pictures in the workarea, and the database is used for input of previously saved originals and for storage of those result pictures that the user considers worth saving. To alleviate the problem of naming the abundance of more or less "volatile" pictures that reside in the workarea during a typical session, these pictures are given names from a fixed set of standard names. In fact, the workarea consists of a set of virtual picture registers and the pictures in them are referenced by the register name. The following figure illustrates the users view of picture I/O, processing and storage:

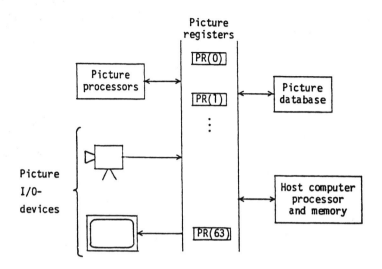

FIGURE 2.   Picture registers.

It should be noted that each user who is logged in at a work-
station is provided with the conceptual structure shown in Fig. 2;
i.e. each user has his own set of 64 picture registers. Further,
the register contents are valid across all modes of operation;
they are lost only when the user terminates the session. As indicated in Fig. 2, pictures can also be brought into the hos computers memory.

What we have is a demand paging system (Denning, 1970) where
the registers constitute the variable size pages. Registers that
are not in PM when they are referenced are brought in from the
disk, possibly after another register has been swapped out to
make room.

The database (Fig. 2) is, as opposed to the picture registers,
intended for storage of pictures beyond the current session. In
its first version the database is simply a set of disk-files under
the filesystem of the host computer. Note that the disk used for
the database is not the disk used to implement the virtual picture
registers. The former one is a standard peripheral of the host
while the latter is a special disk unit connected to the high-
speed bus (Fig 1). A picture file has a header containing a user

supplied name and type and dimensions of the picture. Pictures can
be input to the database either from registers or from magnetic
tape.

PARALLELISM ON THE LOCAL LEVEL

     The overall PICAP architecture is of type MIMD but some of the
individual processors, among them the filter processor FIP, are in
themselves of type SIMD. In FIP, four subprocessors (P_1-P_4) operate in parallel according to a common microprogram, accessing
pixels from a shared local memory. To further increase the processor performance the subprocessors are pipelined. A comparison
with a conventional computer instruction set shows that FIP executes in excess of 10^8 basic instructions per second. An overview
of FIP is shown in Fig. 3. A more detailed description is given in
(Kruse et al, 1980b)

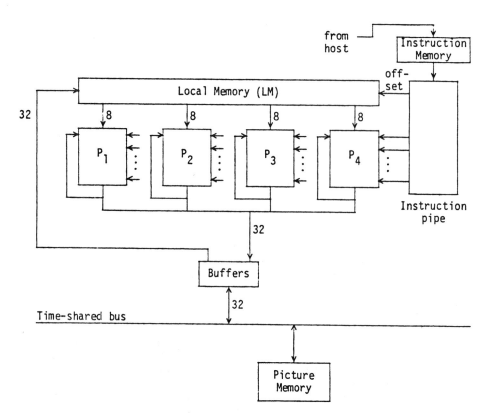

FIGURE 3.   FIP.

The local memory (LM) stores a horizontal strip from a picture
giving access to all neighborhoods along one line, (see Fig. 4).
The operator moves along the line and when ready a new line is
entered into LM while the result is stored back in picture memory.
For 3-D-neighborhood operators a set of strips from the involved
pictures are stored such that corresponding lines are adjacent to
each other. The size of LM is 32 kbyte which for example corresponds to a 64×512 neighborhood operator on a 512×512 picture. The
accessing mechanism that enables the four subprocessors to retrieve data from LM obviously requires a partitioning of LM into
four separate modules. The accessed pixels are horizontally adjacent, one for each module and they are supplied one to each subprocessor after appropriate flipping. A base register holds the
location of the current neighborhood position. The base address is
incremented by four when four neighborhoods have been processed.

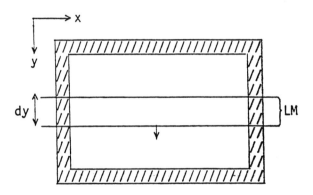

FIGURE 4.   Local memory contents.

The subprocessors of FIP are identically controlled by a
common microprogram. The microinstructions are fed through an instruction pipe from which appropriate controlsignals are derived
(see Fig. 5). A pixel flowing through the pipelined processor is
subjected to the following transformation:

$$ACK^+ = F_2(ACK, F_1(\omega, MAP(p)))$$

p: pixel from LM or a register

ω: constant (immediate operand) from instruction

MAP(·): pixel to pixel mapping

F_1, F_2: linear and non-linear functions

ACK: accumulator in Arithmetic Logic Unit (ALU)

    In the case of for example a linear filter (convolution), F_1 is multiplication and F_2 is addition. Since the resulting pixel can be stored back to a register, it is possible to compose very powerful compound operators involving both linear and non-linear functions.

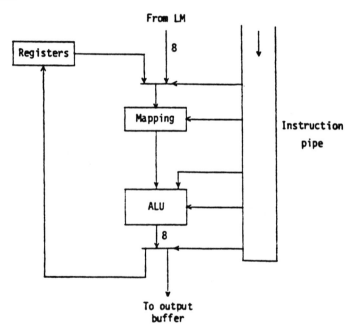

FIGURE 5.   Subprocessor.

CONCLUDING REMARKS

    The structure of the PICAP image processing system has been described. We believe that the modularity of the architecture and the parallelism on both local and global levels is of vital importance for efficient image processing and interactive algorithm development. The basic system has been operational since June 1980 and new facilities are continuously added.

REFERENCES

Danielsson, P.E. (1980a). "Designs for Parallelism in Image Pro-
cessing", Internal Report LiTH-ISY-I-0388.
Danielsson, P.E. (1980b). "The Time-Shared Bus - A Key to Effi-
cient Image Processing, Proc. 5th Int. Conf. on Pattern
Recognition, Miami, Florida.
Danielsson, P.E., Kruse, B. and Gudmundsson, B. (1980). "Memory
Hierarchies in PICAP II", to be published in the Proceedings
of the Workshop on Picture Data Description and Management,
Asilomar.
Denning, (1980). "Virtual Memory", Computing Surveys, Vol. 2, No.
3.
Duff, M.J.B. (1976). "CLIP 4 - A Large Scale Integraded Circuit
Array Parallel Processor", Int. Joint Conference on Pattern
Recognition, Coronado, California.
Gerritzen, F.A. and Aardema, L.G. (1980). "Design and use of a
Fast Flexible and Dynamically Microprogrammable Pipe-lined
Image Processor", First Scandinavian Conference on Image
Analysis, Linköping, Sweden.
Golay, M.J.E. (1971). "Hexagonal Parallel Pattern Transformations",
IEEE Trans. Computers, Vol. C-18, p. 1007.
Granlund, G. (1980). "GOP - A Fast Parallel Processor for Image
Information", First European Signal Processing Conf, Lausanne.
Gray, G.S. (1972). "The Binary Image Processor and its Applica-
tions", Information International Inc., Los Angeles, Calif.
Gudmundsson, B. (1980). "User Level Concepts in an Interactive
Picture Processing System", Proc. 5th Int. Conf. on Pattern
Recognition, Miami, Florida.
Kruse, B. (1973). "A Parallel Picture Processing Machine", IEEE
Trans. on Computers, Vol. C-22, No. 12, p. 1075.
Kruse, B. (1976). "The PICAP Picture Processing Laboratory", Third
Int. Joint Conf. on Pattern Recognition, p. 875.
Kruse, B. (1978). "Experience with a Picture Processing in Pattern
Recognition Processing", Proc. National Computer Conf., Vol.
47, p. 1015.
Kruse, B. (1980a). "System Architecture for Image Analysis",
Structured Computer Vision, Tanimoto and Klinger (eds.), Aca-
demic Press.
Kruse, B. et al (1980b). "FIP - The PICAP II Filter Processor",
Proc. 5th Int. Conf. on Pattern Recognition, Miami.
Kruse, B., Danielsson, P.E. and Gudmundsson, B. (1980). "From
PICAP I to PICAP II", Special Computer Architectures for
Pattern Processing, K.S. Fu and T. Ichikawa, (eds.), CRC Press
Inc.
Luetjen, G.I. (1980). "FLIP: A Flexible Multi Processor System for
Image Processing", Proc. 5th Int. Conf. on Pattern Recognition,
Miami, Florida.

Mori, K., Kidode, M., Shinadao, H., et al. (1978). "Design of
    Local Parallel Pattern Processor for Image Processing", Proc.
    AFIPS, Vol. 47, pp. 1025-1031.
Reddaway, (1975). "DAP - Distributed Array Processor", Academic
    Press.
Sternberg, S. (1979). "Parallel Architectures for Image Pro-
    cessing", Third Int. Ieee COMPSAC, Chicago.

PARALLEL ALGORITHM PERFORMANCE MEASURES

Leah J. Siegel
Howard Jay Siegel
Philip H. Swain

School of Electrical Engineering
Purdue University
West Lafayette, Indiana, USA

I. INTRODUCTION

Parallel processing can provide significant increases in speed for many image and speech processing tasks (e.g., Feather et al., 1980; Mueller et al., 1980; Siegel, 1980, 1981b; Swain et al., 1980). Machines with 64 and 256 processors currently exist (Batcher, 1974; Bouknight et al., 1972); systems with 1K to 16K processors have been proposed (Batcher, 1979; Siegel, 1981a). SIMD (single instruction stream-multiple data stream) machines are parallel systems which are particularly well suited to performing operations on arrays. An SIMD machine typically consists of a control unit (CU), a set of N processing elements (PEs) (each a processor with its own memory), and an interconnection network (Siegel, 1979). The CU broadcasts instructions to all PEs. Each active PE executes each of these instructions on the data in its own memory. Each instruction is executed simultaneously in all active PEs. The network allows data to be transferred among the PEs. Masks are used to enable and disable PEs.

The complexity of SIMD algorithms is, in general, a function of the problem size (number of elements in the data set), machine size (number of PEs), and the network. Performance measures which quantify the relationships among these components can be used in various ways. One is for selecting from alternative algorithms. Second, a measure of the way in which the machine size affects the performance of intended application algorithms will aid in deciding how many PEs to include in a new system design. Lastly, in proposed reconfigurable systems (e.g., Lipovski et al., 1977;

This material is based upon work supported by NSF under Grant ECS-7909016, by AFOSR under Grant No. AFOSR-78-3581, and by Defense Mapping Agency, monitored by the USAF Rome Air Development Center Info. Sci. Div., under Contract No. F30602-78-C-0025 through the Univ. Michigan. The United States Government is authorized to reproduce and distribute reprints for Governmental purposes notwithstanding any copyright notation hereon.

Nutt, 1977; Siegel, 1981a), the machine size can be tailored to
the problem size for execution of a given algorithm if there exist
performance criteria for comparing different choices. The goal of
this paper is to collect measures that have been proposed, define
some new measures, and study the relationships between the measures.

II. PERFORMANCE MEASURES

In this section, measures for evaluating the performance of a
parallel algorithm are defined.

Execution Time: The execution time $T_N(M)$ measures the time to perform the algorithm for a problem of size M on an N-PE system.
$T_N(M) = c_N(M) + o_N(M)$, where $c_N(M)$ is the time spent performing
computations which are actually a part of the task, and $o_N(M)$ is
the "overhead," or time spent performing operations required to
"manage" the parallelism. Two sources of overhead are inter-PE
data transfers and masking operations to enable and disable PEs.
When evaluating the execution time of a parallel algorithm, the
distinction between response time (interval between the request
for execution and the return of results) and throughput (number of
jobs executed per unit time) may be an issue. Consider an algorithm which can use 1 or N PEs and which is to be executed on N
different data sets, each of size M. By assigning a data set to
each PE, N instances of the 1-PE algorithm can be performed in
parallel. Alternatively, the N PEs together can process one data
set at a time, performing the algorithm N times. It can be shown
that the expected value of the response time may be less for the
latter scheme than for the former. The throughput rate will never
be greater for the N-PE algorithm than for the 1-PE algorithm.
However, the 1-PE algorithm requires two conditions that, in general, may not be true: (1) there are N data sets to be processed
and (2) each PE has sufficient memory space (and/or a memory
management system) to allow it to handle an entire data set.

Speed: The speed $V_N(M) = M/T_N(M)$ is the number of data points processed per unit time. Assuming at least one instruction is needed
to process a data point and that a unit of time equals the instruction cycle time of a PE, the maximum attainable speed is N.
The reciprocal of speed is execution time per data point.

Speedup: The speedup $S_N(M) = T_1(M)/T_N(M)$ (Kuck, 1977). For a
given algorithm and fixed M, $S_N(M) = \text{constant} * V_N(M)$ for all N.

Ideal speedup is N, which implies $o_N(M) = 0$. Speedup on computations alone is $T_1(M)/c_N(M)$ and is ideal if it is N.

Efficiency: The efficiency $E_N(M) = S_N(M)/N = T_1(M)/(N*T_N(M))$ (Kuck, 1977). For SIMD algorithms, $E_N(M) \leq 1$. $E_N(M) = 1$ for N=1; for N>1, $E_N(M) = 1$ if and only if the speedup on computations is ideal and $o_N(M) = 0$. Execution time together with $dE_N(M)/dN$ may be of interest. Intuitively, efficiency is a measure of how the achieved speedup compares to the ideal speedup.

Overhead Ratio: The overhead ratio $O_N(M) = o_N(M)/T_N(M)$. If the speedup on computations is ideal, then $O_N(M) = 1 - E_N(M)$.

Utilization: The utilization $U_N(M)$ is the fraction of time during which the PEs are executing computations of the algorithm. If there are X steps in the N-PE algorithm, with each step x, $0 \leq x < X$, can be associated a time t_x to perform the step and a count p_x of the number of active PEs. Then

$$U_N(M) = \sum_{x=0}^{X-1} t_x p_x/(N*T_N(M)).$$

When the speedup on computations is ideal, $p_x = N$ for all x, so $U_N(M) = c_N(M)/T_N(M)$, and $U_N(M) = 1 - O_N(M) = E_N(M)$. Utilization is 1 if and only if the speedup on computations is ideal and the overhead is 0.

Redundancy: The redundancy $R_N(M) = \sum_{x=0}^{X-1} t_x p_x/T_1(M)$. If the simplifying assumption is made that all operations take 1 time unit to execute, then this is equivalent to the redundancy in (Lee, 1980). $R_N(M) \geq 1$, and $R_N(M) = 1$ if and only if N=1 or the speedup on computations is ideal.

Cost Effectiveness: The cost effectiveness $CE_N(M) = V_N(M)/C_N$, where $V_N(M)$ is the speed and C_N is the "cost" of the system. The cost is approximately the cost of the CU, N*(the cost per PE), and the cost of the network. For large N, the CU cost will be negligible. In general, the cost for an n-stage network (e.g., omega, generalized cube, STARAN flip, and ADM) can be represented as $\mu n N p_s$, where $n = \log N$, μ is a constant depending on the number of switches in the particular network, and p_s is the cost of a switch. (All logs are base 2.) Within a network, all switches cost the same; switches for two different networks may differ in cost. For a nearest neighbor network, the cost will be $\sigma N p_s$, where σ is a constant reflecting the number of neighbors.

By a (possibly unwieldy) change of variables, for a fixed cost effectiveness, the speed could be written as a function of C_N and M. It would then be possible to obtain an expression for d(speed)/d(cost), and answer two key types of questions: (1) To increase the speed by a factor of f, what will be the increase in cost? (2) If the cost is decreased by a factor of g, what will be the resulting decrease in speed?

Price: Like cost-effectiveness, price integrates considerations of both speed and system cost. Let p_t be the cost of using a unit of time, reflecting factors related to the variable costs of operating the system. The "cost" of the computation will be $p_t * T_N(M)$ and in general will decrease as N increases. However, increasing N increases the hardware and implementation costs. Let p_{PE} be the cost of a PE and p_i a proportionality constant relating the hardware cost to the total cost of implementing the system. The total implementation cost will be $p_i * (N p_{PE} + \mu n N p_s)$.

The cost associated with implementing and using a system having N PEs and an n-stage network for a problem of size M is the price of the computation: $P_N(M) = p_t * T_N(M) + p_i * (N p_{PE} + \mu n N p_s)$. The optimal N can be found by minimizing price with respect to N.

In some applications, the cost related to execution time and the cost due to system implementation may not be of equal importance. For example, the system designer may be prepared to pay a several-fold increase in implementation cost for even a modest improvement in speed. Therefore, it is useful to define a generalized price:

$$P'_N(M) = \frac{\alpha}{\alpha+1} * p_t * T_N(M) + \frac{1}{\alpha+1} * p_i * (N p_{PE} + \mu n N p_s),$$

where the parameter α ($\alpha > 0$) reflects the relative importance of speed and time-variable costs versus system costs. In many cases, α will be a subjectively chosen factor based on cost and speed constraints for a particular application.

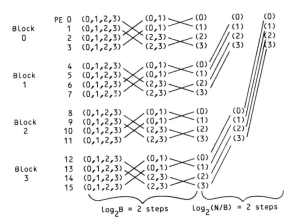

Fig. 1. Histogram calculation for N=16, B=4. (w,...,z) denotes
that bins w through z of the partial histogram are in the PE.

III. PERFORMANCE OF AN EXAMPLE ALGORITHM

In order to discuss and compare performance measures, an example SIMD algorithm is examined. Conclusions drawn from the individual measures and comparisons among them are discussed.

A. Algorithm

Consider an SIMD algorithm to compute the B-bin histogram of an
m-by-m image using $N \geq B$ PEs (Siegel, 1981b), where $m^2 = M$ and $B = 2^b$.
Each picture element (pixel) of the image is represented by a
"gray level," an integer from 0 to B-1. The histogram contains j
in bin i if exactly j pixels have a gray level of i, $0 \leq i < B$. Each

PE is assigned a $(m/\sqrt{N})$-by-$(m/\sqrt{N})$ subimage. The "local" histograms for the subimages are computed in parallel, so that each PE
contains a B-bin histogram for its subimage. These local histograms are combined using a form of overlapped recursive doubling.
(See Fig. 1.) In the first b steps, each block of B PEs performs B
simultaneous recursive doublings (Stone, 1975) to compute the histogram for the portion of the image contained in the block. At
the end of the b steps, each PE has one bin of this partial histogram. The next n-b steps use recursive doubling to combine the
results for these blocks to yield the histogram of the entire image, with the sum for bin i in processor i, $0 \leq i < B$.
A sequential algorithm to compute the histogram of an M pixel
image requires M additions. The SIMD algorithm uses M/N additions
for each PE to compute its local histogram and B-1+n-b steps
(transfer-and-add) to merge the histograms into the first B PEs.
2(B-1 + n-b) masking operations are used: in the first b steps to

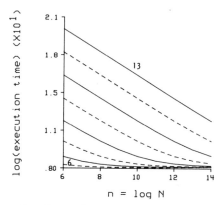

Fig. 2.   Log of execution time. $t_a=1$.

control which direction ("up" or "down") a PE sends its partial
results, and in the final n-b steps to select the PEs which are to
send data and to activate the PEs which received data.

B. Performance

    For the performance measures comparison, assume: t_a = time for
1 integer addition, t_m = time for 1 mask operation, and $t_t(N)$ =
time for 1 inter-PE data transfer. In order for the network to
interface with the PEs, the effective transfer time $t_t(N)$ must be
a multiple of the PE cycle time. For simplicity, assume
$t_a = t_t(N)/2 = 2t_m$. Since the purpose of the example is to demonstrate performance measures rather that to evaluate the absolute
performance of the algorithm, these simplifying assumptions can be
made. In general, program control operations can be overlapped
with the PE operations. The performance of the histogram algorithm will be evaluated as a function of N for m-by-m images,
$64 \leq N \leq 16,384$ and $64 \leq m \leq 8192$. B is fixed at 64.

Execution Time: Fig. 2 shows the log of the execution time as a
function of N and M, normalized to $t_a=1$. (From bottom to top, the
plots are for $m=2^k$, $6 \leq k \leq 13$). For large images, the execution
time decreases as N increases. For small images, an increase in
the number of PEs has little effect on the execution time, and
there is little or no advantage to using a large machine. For
each m in the range 256 to 1024, the execution time decreases as N
increases. However, the rate at which $T_N(M)$ decreases ($dT_N(M)/dN$)
also decreases as N increases. Therefore, it may also be appropriate to consider $dT_N(M)/dN$.

Speed: Fig. 3 shows the log of speed, with t_a the basic time unit.

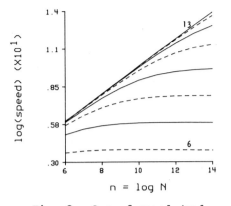

Fig. 3. Log of speed (and speedup).

Fig. 4. Efficiency.

For given N, speed increases as m increases. This happens because, in this case, the effect of the overhead is reduced. For large m, the speed increases almost linearly with N, and the number of pixels processed per unit time ≃ N, the maximum attainable speed. For small m, speed increases to a maximum then decreases for larger N. However here, the total range of speed as a function of N is small so the penalty is not great for choosing a non-optimal N. For m in the middle range, the speed increases at a rate that levels off as a function of N.

Speedup: In this case, for any given m, speedup equals speed (Fig. 3). Using a combination of speedup with d(speedup)/dN (and/or a measure such as utilization or efficiency) may yield a more practical criterion than speedup alone.

Efficiency: As shown in Fig. 4, for all m the efficiency decreases as N increases, but the rate of decrease is slower for large m. For very large m, efficiency is high for all N considered, and the choice of N does not significantly affect the efficiency. This is different from the information provided by execution time, which indicates that, for large m, N should be as large as possible. For small m, efficiency is small for all N considered. Here, the conclusion that efficiency is poor regardless of the choice of N is consistent with the information from execution time. For m in the middle range, the choice of N has a great impact on efficiency. For such m, the value of N which would be chosen to give high efficiency is smaller than the value which would be chosen to minimize execution time or maximize speedup.

    From the observation that efficiency is higher for large m, it can be concluded that for a fixed N there is no advantage to decomposing a size m problem into smaller subproblems, even if the result can later be recombined at low cost. However, given a reconfigurable system such as PASM (Siegel, 1981a) in which N may vary, a somewhat different conclusion can be drawn. For example, in such a system, using efficiency as the only measure, for M=64K (m=256) it would be more efficient to process 4 images simultaneously, each on machines of size 512 than 4 images sequentially, each on a machine of size 2048. The execution time for processing each image would increase, but the total processing time for all four images would be less in the 4-machine case.

Overhead Ratio: Fig. 5 shows the overhead ratio. The conclusions obtainable are consistent with those from efficiency.

Utilization: Fig. 6 shows the utilization. The patterns and conclusions obtainable are consistent with those for efficiency and overhead ratio. In this example, because most steps use all N PEs, $\sum t_x p_x \simeq N*c_N(M)$, and $U_N(M) \simeq c_N(M)/T_N(M) = 1-O_N(M)$.

Redundancy: Fig. 7 shows the log of redundancy. For all m, redundancy increases as N increases, although the rate at which redundancy increases is less for large m than for small m. The redundant computations arise from the additions performed to combine the local and block histograms. When the number of additions to compute the local histograms is much greater than the number of additions to combine them, redundancy is small. For small images, however, the additions to combine the local histograms far outweigh the additions performed to compute them. As in the case of overhead, a high redundancy will be reflected as a decrease in efficiency. From efficiency alone, however, it is not possible to separate the losses due to overhead, utilization, and redundancy. In this example, all three contribute to the loss in efficiency.

Cost Effectiveness: Fig. 8 shows the log of cost effectiveness for an N-PE system having an n-stage network, $\mu=1$, and $p_{PE}=\delta*p_s=1$, with $\delta=32$. The cost effectiveness curves for $\delta=128$ are similar, with corresponding points having slightly higher values for $\delta=128$. Intuitively, the relation between cost effectiveness and speed can be considered by observing, for a fixed m, whether speed increases with N. If it does, cost effectiveness stays high. If it does not, cost effectives decreases. For each value of m in this example, cost effectiveness decreases as n increases. This occurs because doubling N will increase speed by less than a factor of two, but will more than double the cost.

Price: Fig. 9 shows the log of the price for $p_t=p_i=p_{PE}=\delta*p_s=1$, $\mu=1$, and $\delta=32$. Each of the curves has a minimum. The optimal N is greater for large images than for small because for large images execution time continues to decrease significantly as N increases, while for small images the rate at which the execution time decreases falls off for large N. For this example under the simplifying assumptions, each of the minima occurs at N=m.

Using the same assumptions, Fig. 10 shows generalized price as a function of N and α for a 1024-by-1024 image. As expected, the

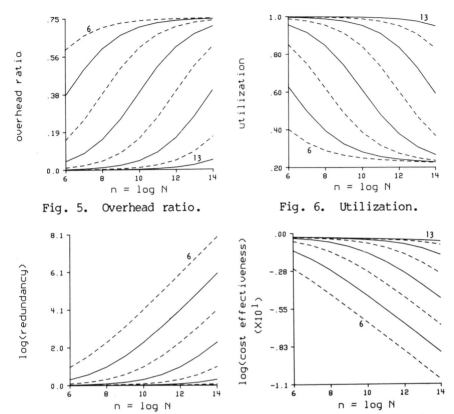

Fig. 5. Overhead ratio.

Fig. 6. Utilization.

Fig. 7. Log of redundancy. The values for m=8192 coincide with the horizontal axis.

Fig. 8. Log of cost effectiveness; t_a = basic time unit.

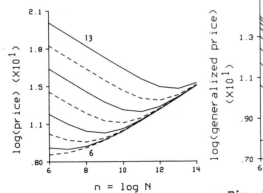

Fig. 9. Log of price, where $P_t=P_i=P_{PE}=32*P_s=1$.

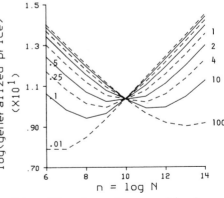

Fig. 10. Log of generalized price, where $P_t=P_i=P_{PE}=32*P_s=1$, m=1024. Labels show α values.

optimal value of N shifts to the right when execution time is more
critical than system cost ($\alpha > 1$) and to the left when system cost
dominates ($\alpha < 1$). The curves intersect at n=10 because under the
assumptions made, for fixed M, $T_{N=1024}(M) \simeq C_{N=1024}$, so at this
point the generalized price is independent of α.

C. Discussion

From the plots for the example, the measures can be grouped according to which measures give similar answers to two types of
questions: (1) Based only on the performance measure being examined, what value(s) for N are "good" for a given problem size?
(2) What is the effect of choosing the "wrong" N?

Similar patterns are observed for execution time, speed, and
speedup. For large m, N should be chosen as large as possible
(over the range considered); there is no disadvantage to such a
choice. Smaller N gives significantly poorer performance. For
small m, N has little effect on performance. For m in the middle
range, performance improves as N increases, but criteria specifying a desired rate of improvement may dictate choice of an intermediate N. These three will be the most useful measures when the
operating constraints are in terms of required execution time, as,
for example, in systems aimed at real-time processing.

Efficiency, overhead ratio, and utilization exhibit similar
characteristics for the example. For very large m, N has little
effect on these measures. This conclusion differs significantly
from that drawn from the first three measures. For small m, N
should be as small as possible. For m in the middle range, there
exist (small) N for which performance is very high, and (large) N
for which performance is very low: the choice of N is critical and
is smaller than would be dictated by the execution time, speed, or
speedup alone.

Together with redundancy, these three measures quantify the
various ways in which the system resources are used by the algorithm. An example of when such measures might profitably be applied is in deciding how to partition a reconfigurable SIMD system
to perform the same algorithm on a number of different data sets,
as was discussed above.

For the histogram example, the choices for N dictated by cost
effectiveness, which takes account of both speed and system cost,
are consistent with the choices indicated by the four resource
utilization measures. For the considered range of N, however, the
choices appear to run counter to the choices suggested by the
speed criterion. Over this range, cost is increasing faster than
speed. Cost effectiveness can be related to the resource utilization measures by observing that high cost effectiveness requires
efficient use of the resources.

Price and generalized price are the only measures which, for
the parameters chosen, indicate clear choices of system size. The

N value indicated by price appears in most cases to be a compromise between the values suggested by the time/speed measures and
those indicated by the resource utilization measures. Generalized
price, as shown for the m=1024 example, makes explicit the tradeoffs between system cost and execution time. For large α (execution time critical), the optimal values for N are consistent
with the N values for high performance under the time/speed measures. Conversely, for small α (system cost critical), the optimal
values of N are consistent with the N values for high performance
under the efficiency, overhead ratio, utilization, and cost effectiveness measures. However, the apparently optimal values of N
are optimal in a meaningful sense only if the various parameters
are determined in a meaningful and realistic way. The determination of the weights will most likely be nontrivial. From a practical viewpoint, if appropriate values can be assigned to these
parameters, price and generalized price may be the most useful of
the measures discussed here because they provide the opportunity
to incorporate real cost factors and make explicit tradeoffs
between system cost and raw computational speed.

IV. CONCLUSIONS

The goal of this paper was to collect, create, and compare different criteria for determining the "goodness" of an SIMD-
machine/algorithm/problem-size combination. As a result of this
study, one may conclude that no single measure is sufficient for
all processing environments. A number of the performance measures
quantify computational speed but do not account for how effectively the resources of the parallel processing system are being used.
For other measures the reverse is true. The cost effectiveness
and price criteria attempt to provide an integrated picture of
speed and resource utilization. Generalized price offers the most
flexible approach for "tuning" the system size based on the relative importance of processing speed versus system cost, but it
also requires the most knowledgeable specification of system and
application parameters in order to obtain meaningful results.

REFERENCES

Batcher, K. (1974). "STARAN parallel processor system hardware,"
    AFIPS Conf. Proc. 1974 NCC, 43, 405.
Batcher, K. (1979). "MPP—a massively parallel processor," 1979
    Int. Conf. Parallel Processing, 249.
Bouknight, W., et al., (1972). "The Illiac IV system," Proc.
    IEEE, 60, 369.

Feather, A., Siegel, L., and Siegel, H. J. (1980). "Image corre-
    lation using parallel processing," 5th Int. Conf. Pattern
    Recognition, 503.
Kuck, D. (1977). "A survey of parallel machine organization and
    programming," ACM Computing Surveys, 9, 29.
Lee, R. (1980). "Empirical results on the speed, efficiency,
    redundancy and quality of parallel computations," 1980 Int.
    Conf. Parallel Processing, 91.
Lipovski, G., Tripathi, A. (1977). "A reconfigurable varistruc-
    ture array processor," 1977 Int. Conf. Parallel Processing,
    165.
Mueller, P., Siegel, L., and Siegel, H. J. (1980). "Parallel al-
    gorithms for the two-dimensional FFT," 5th Int. Conf. Pattern
    Recognition, 497.
Nutt, G. (1977). "Microprocessor implementation of a parallel
    processor," 4th Symp. Computer Architecture, 147.
Siegel, H. J. (1979). "A model of SIMD machines and a comparison
    of various interconnection networks," IEEE Trans. Comp., C-28,
    907.
Siegel, H. J. (1981a). "PASM: a reconfigurable multimicrocomput-
    er system for image processing," in "Languages and Architec-
    tures for Image Processing" (M. Duff and S. Levialdi, ed.),
    Academic Press, London.
Siegel, L. (1980). "Parallel processing algorithms for linear
    predictive coding," 1980 IEEE Int. Conf. Acoustics, Speech, and
    Signal Processing, 960.
Siegel, L. (1981b). "Image processing on a partitionable SIMD
    machine," in "Languages and Architectures for Image Processing"
    (M. Duff and S. Levialdi, ed.), Academic Press, London.
Stone, H. (1975). "Introduction to Computer Architecture," (H.
    Stone, ed.), S.R.A., Chicago.
Swain, P., Siegel, H. J., and El-Achkar (1980). "Multiprocessor
    implementation of image pattern recognition: a general ap-
    proach," 5th Int. Conf. Pattern Recognition, 309.

CELLULAR ARCHITECTURES: FROM AUTOMATA TO HARDWARE[1]

Azriel Rosenfeld

Computer Vision Laboratory
Computer Science Center
University of Maryland
College Park, Maryland

Cellular automata have been studied for many years both as pattern generators and as acceptors for pattern languages. At the same time, computer architectures analogous to two-dimensional cellular automata have been proposed and used for image processing and recognition. In recent years, various extensions to the basic cellular automaton concept have been proposed. These extensions should be of interest in connection with the design of future hardware systems for image processing and analysis.

I. INTRODUCTION

Two-dimensional cellular automata--arrays of finite-state machines that operate in parallel and each of which can communicate with its neighbors--were introduced nearly 30 years ago as models for pattern generation and self-reproduction. Over 20 years ago, a two-dimensional cellular architecture (i.e., an array of processing elements) was proposed for image processing applications (Unger, 1958,1959). More recently, some attention was given to the use of cellular automata as acceptors for two-dimensional formal languages (Smith, 1970,1971a). Three monographs dealing primarily with cellular automata were published

---

[1]*Supported by the U.S. Air Force Office of Scientific Research under Grant AFOSR-77-3271.*

in the USA, Germany, and the USSR at the end of the 1970's
(Rosenfeld, 1979; Vollmar, 1979; Aladyev, 1980).

Since Unger's work, a succession of hardware systems incorporating cellular parallelism have been proposed or built.
Notable among these are ILLIAC III, involving a 36 by 36 array
(McCormick, 1963) (in contrast, the later ILLIAC IV has only 64
processors); CLIP IV, 96 by 128 (Duff, 1973); ICL's DAP, 128 by
128 (Marks, 1980); and NASA's MPP, 128 by 128 (Batcher, 1980).

In recent years, the classical cellular automaton model has
been generalized in a number of ways. The assumption that the
individual cells are finite-state machines (i.e., each cell has
a bounded amount of memory, no matter how many cells there are)
is overly restrictive in practice; by assuming, e.g., that a
cell's memory grows logarithmically with the number of cells, the
power and speed of the system can be increased in a variety of
ways. More interestingly, the topology of the system can be
generalized, e.g., by permitting communication over power-of-2
distances, or by building a "pyramid" of arrays and allowing
communication vertically as well as (or instead of) horizontally.
Still more generally, one can consider cellular automata in which
the "neighbor" relation defines an arbitrary graph of bounded
degree (or perhaps degree logarithmic in the number of cells),
rather than a regular array. Finally, one can allow the neighbor
relation to change during the course of the computation.

This paper reviews basic concepts about cellular automata
and their generalizations, and suggests that some of these generalizations may be of interest to the designers of future hardware systems for image processing and recognition.

II.   BOUNDED CELLULAR AUTOMATA (BCAs)

Traditionally, a two-dimensional bounded cellular automaton
(BCA) is an array (say rectangular) of finite-state machines
("cells") all having the same transition function, where the new
state of a cell depends on the current states of itself and its
neighbors in the array. In more practical terms, we can think
of a BCA as an array of processing elements each of which has a
finite amount of memory, and all of which operate synchronously,
in discrete time steps, in accordance with the same stored program. Each processor initially receives a piece of input data,
and at subsequent time steps, each processor accepts inputs from
its neighbors.

When a BCA is used for image processing or analysis, the input data given to each processor is the gray level of a pixel
(or block of pixels), with neighboring processors getting data
from neighboring pixels or blocks. For simplicity, we will usually assume a single pixel per processor from now on. We will
not consider here how images are input to or output from the BCA;

to avoid the need to consider I/O time, we may suppose that the
input images are sensed directly by an array of sensing elements,
one per processor, and that output images are directly displayed
by an array of light-emitting elements, one per processor.

The greatest advantage of BCAs over conventional computers
is that BCAs can perform local operations on the input image in
parallel, so that the computation time required grows only with
the complexity of the operation, not with the image size (at the
price, of course, of requiring the number of processors to grow
with the image size). Their advantage is somewhat reduced when
we use them to compute properties of the image, or to make decisions on the basis of such properties ("acceptance", in automata
terminology. Here the fact that only neighbors can communicate
causes the computation time to grow with the image diameter. As
a simple example, if we want to compute the histogram of the
image (i.e., count how often each gray level occurs in it), we
must send signals representing each value to a common location so
they can be summed, and some of these signals must travel a distance on the order of the image diameter.

Some of the BCA algorithms for efficient property measurement or recognition are quite nonobvious (see, e.g., (Rosenfeld,
1979) for a collection of such algorithms). An example is the
shrinking-and-shifting process that can be used to determine,
for a given binary-valued input image, whether the 1's (or 0's)
are connected, in time on the order of the image diameter.
Labelling the connected components can also be done in O(diameter)
time, but the algorithm is even less obvious (Kosaraju, 1979).
Automata theorists have made important contributions to our
understanding of how arrays of processors can be used efficiently
for basic image processing tasks.

III. EXTENSIONS

In a classical BCA, each cell has a bounded amount of memory,
no matter how many cells there are. This implies, in particular,
that a cell cannot know its address in the array, since its
memory may not be large enough to hold that address; and a cell
cannot explicitly address cells more than a bounded distance
away. Historically, when the pattern generation capabilities
of BCAs were studied, efforts were even made to minimize the
number of states (i.e., the amount of memory) in each cell.
Keeping the amount of memory per cell low does indeed reduce
the hardware cost of a BCA; but from a practical viewpoint, there
is no reason not to allow the memory per cell to grow, e.g.,
logarithmically, with the number of cells. Logarithmic growth
makes operations such as histogramming easier (the values can be
summed by a single cell, rather than using a set of cells as a
counter), and also makes it easy to compute such properties as

moments and the autocorrelation. It also facilitates connected
component labelling (e.g., use the coordinates of a distinguished
cell in each component as a label), run length coding, border
coding, medial axis transformation, and quadtree construction,
as well as computation of region properties such as area, perimeter, height, width, diameter, compactness, elongatedness, and
convexity. For further details on algorithms for these tasks
see (Dyer and Rosenfeld, 1981).

The speed of many BCA operations can be increased by allowing cells to communicate not just with their immediate neighbors,
but with cells at distances 2, 4, 8,...(Klette, 1979,1980).
(On neighborhood size tradeoffs in BCAs see (Smith, 1971b)).
This allows information to be sent to a common destination in
time proportional to the log of the BCA's diameter. An alternative idea (Rosenfeld, 1979, Ch. 6; Dyer and Rosenfeld, in press)
is to use a "pyramid" of BCAs, each half of the size of the pre-
ceding (e.g., the sizes are $2^n \times 2^n$, $2^{n-1} \times 2^{n-1}$,...,$2 \times 2, 1 \times 1$, where
each cell communicates not only with its "brother" neighbors on
its own level of the pyramid, but also with four "sons" on the
level below and with a "father" on the level above. Note that
the total number of cells is

$$2^n \times 2^n (1 + \frac{1}{4} + \frac{1}{16} \ldots) < 2^n \times 2^n \times 1\frac{1}{3},$$

not much greater than the number of cells in the base of the pyramid alone. Here the height of the pyramid is the log of the
BCA's diameter, and tasks such as histogramming can be performed
in O(log diameter) time using the apex node of the pyramid as the
counter. Many of the algorithms for such pyramid BCAs require
communication upward only, and require the memory in a cell to
grow (at most) with its level in the pyramid, not (otherwise) with
the number of cells, so that the total amount of memory is proportional to that of an ordinary BCA.

IV. ONE-DIMENSIONAL BCAs

Two-dimensional BCAs are rather expensive to build, and the
largest ones now in existence are 128×128 arrays. For the same
cost one could build very large one-dimensional BCAs, consisting of tens of thousands of cells. (Indeed, the MPP has the
option of operating as a 16,000-cell one-dimensional BCA.) Such
BCAs could be used for fast parallel processing of various types
of waveforms; one might, for example, use two of them alterna-
tingly, so that one processes the previous waveform segment
while the other loads the current segment. Further speedup is
possible by using a "triangle" of such BCAs, each half of the
size of the preceding, where each cell communicates with its
two brothers on its own level, two sons on the level below, and
one father on the level above; here the total number of cells

is $2^n+2^{n-1}+\ldots<2^{n+1}$, and the processing time for many tasks is
O(n). In this section we briefly mention two other possibilities
for using one-dimensional BCAs in image-related tasks.

A one-dimensional BCA can be used to scan an image one row
at a time, operating in parallel on each row and "moving" sequentially from row to row. (Again, imagine two of them operating
alternately on video data, one processing the previous row while
the other loads the current row.) Algorithms for such "parallel-
sequential" BCAs are given in (Rosenfeld, 1979, Ch. 7; Rosenfeld
and Milgram, 1973). As an example, if the cells have memory
proportional to the log of the row length, it is easy to do
histogramming in time proprotional to the image diameter. Local
operations, however, now also take O(diameter) time, rather than
O(constant) time.

One-dimensional BCAs can also be used to process border or
curve information represented by chain codes. For example, such
a BCA can determine, in O(length) time, the intersections (if
any) of two given codes, and can determine the code(s) of the
border(s) of the union or intersection of the regions having the
given codes as borders. Various algorithms for chain code analysis using one-dimensional BCAs are given in (Dubitzki et al.,
in press (a)).

V.   GRAPH-STRUCTURED BCAs

One-and-two-dimensional BCAs, pyramids, etc. are all composed
of cells each of which communicates with a fixed number of neighbors (ignoring border effects). More generally (Rosenstiehl
et al., 1972; Wu and Rosenfeld, 1979,1980), one can consider
graph-structured BCAs in which the neighbor relation defines an
arbitrary graph of bounded degree. If the amount of memory per
cell is allowed to grow logarithmically with the number of cells,
one can also allow the degree of the graph to grow in the same
way.

A BCA that has a fixed graph structure is of limited interest
unless there are many sets of input data to be analyzed that all
have the same graph structure (e.g., all images are arrays). In
image analysis, various types of graph structures do arise (e.g.,
the adjacency graph of a segmentation of an image into regions),
but they differ from image to image, and even vary in the course
of processing a single image (e.g., if regions merge or split).
Thus it is of greater interest to study graph-structured BCAs
in which the initial graph structure can be defined arbitrarily
and can then modify itself in the course of a computation.

A class of self-modifying graph-structured BCAs is defined
in (Wu and Rosenfeld, 1979b; Dubitzki et al., 1979; Rosenfeld
and Wu, 1980). It is shown in (Dubitzki et al., 1979; Dubitzki
et al., in press(b); Rosenfeld and Wu, in press) how such BCAs

can be initially configured to represent a given segmentation of
a given image, e.g., in terms of its region boundary segment
graph or its quadtree. (The representation in terms of the
region adjacency graph is simpler, but requires graphs of degree that can grow with the image size.) It is also shown how
such a BCA can modify its configuration as the image representation changes, e.g., as regions split or merge, and can perform
subgraph matching in parallel, avoiding the need for combinatorial search.

Arbitrarily graph-structured BCAs may not be easy to implement in hard-wired form unless they have regular structures
(array, tree, etc.). On the other hand, conventional multiprocessor communication systems can be used to simulate BCAs
that have either fixed or variable graph structures. For example,
the ZMOB system (Rieger *et al.*, 1980), a collection of 256 Z80A
microprocessors that communicate via a fast bus, can be used to
simulate a reconfigurable graph-structured BCA having up to 256
cells. Such a simulation would be adequate for many real-world
tasks involving region-level image processing.

VI. CONCLUDING REMARKS

Two-dimensional BCAs are a classical model for image processing and analysis at the pixel level. Such BCAs are beginning
to be built in reasonable sizes, but are still quite costly,
and are limited in speed for some tasks due to communcation
delays. Their performance can be speeded up by extending them in
various ways: cell memory that grows logarithmically with the
number of cells; connections to cells at power-of-2 distances,
or the use of "pyramids" of BCAs. At the same time, there exist
types of BCAs that can be built today at reasonable cost and that
can solve practical problems--e.g., one-dimensional BCAs for
processing waveforms or chain codes, or for row-by-row processing
of images. Graph-structured BCAs whose structure varies in the
course of a computation can be used for image analysis at the
region level; here again, the number of cells required is not
very great, and a variable-structure BCA can be simulated by a
multi-microprocessor system that allows sufficiently flexible
interprocessor communication.

REFERENCES

Aladyev, V. (1980). "Mathematical Theory of Homogeneous Structures and their Applications." Valgus, Tallinn.

Batcher, K. E. (1980). Design of a massively parallel processor. *IEEE Trans. C-28*, 836-840.

Dubitzki, T., Wu, A., and Rosenfeld, A. (1979). Local reconfiguration of networks of processors: arrays, trees, and graphs. TR-790, Computer Vision Laboratory, Computer Science Center, University of Maryland, College Park, MD.

Dubitzki, T., Wu, A., and Rosenfeld, A. (in press(a)). Parallel computation of contour properties. *IEEE Trans. PAMI*.

Dubitzki, T., Wu, A., and Rosenfeld, A. (in press (b)). Region property computation by active quadtree networks. *IEEE Trans. PAMI*.

Duff, M. J. B. (1973). A cellular logic array for image processing. *Pattern Recognition 5*, 229-247.

Dyer, C. R. and Rosenfeld, A. (1981). Parallel image processing by memory-augmented cellular automata. *IEEE Trans. PAMI-3*, 29-41.

Dyer, C. R. and Rosenfeld, A. (in press). Triangle cellular automata. *Info. Control*.

Klette, R. (1979). A parallel computer for digital image processing. *EIK 15*, 237-263.

Klette, R. (1980). Parallel operations on binary images. *Computer Graphics Image Processing 14*, 145-158.

Kosaraju, S. R. (1979). Fast parallel processing array algorithms for some graph problems. Proc. 11th STOC, 231-236.

Marks, P. (1980). Low-level vision using an array processor. *Computer Graphics Image Processing 14*, 281-292.

McCormick, B. H. (1963). The Illinois pattern recognition computer - ILLIAC III. *IEEE Trans. EC-12*, 791-813.

Rieger, C., Bane, J., and Trigg, R. (1980). ZMOB: a highly parallel multiprocessor. Proc. IEEE Workshop on Picture Data Description and Management, 298-304.

Rosenfeld, A. (1979). "Picture Languages: Formal Models for Picture Recognition." Academic Press, New York.

Rosenfeld, A., and Milgram, D. L. (1973). Parallel/sequential array automata. *Info. Proc. Letters 2*, 43-46.

Rosenfeld, A., and Wu, A. (1980). Reconfigurable cellular computers. TR-963, Computer Vision Laboratory, Computer Science Center, University of Maryland, College Park, MD.

Rosenfeld, A., and Wu, A. (in press). Parallel computers for region-level image processing. *Pattern Recognition*.

Rosenstiehl, P., Fiksel, J. R., and Holliger, A. (1972). Intelligent graphs: networks of finite automata capable of solving graph problems. *In* "Graph Theory and Computing" (R. C. Read, ed.), pp. 219-265. Academic Press, New York.

Smith, A. R. III (1970). Cellular automata and formal languages. Proc. 11th SWAT, 216-224.

Smith, A. R. III (1971a). Two-dimensional formal languages and pattern recognition by cellular automata. Proc. 12th SWAT, 144-152.

Smith, A. R. III (1971b). Cellular automata complexity trade-offs. *Info. Control 18*, 466-482.

Unger, S. H. (1958). A computer oriented toward spatial problems. *Proc. IRE 46*, 1744-1750.

Unger, S. H. (1959). Pattern detection and recognition. *Proc. IRE 47*, 1737-1752.

Vollmar, R. (1979). "Algorithem in Zellularautomaten." Teubner, Stuttgart.

Wu, A. and Rosenfeld, A. (1979a). Cellular graph automata (I and II). *Info. Control 42*, 305-353.

Wu, A. and Rosenfeld, A. (1979b). Local reconfiguration of networks of processors. TR-730, Computer Vision Laboratory, Computer Science Center, University of Maryland, College Park, MD.

Wu, A. and Rosenfeld, A. (1980). Sequential and cellular graph automata. *Info. Sciences 20*, 57-68.

PARALLEL ALGORITHMS AND THEIR INFLUENCE
ON THE SPECIFICATION OF APPLICATION PROBLEMS

M J B Duff

Department of Physics and Astronomy
University College London
England

I. INTRODUCTION

The recent sudden growth in interest in novel computer architectures, which has been stimulated by an unprecedented availability of cheap computers, microprocessors, and special logic
circuits, has, in its turn, stimulated a related interest in the
design of algorithms. To a generation brought up on batch processing in mainframe computers, in which the computer architecture is almost completely hidden from the high level programmer
by sophisticated software packages and efficient compilers, it
is not immediately obvious that algorithms should be tailored to
architectures. In fact, it has been argued that a need for the
programmer to know anything at all about how his computer 'works'
is a sign of weakness in the language he is using. Be that as it
may, it is the purpose of this paper to explore the relationships
between tasks, algorithms and architectures, with particular
reference to the properties of processing arrays. Although the
arguments presented will usually have some general applicability,
the CLIP4 architecture (Duff, 1978) will be used to provide illustrations, thus avoiding the temptation to lapse into an over-
generalising vagueness.

II. GENERAL PURPOSE COMPUTERS

Computer users do not expect to be limited by their computer
as to what can be computed, except in so far as available memory,
word lengths and computing speed may place bounds on the size of
the problem or the accuracy to which it may be computed within a
given maximum time span. It would be unthinkable for any computer worthy of the name to reject a program because it could not
calculate square roots, for example. The reason for this expected flexibility is that all arithmetic problems can be decomposed
into sequences of simple, fundamental arithmetic operations such
as addition, inversion and overflow detection. Similarly, logic

operations can be reduced to sequences of binary boolean operations. After adding input-output capability and an adequate
amount of storage, the computer so constructed should be applicable to problems of all types and complexities, flexibility and
programmability now depending primarily on the range and efficiency of the system's software.

Given that a computer possesses all these characteristics,
then it can be said to be 'general purpose' in that by sufficiently ingenious programming it could be used to solve all sol-
vable problems, i.e., problems which have been or could be solved
on at least one other computer and for which an algorithm is
known.

CLIP4 is certainly a 'general purpose' computer as defined in
this manner but in common with many other computers, it is also
describable as 'special purpose'. The term 'special purpose'
should not be taken here to imply a limitation but rather to point
to the optimisation of CLIP4 for computations on two dimensional
arrays of data, particularly binary data. The same is true for
many 'special purpose' computers whose main areas of application
are in image processing or pattern recognition. At the same time,
it should be noted that this field of application has spawned many
processors which are capable only of a well-defined range of operations of particular value in that field. It is not the purpose
of this paper to discuss processors of this very restricted type
since the choice of algorithms to be used with them will not, in
most cases, be in the hands of a programmer but will be wired
into the system.

III. LINE-BY-LINE TRANSLATION

Let us assume that a conventional serial computer is being
used as the computing component in an image analysis system being
applied to an inspection problem: the classification of a stream
of manufactured components on a conveyor belt, so that they can
be sorted into bins.

(It is interesting to note in passing that certain industrial
inspection problems which arise would appear at first to be completely artificial and merely invented to provide a vehicle for
students of image analysis techniques requiring a gratifyingly
simple problem to solve. Although, on the face of it, it would
seem to be absurd, having manufactured a range of components, to
throw them into a heap, and then to try to devise an automatic
sorting system to re-identify the components, nevertheless,
exactly this procedure is practised in certain industries. The
best advice, i.e., to refrain from mixing the components, may
turn out to be unacceptable, for logistic reasons which may not be
immediately apparent, so the problem must be seen as 'real' by
the image analyst).

A typical procedure would be as follows:
1. The opaque objects, as they pass on a conveyor belt, are viewed by transmitted light and their silhouettes recorded in a television camera.
2. The video signal is thresholded and the resultant binary data entered directly into computer memory as an array of ones and zeros (assume ones represent the white background and zeros the black silhouettes of the objects). The data might appear as shown in Fig. 1. The small regions labelled (a) represent noise due to imperfect illumination or thresholding; (b) are two parts of an object just about to appear in view whereas (c) is part of an object just passing out of view. Finally, (d) is the object currently to be classified.
3. The data is raster scanned, starting at the top left hand corner, until a zero element is found. The co-ordinates of this element are recorded and a border following routine entered.
4. The border following routine is used to trace once round the object (we will assume all the objects are simply connected), and the chain code for the edge is extracted and recorded. The edge co-ordinates are also recorded.
5. The raster scan is continued to find further objects, ignoring those one elements whose co-ordinates lay between pairs of co-ordinates in the edges of objects already detected.
6. The scan ceases when the bottom right hand corner of the array is reached.
7. Edges (and their chain codes) which have co-ordinates in the array sides are ignored and assumed to be parts of objects already analysed or to be analysed later.
8. Very small objects are rejected as noise.
9. Chain code analysis is applied to the remaining chain to identify the object (d).

Despite the complexity of this algorithm, it is not sufficiently sophisticated. For example in Fig. 2, which shows two objects, (a) and (b), the first object (a) could be border followed, passing through points a_1, a_2, a_3 and a_4. However, when the raster scan is resumed, points such as b_1 and b_2 would be ignored as they lie between a_2 and a_3. In fact, object (b) would not be detected at all. It would seem that a more detailed examination of the co-ordinates a_1 to a_4 is needed before others can be rejected.

Ignoring these refinements, the broad structure of the serial algorithm is described in the nine steps listed. Suppose now that a parallel processor is introduced with a view to speeding up the operation. Then one possible procedure to adopt would be to take the nine steps one by one and to write a line-by-line translation of the existing serial program into code to be run on the parallel processor. If this were to be done, in most cases it would be found that the new program would run more slowly on the 'fast' parallel processor than on the 'slow' conventional serial computer. The parallel machine would be programmed

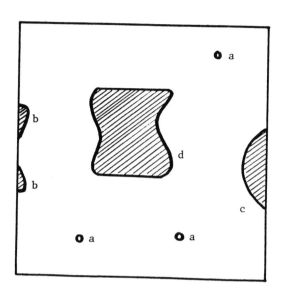

FIGURE 1.   Binary view of objects to be recognised

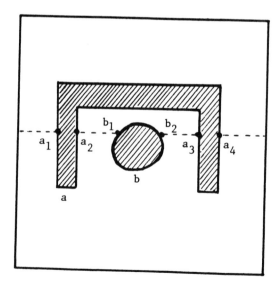

FIGURE 2. Objects requiring special
treatment for segmentation

to carry out tasks in which it had no additional competence, whereas no advantage would be taken of its parallel processing capability. Yet the algorithm has been shown, presumably, to work. Is it therefore absolutely necessary to discard a proven algorithm in order to take advantage of a faster processor?

IV. HIGH-LEVEL DESCRIPTIONS

A little thought quickly reveals the reason for the failure of the line-by-line approach in any specific example. In CLIP4, each of the 9216 processors is extremely simple in construction having only a single-bit processing capability. Furthermore, the semiconductor technology used (metal gate NMOS) is slow compared with that to be expected in contemporary minicomputers. If only one of these processors is in use at a particular stage in the calculation, with the other 9215 not performing any useful function, then the overall performance of the system will compare extremely unfavourably with that of the minicomputer. What is needed is for the algorithm to be restructured so that the bulk of the processors can be gainfully employed during the greater part of the processing time. The nine steps of the working algorithm must be re-examined to see if such a restructuring is possible.

As a first approach, it is worthwhile to restate the algorithm in less detail, i.e., at a higher level. It is important here to make a clear distinction between a _task_ and a _procedure_. If a task is broken up into a set of sub-tasks, then this set might eventually be seen as a procedure. The danger is that it may not be appreciated that the set of sub-tasks is not unique; alternative sets may be equally effective. Turning attention again to the inspection problem, a higher level description of the procedure could be stated thus:

A. (Steps 1,2) Represent silhouettes of the objects as binary arrays in computer memory
B. (Steps 7,8) Reject image noise and all objects touching the array sides
C. (Steps 3-6) Extract and store edges of all binary objects in view
D. (Step 9) Analyse the remaining edge(s) to identify the object(s)

This description is obviously less restricting to the programmer but, at the same time, sufficiently less precise to lead to the danger that a new program fitting the description might not work in practice. Disregarding this objection for the present, a parallel processor now looks very suitable for the task described. Thus for CLIP4, the first part of the program becomes (using the assembly level language CAP4 (Wood, 1977) which is used to program CLIP4):

```
SCAN                           )
HTH L                          )   STEP A

PST 0                          )
SET  P+A, [1-8] A              )
PST 1                          )
LDA 1                          )
SET  -P.A, [1-8] -A            )   STEP B
PST 1                          )
LDA 1                          )
SET  P+A, [1-8] P.-A,E         )
PST 0                          )

LDA 0                          )
SET  P.-A, [1-8] A             )   STEP C
PST 1                          )
```

As a result of these three steps, edges of the centrally
placed objects are stored in memory address 1. This is the same
stage as that reached having executed steps 1-8 in the fuller
description of the algorithm, except that the chain code information has not been derived. Now a chain code does, by its very
nature, represent a spatially serial concept. Even so, it is
possible to derive a chain code with a fair degree of parallel
operation in the following way:
1. At every edge element E, examine neighbouring elements N_1 in
   the corresponding array containing the original object. Continue in a clockwise direction, examining elements N_2, N_3 etc
   up to N_8 (see Fig. 3).
2. When N_x is a one element and N_{x+1} is a zero element, store the
   label $x+1$ at E (taking 8+1 as equivalent to 1).
   Having labelled every edge pixel with its appropriate chain
code, it remains only to match the code with stored prototypes,
making due allowance for rotation and for digitizing variability.
At this stage in the procedure, it is almost certain that the
process would become almost entirely serial so that the parallel
processor would be at a disadvantage.
   Summarising this approach, it can be seen that what is essentially the same process might be restructured so as to take advantage of a widely differing computer architecture, so that
some intermediate results during the process are the same in both
procedures (extracted edges, for example), but with the inherent
difficulty that certain parts of the procedure (such as chain
code analysis) remain unsuitable for implementation on the new
computer. It would seem that this approach is still not completely satisfactory.

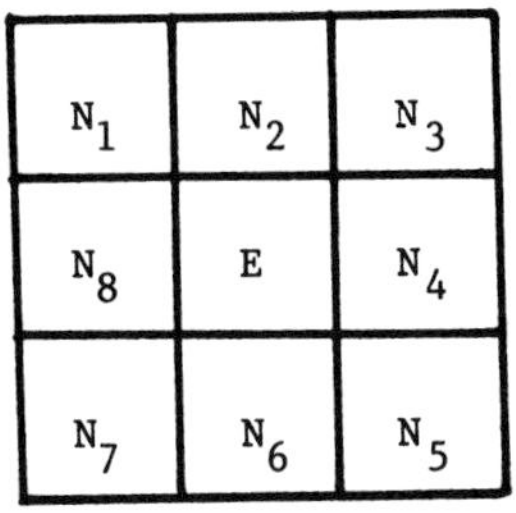

N_1	N_2	N_3
N_8	E	N_4
N_7	N_6	N_5

FIGURE 3.   Elements for deriving chain codes

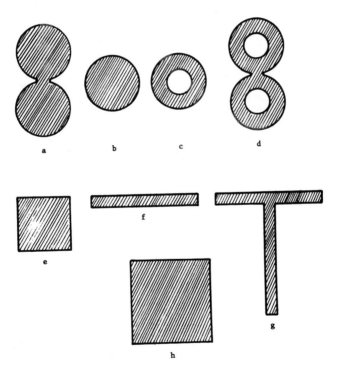

FIGURE 4.   Eight test objects

V. REDEFINITION OF THE ALGORITHM

Taking the example used in previous sections, it is obvious that making use of a chain code to describe edge shapes is likely to be inefficient for a parallel processor with an array structure. At the very least, the number of edge elements is small compared with the number of array elements so that it would be difficult to make effective and simultaneous use of all the processors in the array during the chain code analysis phase of the operation. Following the previous line of thought, the next step would be to seek an even higher level definition of the procedure:

A. As before

B. As before

C. Recognise the shape of the remaining object(s)

By now, so little information is given in these three statements that they constitute little more than the definition of the task. There can consequently be no guarantee that programs designed to implement these steps will be successful. At the same time, the programmer is further freed to explore what are, essentially, completely new algorithms which take full advantage of the available architecture. The task remains the same: inspect the object silhouette and recognise its shape.

By way of illustration, consider Fig. 4 which shows eight object silhouettes labelled (a) to (h). Let us suppose that these objects may appear in the field of view in any orientation. Having a parallel processor, such as CLIP4, certain operations are easily performed with high efficiency. The problem is to define procedures for recognising these objects making use primarily of efficient operations. Once again, we assume the objects to be binary, i.e., represented by clusters of zero (black) elements on a ones (white) background. The measures proposed are:

TOTAL OBJECT AREA	A
TOTAL HOLES AREA	H
EXTERNAL PERIMETER LENGTH	P
NUMBER OF 'SHRINKS' TO ERASE THE OBJECT	S
NUMBER OF DISCONNECTED OBJECTS PRODUCED DURING SHRINKING	N

A, H and P are self-explanatory. S is obtained by repeatedly 'shrinking' the object (after filling in any holes) until it vanishes. In this case, 'shrink' implies changing a zero element to a one element if at least one of the elements in its three by three neighbourhood is a one element. S is the number of times this operation must be repeated to cause the object to vanish. As the operation is repeated, objects with concavities (such as (a) and (d) in Fig. 4) will break up to form more than one

smaller object. N is the number of objects so formed.
A table can be prepared to describe every object in terms of
these parameters:

TABLE I.   *Parameter Values*

Object	Parameter				
	A	H	P	S	N
a	2π	0	11.1	3	2
b	π	0	2π	3	1
c	0.84π	0.16π	2π	3	1
d	1.68π	0.32π	11.1	3	2
e	4	0	8	3	1
f	1.6	0	8.8	1	1
g	3.2	0	15.8	1	1
h	9	0	12	4	1

It is then a simple step to construct a decision tree based on
the table (see Fig. 5). The details of this method are unimpor-
tant; the significant factor is that the same overall result is
obtained (sorting the objects) without making use of conceptually
serial measures such as the chain code. It may at first sight
seem that this method is not sufficiently general to deal with
all possible object shapes. Whilst this is true, similar objec-
tions can be raised with respect to chain code analysis. This
only goes to underline the truth of the statement that all pattern recognition algorithms are likely to be data dependent in
that they are in danger of failing when unexpected shapes are
seen for the first time.

VI. REDEFINITION OF THE TASK

It is a common complaint from those who spend their time
devising automatic machinery for others to use, that the task
itself is often in need of redefinition. A task which is being
performed by a human being will most usually be shaped by the
needs and capabilities of a human being. If the task is then to
be automated, it is more than likely that some absurdities will

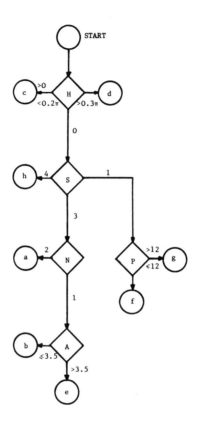

FIGURE 5.   Flow diagram for object recognition

be built into the system so produced. For example, it would be
ludicrous to try to build a robot capable of walking up and down
stairs in order to carry loads from one floor to the next; surely
a lift would be a more sensible solution? Yet image analysts are
often faced with just this situation. When diseased cells are
detected by a pathologist, the observed features are those most
readily detectable by the human eye and not necessarily optical
features most characteristic of the cell's diseased condition.
It seems natural, even inevitable, that an automatic analysing
microscope will be expected to use these same features to diag-
nose disease, regardless of whether a much simpler and potentially more reliable device might be made to work using quite
different criteria.

The mismatch between human and non-human methodology stems
largely from the human eye's excellence in recognising patterns
coupled with its almost complete inability to make precise and
absolute measurements of length, area, intensity, etc.
    Some of this disparity is reflected in the differences be-
tween serial and parallel processors. In a serial system, every
picture element is handled individually by the one processor so
that counting operations can often be carried out with almost no
additional computational cost. On the other hand, in an array
processor, the individual processors will go through the motions
of, say, shrinking the black regions of an image even in parts of
the array where no black regions of the image exist. Again, in
an array processor but not in a serial computer, it is far easier
to find the edges of an object than to measure its width. Broadly
speaking, in a discrete image, most measurements become counting
operations and counting is conceptually more serial than parallel.
    This difference is further accentuated by the limited arith-
metic capability of individual processors in most arrays, which
can be compared with the almost non-existent calculating ability
of the human retina and its associated visual cortex.
    Taking all these and other related points into consideration,
it becomes clear that successful image analysis is not likely to
be achieved unless:
a) the programmer interacts with those posing the problem, so as
   to avoid tacit assumptions as to the choice of procedures
   which may distort the definition of the task,
b) the programmer is aware of the inherent strengths and weaknesses of the computer which will be used, to implement the
   procedures.

VII. COMPILERS

    It is a matter for discussion as to how far the ingenuity of
a compiler designer might be expected to protect the programmer
from the need to know the details of the architecture of his computer. At the simplest level, a well written compiler would
generate efficient code to compute small program fragments. At
this level, it is usually true that almost all possible algorithms have been invented (or discovered) so it is possible to
select the one which runs most efficiently on the computer being
used. At the next level of complexity, an efficient compiler
would be required to check that the sequences of fragments remain optimal, which is a much more difficult task involving some
reselections of the fragmentary algorithms. In the limit, there
would seem to be no way in which a computer, unsupported by a
vast knowledge base, could produce maximally efficient code from
ultra high-level instructions which do little more than specify
the task.

VIII. THE ROLE OF THE PROGRAMMER

Parallel machines have acquired a largely undeserved reputation for being 'hard to program'. There are many possible explanations for this, of which the most likely is the fact that most such machines are fairly new, as a consequence of which the customary software support is not yet available. It is as though large mainframe and minicomputers were always to be programmed in assembly level languages.

Most programmers have learned to program in one of the more popular languages (BASIC, FORTRAN, ALGOL, etc) using a conventional serial computer. Inevitably, the general approach to program design has been pointed towards serial machines, to their advantages and to their limitations. Certain fundamental 'rules of thumb' have to be rejected and others adopted if efficient use is to be made of parallel machines. Some of these rules are as follows.

A. Order of the Computation

It is customary to evaluate a computation on an n by n pixel picture in terms of the order of the number of operations to be performed: $\log n$, n, n^2 etc. Since a processing array performs n^2 operations each cycle, this approach is usually unhelpful, unless the order significantly exceeds n^2.

B. Reducing the Dimensionality of the Problem

Relating to the previous point, in a serial machine it is advantageous to reduce the dimensionality of the data as soon as possible, even though this may increase the complexity of the computation. In other words, a more complex calculation on fewer data points may be faster than a simple calculation on the original number of data points (pixels). This is the reverse of the situation in processing arrays where increasing the dimensionality might well lead to higher efficiency.

C. Time Ordered Computation

Since pixels are visited sequentially in a serial machine, it naturally follows that certain parameters such as the ordering of features and directions of edges can be determined without computational cost. This will not generally be true in parallel machines.

D. *Precision*

Most serial machines will be structured to provide a fixed precision in arithmetic calculations (with double precision extensions). The fact that less precision may be adequate for a particular calculation cannot usually be turned to advantage. On the other hand, the typical single bit structure of an array processor does imply an increasing penalty as higher precisions are demanded; it follows that efficient parallel programming will imply a careful estimation of the minimum precision required at many stages of a calculation.

E. *Local Neighbourhoods*

In an array processor, it is easy to combine data in neighbouring pixels since neighbouring processors will have data paths between them. It is much less easy to explore more widely spread neighbourhoods. However, in a serial machine, data from other array locations is obtained by offset address calculation and memory fetching, neither of which are less efficient when larger array distances are involved. The strong pressure to confine algorithms to immediate neighbourhoods is therefore felt only in algorithms designed for processing arrays.

F. *Input-output*

The balance between I/O time and processing time is very different when processing times are so greatly diminished, in processing arrays. This sets a higher low limit to problem complexity before I/O time predominates. In terms of the acceptable task specification, it is implied that more complex tasks should be attempted in order to make effective use of the fast computation available in a parallel machine, particularly if techniques are not available to speed up the I/O process. This does point to the dangers inherent in benchmarking new processors in terms of already existing problem specifications.

G. *Memory Limitations*

In a similar way to the previous argument concerning I/O, it can also be argued that programs to run on processing arrays should, whenever possible, not exceed the in-array memory provision. Although overflow into a host computer memory (or onto disk or magnetic tape) is a possibility, once again I/O times will begin to predominate. This limitation points to the need for sound approximations and efficient memory management.

IX. SUMMARY

It can be concluded that, in the present state-of-the-art in machine and compiler design, it is not possible to adequately protect the programmer from the need to have some knowledge of machine architecture if
a) efficient programs are to be written and
b) tasks are to be interpreted intelligently in the interests both of those specifying the tasks and of those attempting to carry them out.
A few guide lines can be followed to aid the writing of programs which will run efficiently on processing arrays. Nevertheless, the evidence points to the need for parallel machine programmers to dig deeper into the problem poser's ultimate requirements and to resist the temptation to produce line-by-line translations of existing serial programs.

REFERENCES

Duff, M.J.B. (1978). Proc. 1978 National Computer Conference, 1055.
Wood, A. (1977). CAP4 programmers manual, University College London Internal Report.

MPP ARCHITECTURE AND PROGRAMMING

J. L. Potter

DIGITAL TECHNOLOGY
GOODYEAR AEROSPACE CORPORATION
AKRON, OHIO

I. INTRODUCTION

Because of the large volume of data handled during image processing, parallel processors are required if execution times are
to be achieved in real time. The Massively Parallel Processor
(MPP) is a parallel processor being built for NASA by Goodyear
Aerospace Corporation for image processing applications. The MPP
is completely programmable and extremely flexible. Consequently,
it can be applied to any task requiring large amounts of computing power. This paper will describe the basic concepts behind
programming a parallel computer like the MPP. It will summarize
the basic MPP instruction set. Also, it will briefly describe a
structured parallel FORTRAN language for the MPP. The final
section will describe some proposed language extensions that would
be particularly valuable for image processing applications.

II. SIMD PROGRAMMING

The MPP is a member of the SIMD class of computers. This
class is quite important for two reasons. First, the single instruction stream aspect makes many of the techniques and practices
developed for programming conventional serial computers directly
applicable. Second, the parallelism is in the execution of a
single task, not in multi-tasking. Consequently, the parallelism
closely models the thought process. For example, the single
thought "add 5 to every pixel" is expressed in an SIMD machine
by the single statement: PIXEL = PIXEL + 5 instead of the
sequence of statements: DO 100 I=1,ALL
                        100 PIXEL(I)=PIXEL(I)+5

MULTICOMPUTERS AND IMAGE PROCESSING
ALGORITHMS AND PROGRAMS

275

Since a common ground of knowledge exists in serial computers, it provides the most reasonable basis for discussion of SIMD programming. Thus, approaches to converting ordinary serial constructs into parallel ones will be explored. In order to keep the discussion simple and to avoid specific hardware details at this level, the generalized SIMD architecture of Figure 1 will be assumed.[1] The processing units (PU's) and associated memories are organized into a special unit called an array. The sequential program is stored and executed out of the Sequential Control Unit and its associated memory.

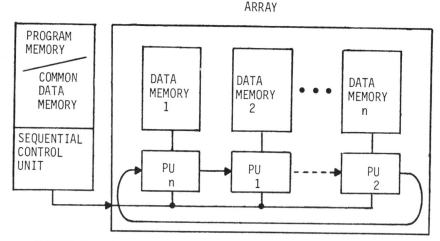

FIGURE 1.   Generalized SIMD model

Loops are one of the most fundamental constructs in a serial computer. They are used to make "inherently parallel" algorithms serial. Take for example the parallel concept, "clear array DATA to zero." In a parallel machine, every element of the DATA array is cleared simultaneously. In a serial machine, loop control devices and parameters are required to assure that every element of DATA is cleared one location at a time. The key conceptual relationship between the serial and parallel implementation of this construct is memory layout. In a serial environment, the array D is assigned to a physically contiguous block of memory. In a parallel environment, each element of D is assigned to the same memory location (address X for example) in each of the data memories (see Figure 2). Thus, the single instruction "clear X" broadcast from the Sequential Control Unit to all of the PU's causes all elements of the array D to be cleared simultaneously and therefore, is equivalent to a loop in a conventional computer.

[1]Note this architecture has a linear array of processors while the MPP has a two-dimensional array. The MPP architecture is discussed in more detail later.

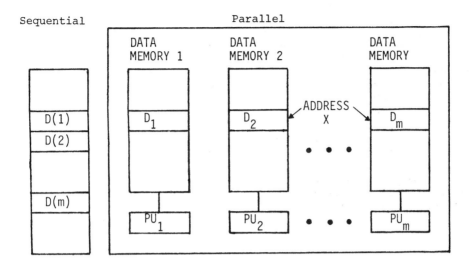

FIGURE 2. Memory assignments in serial and parallel computers for an array of size (m) called D.

Frequently, algorithms involve intra-array operations. The array organization shown in Figure 1 for the parallel machine appears to preclude this type of operation. However, all SIMD computers have some degree of communication between processing units. This intra-module communication capability can be used for intra array operations. If, for example, the computation B(J) = D(J) + D(J+C) where C is a constant, were to be performed, a common procedure in a conventional computer would be to use an index register as shown in Figure 3a.

In an SIMD computer, the same computation would be performed simply by moving (shifting) the data C processing units. Conceptually, the assembly language instructions would be as shown in Figure 3b.

```
        R0      address of D
        R1      address of B
        R2      loop count
LOOP  LOAD   0(R0)                    LOAD    D
        ADD    C(R0)                    ADD     D(SHIFTED C UNITS)
        STORE  0(R1)                    STORE   B
        INC    R0,R1
        DEC    R2
        BGZ    LOOP
```

a.   Sequential Code   b.   Parallel (SIMD) Code

FIGURE 3. Displacement indexing.

In all computers, there is a fundamental relationship between memory size and the amount of data to be processed. To
this point it has been assumed that all of the data would fit
into one memory array. That is, there are more PU's than data
elements. If this is not true, techniques used in conventional
serial computers must be used to reconcile the amount of data
with the memory capacity available.

Since data is normally broken into records, this discussion
will center on record size in relation to the number of PU's. If
there are n PU's in a machine, any record of size n or less is
easily accommodated as discussed above. If a record has more
than n elements, then the data can be "folded" over and processed
as shown in Figure 4. The code shown is the equivalent of a
nested loop in a sequential computer where the "inner loop" is
performed in parallel. Thus, it is clear that the solution to

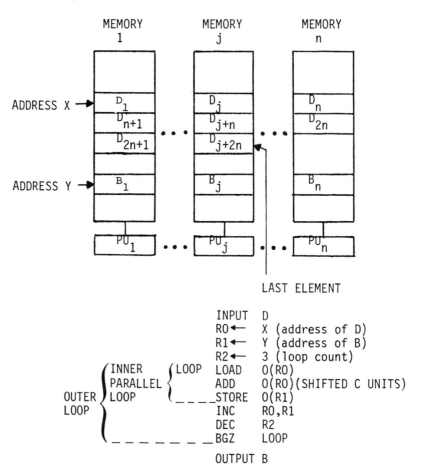

FIGURE 4. Folded data

the limited memory problem is basically the same whether a
serial or a SIMD computer is involved.

To this point, figures have been used to illustrate the
concepts of SIMD processing. However, in order to conserve space
a vector notation will now be introduced. Thus, if the contents
of address D of the memory associated with PU_i is written as D_i,
then the entire contents of the array of PU's at a memory address
D can be denoted by $D = (D_1, D_2, ---D_n)$ where n is the number of
processing units in the SIMD array. Folded data is represented
by a second index indicating the "row" number.

Occasionally loops contain statements which are dependent on
the loop variable. In these cases, a parallel machine may require
an auxiliary data array such as $I = (1,2,3,---n)$. Then, calculations which are a function of the loop index are obtained simply
by using the index array. For example, if $K = (5,20,10,7,12,15)$;
then the SIMD statement $D = K*I$ yields $D = (5,40,30,28,60,90)$ and
is equivalent to the sequential code: DO 100 I = 1,6
$$100 \ D(I)=I*K(I).$$
Frequently, loops have conditional statements in them. In order
to accommodate this situation every processing unit in a SIMD
array must have its own dedicated condition code memory (CCM).
The CCM for each individual PU can be set to true or false as a
function of any logical combination of comparisons between
constants and variables. During the execution of a conditional
statement, the operations are executed normally except only those
PU's which have a "true" CCM participate in the operation. The
PU's whose CCMs are not true perform no-ops during the conditional
operation. As can be seen in Figure 5a and 5b the code for this
type of loop is nearly identical for both serial and parallel
machines, the loop statements and CCM initialization being the
only difference.

```
    DO 100 I=1,5                        CCM=.TRUE.
    B(I)=2*K(I)                         B=2*K
    IF(D(I).LE.20)B(I)=D(I)+K(I)        IF(D.LE.20)B=D+K
100 CONTINUE
```

      a. Serial Code   b. Parallel Code

FIGURE 5.   Conditional execution.

If two identical serial computers are programmed with an
identical algorithm but are presented different data, at any
given time they will, in general, be at different program
locations as a function of the input data and the decision nodes
in the algorithm. In a SIMD computer each processing unit in
the array will be processing a different set of data. Therefore,
all paths of the algorithm must be traversed by the sequential
control unit to assure that every PU executes the portions of

the algorithm that the data it was presented requires. Yet, a
method must be provided to prevent portions of the algorithm from
being executed by PU's whose data does not require them. This
can be accomplished quite simply by requiring each PU to perform
a no-op when an instruction it should not be executing is
presented to it. This is exactly the function the CCM performs.
The CCM mechanism and the requirement that all algorithm paths
must be executed causes SIMD computers to be very well suited
to structured programming. (See Figure 6)

```
            EXECUTE A
            IF (I)
                EXECUTE B
            ELSE
                EXECUTE C
                IF (II)
                    EXECUTE D
                ENDIF
                EXECUTE E
            ENDIF
```

FIGURE 6.   Parallel structured code.

III.   MASSIVELY PARALLEL PROCESSOR

The Massively Parallel Processor has been described in
detail elsewhere (Batcher 1980). Therefore, this discussion will
be limited to those design aspects which most directly affect
program generation. The MPP has a two-dimensional array of 128 x
128 processing units. Each unit has a thousand bits of memory
and is connected to its East, West, North and South neighbor by
a communications path. The left (West) most column of processing
units can under program control be connected to the right most
column to form a cylinder. In a similar manner, the top and
bottom rows can be connected.

Figure 7 diagrams the basic structure of the MPP. The data
flows into the system from the top into the staging memory,
from the staging memory into the I/O plane, from the I/O plane
into the Processor Memories. The data output flows from the
processor memories to the I/O plane, from the I/O plane to the
staging memory, from the staging memory, out of the system.

This I/O system performs two very important functions. First,
the data format into and out of the I/O plane must be organized
into 128 rows of 128 bits; each bit from a different pixel.
Input to the system will be conventional sequential computer
format where all bits of a pixel are organized into a byte or
word. The staging memory performs this data transformation.

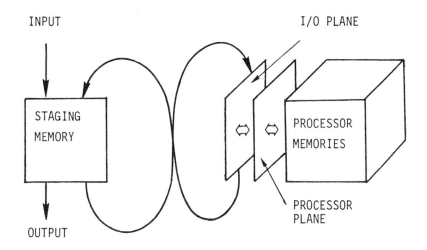

FIGURE 7.   MPP configuration

     The second major function of the Staging memory is to allow
subimage overlap. If a 3 x 3 convolution is being performed on
a 512 x 512 image, since the MPP array is only 128 x 128, the
calculations of the output values for the 128th row and column
require values from the 129th row and column respectively which
may not be in the array memory.

     To resolve this problem, the programmer can specify that the
next 128 x 128 subimage start at column 127 and the next row of
subimages start at row 127. The staging memory is capable of
addressing a 128 x 128 subimage starting at any offset from the
origin of the original image as long as the original image can
fit into the staging memory. The address space of the staging
memory is 64m bytes which could handle an image of over 8000 x
8000 eight bit pixels.

A. Main Control Language

     The MPP instruction set like all SIMD machines can be divided
into two classes - sequential and parallel. The sequential instruction set is much like a conventional sequential computer.
The parallel instruction set is basically a triple address
language with memory-to-memory operations. It can be subdivided into two subclasses, array-to-array operations and
scalar-to-array operations. Table I very briefly summarizes the
instruction set. It in no way conveys the power of the complete
instruction set. The table is intended only to provide the
general concepts of the basic programming language of the MPP.

TABLE I.   MPP Instruction Summary

Sequential Instructions

Load/Store Register from Memory/Register	LR Rn,address
Conditional/Unconditional Branches	Bcc address
Shift and Rotates	ASL bits
Arithmetic Functions (ADD,SUBTRACT,MULT)	ADD Rn,address
Comparisons	CMP Rn,address
Logical Operations	AND Rn,address

Array-to-Array Parallel Instructions

Move Array of Data	MVA	address1,address2
Logical Operations	ANDAA	address1,address2,address3
Compare Operations	GEAA	address1,address2,address3
Arithmetic Operations	ADAA	address1,address2,address3
Miscellaneous Operations	MAXA	address1,address2

Scalar-to-Array Parallel Instructions

Move Scalar-to-Array	STORSA	address1,address2
Move Array Element to Scalar	LOADSA	address1,address2
Logical Operations	ANDSA	address1,address2,address3
Comparison Operations	GESA	address1,address2,address3
Arithmetic Operations	ADSA	address1,address2,address3

B.   Structured Parallel FORTRAN

The Structured Parallel FORTRAN (SPFOR) compiler is intended to be the nucleus of a parallel FORTRAN language which can be easily expanded and modified to meet user specific needs such as image processing, data base management, etc.

The design philosophy of SPFOR is to require the programmer to partition the program into modules which are exclusively to be executed on the MPP and modules which are exclusively intended for the host. SPFOR and/or the host HOL will provide the facilities to allow programs composed of both types of modules to be compiled, linked, loaded and executed on the MPP/HOST system in a parallel multi-tasking fashion.

In addition to extending conventional (scalar) arithmetic and logical concepts to vectors and matrices, several new concepts must be addressed.

One concept is concerned with typing. Variables may be typed as SCALAR, VECTOR or MATRIX,[2] but the typing of a variable may change depending on its mode of reference. Thus, a VECTOR variable "indexed" with a SCALAR is a scalar.

For example, in the right-hand side of an assignment statement or in any expression, a VECTOR(SCALAR) is a scalar variable whose value is the value of the SCALARth element of VECTOR. On the left-hand side of an assignment statement, the SCALARth element of VECTOR is replaced by the value of the right-hand side. This nomenclature and its extension to MATRICES is inherent in the arithmetic/logical grammars discussed in this subsection.

Two special functions are desirable in a SIMD HOL. One is a unit array generator which generates an array where a single element is a one and all other elements are zero. The SPFOR language will have such a capability provided by the special functions UNITVECTOR (n) and UNITMATRIX (n,m).

Another special function is the vector shift index operator. If the notation VECTOR ($+5)[3] is used, the meaning would be that the elements of the vector VECTOR are index shifted right five places with zero fill. Thus, if VECTOR = (1,2,3,4,5), then VECTOR($+2) = (0,0,1,2,3). If a matrix is shifted, both a row and column shift may be specified.

The arithmetic statement capability of SPFOR is summarized in Table II. SPFOR extends conventional HOL operators to vectors and matrices on an element by element basis. Operands must agree in size (dimension) in arithmetic expressions. In an analogous manner, the conventional scalar operators are extended to vectors, matrices and combinations thereof on an element-by-element basis.

Table III lists the control statements in SPFOR. Examples of the use of these block-oriented statements are discussed below. In addition to these statements, SPFOR contains typing statements, precision statements, and subroutine call and return statements.

Any statement with an '=' is an assignment statement. The value on the right of the equal sign is assigned to the variable on the left of the equal sign. Both arithmetic and logical assignments are allowed. If a vector (matrix) is specified with an index operator on the left of the equal sign, then an insert element (row, column or element for matrix) is indicated.

The IF statement causes conditional execution. There are several different cases depending on the mode of the conditional and the statement bodies.

[2]It is important to realize that the terms VECTOR and MATRIX are used loosely here and refer to the logical organization of the data only not to any inherent mathematical properties of these data types.

[3]The $ notation as in VECTOR($), is used to denote all elements of a vector or matrix

TABLE II. Arithmetic Statements/Expressions

```
<in>      ::= <int>|<cons>

<scalv>   ::= <cons>|<scd>|<vec(in)>|<mat(in,in)>
                                $,in
<vectv>   ::= <vec($)>|<mat({    })>|UNITVECTOR(in)>
                                in,$
<matv>    ::= <mat($,$)>|<UNITMATRIX(in,in)>

<ave>     ::= <vectv><op>{<ave>}|<vectv>
                        {<are>}|

<are>     ::= <scalv><op><are>|<scalv>

<ame>     ::= <matv><op><ame>|<matv>

<op>      ::= +|-|÷|*

 as       ::= <vectv>=<ave>|<matv>=<ame>|<scalv>=<are>
```

in   ≡ index·   vectv ≡ vector variable

scalv ≡ scalar variable   matv ≡ matrix variable

int ≡ integer   ave   ≡ arithmetic vector expression

cons ≡ constant   are   ≡ arithmetic scalar expression

scal ≡ scalar   ame   ≡ arithmetic matrix expression

vec ≡ vector   as   ≡ arithmetic statement

mat ≡ matrix·

If the statement body and the conditional statements are both vectors (matrices) then only those elements of the vectors (matrices) in the statement body which correspond to true elements of the condition statement are affected. For example, given A = (3,4,5,6,7), B = (7,6,5,4,3), C = (1,1,1,1,1) and D = (2,2,2,2,2) then IF (A.GT.B) C = D results in A,B,D unchanged and C = (1,1,1,2,2). Table IV shows the results if scalar and vector (or scalar and matrix) statements are intermixed in the IF statement. Note that in the conditional part, the final expression type is the determining factor, but that in the statement body the effect is determined on a statement-by-statement basis. Comparison of VECTOR and MATRIX types are not allowed in SPFOR.

The ANY statement will cause execution of the statement block if the logical vector (matrix) expression is true for any (at least one) of the elements of the resultant vector (matrix) expression. A scalar conditional expression is not allowed.

TABLE III. Control Statements

```
FOR scalar IN (logical vector expression)
    statement block - scalar may be referenced
ENDFOR

IF (logical expression)
    statement block 1
ELSEIF (logical expression)
    statement block 2
ELSE
    statement block 3
ENDIF

ANY (logical vector expression)
    statement block
ENDANY

NEXT X IN (logical vector expression)
    statement block
ENDNEXT

FIRST
    initialization statement block
LOOP
    statement block 1
WHILE (logical expression)
EXITON (logical expression)
EXITAFTER (scalar expression) TIMES
    statement block 2
ENDLOOP
```

TABLE IV. Results of Mixed Typing in IF

		Type of Body Statements	
		Scalar Statements	Vector or Matrix Statements
Type of Conditional Expressions	Scalar	all statements executed if condition is true	all elements of all statements processed if condition is true
	Vector or Matrix	all statements executed if any element is true	all statements processed element by element if condition element is true

The FOR scalar IN statement will assign to scalar the ordinal
value of each element for which the conditional statement is true.
If X is a set and x are elements of the set, the FOR scalar IN
statement allows the user to identify a subset of X by the
conditional statement and to operate on each element of that subset in turn as a scalar. Figure 8 gives an example of a FOR
statement inside a LOOP statement.

GIVEN: SON1 = (2,4,6,8,-1,-1,10,-1,-1,-1,-1)
         SON2 = (3,5,7,9,-1,=1,11,-1,-1,-1,-1)

Then

              FIRST
                LEVELS = 0
                LEVEL(1) = 0
              LOOP
                FOR X IN (LEVEL($) .EQ. LEVELS)
                  LEVEL(SON1(X)) = LEVELS + 1
                  LEVEL(SON2(X)) = LEVELS + 1
                ENDFOR
                EXITON (SON1($) .EQ. -1 .AND. LEVEL($) .EQ. LEVELS)
                  LEVELS = LEVELS + 1
              ENDLOOP

Results in

              LEVEL = (0,1,1,2,2,2,2,3,3,3,3)

          FIGURE 8.   For/Loop Example

The LOOP statement (see Table III) allows an optional
initialization preamble and allows the exit condition to be
evaluated at: 1) the beginning (therefore if the condition is
false the code is not executed), 2) at the end (so that at least
one iteration is made regardless of the condition) and 3) in
the middle (to allow exit on input EOF, etc.). Multiple exit
statements are allowed. If the logical expression is a vector
or matrix expression, EXITON acts like ANY; i.e., the loop will
be exited if the condition is true for any one of the elements.
The WHILE statement can be used instead of EXITON for readability
because EXITON (.NOT. A) is equivalent to WHILE (A).
    The NEXT scalar statement allows the user to sequence through
the elements of a vector or matrix. The ordinal value of the
"next" element of a vector or matrix is saved in scalar. Figure
9 shows an example of NEXT and LOOP.

```
          VECTOR IDLE, NAME, ADDRESS
          SCALAR X,EOF
          FIRST
          IDLE($)=1

          LOOP
          INPUT INPUTNAME,INPUTADDRESS
          EXITON (EOF  .EQ.1)
          NEXT X IN IDLE($)

          NAME(X) = INPUTNAME
          ADDRESS(X) = INPUTADDRESS

          ENDNEXT
          ENDLOOP
```

FIGURE 9.   Next scalar example.

C.   Image Processing Functions

One of the application extensions of SPFOR would be in the
area of image processing. Four types of extensions are being
considered, exemplification, scene analysis comparative primitives,
associative data structures and production statements.
Exemplification or "windowing" allows the programmer to
simply state the mathematical function to be executed. The
function is then applied to the entire image automatically with no
need for loop control or other administrative overhead.
Typically, these functions are area functions such as convolution
or median filtering. Since the center of the "window" is thought
of as the origin, an ORIGIN statement allows this type of biasing.
Figure 10 shows an example of exemplification.


```
MATRIX    I1(512,512),I2(508,508)
SCALAR    W(5,5)
ORIGIN    W(3,3)

OPEN
INPUT     I1
I2($,$) = I1($,$)*W(0,0)+I1($+I,$)*W(1,0)+I1($,$+1)*W(0,1) ETC
OUTPUT    I2
CLOSE
```

FIGURE 10.   Exemplification

In scene analysis, it is desirable to be able to make relative positional comparisons. SPFOR will be extended to provide these capabilities. Table V lists the comparatives.

TABLE V. Comparative Primitives

EXPRESSION	EVALUATION
A LEFTOF B	HIGHX(A) .LE. LOWX(B)
A RIGHTOF B	LOWX(A) .GE. HIGHX(B)
A TOPOF B	LOWY(A) .GE. HIGHY(B)
A BOTTOMOF B	HIGHY(A) .LE. LOWY(B)
A INSIDEOF B	LOWX(A) .GE. LOWX(B) .AND.
	HIGHX(A) .LE. HIGHX(B) .AND.
	LOWY(A) .GE. LOWY(B) .AND.
	HIGHY(A) .LE. HIGHY(B)
A OUTSIDEOF B	HIGHX(A) .LE. LOWX(B) .OR.
	LOWX(A) .GE. HIGHX(B) .OR.
	HIGHY(A) .LE. LOWY(B) .OR.
	LOWY(A) .GE. HIGHY(B)
A AT(T) B	T .GE. SQRT ((X1(A)-X1(B))**2+(Y1(A)-Y1(B))**2).OR.
	T .GE. SQRT ((X2(A)-X2(B))**2+(Y2(A)-Y2(B))**2).OR.
	T .GE. SQRT ((X2(A)-X1(B))**2+(Y2(A)-Y1(B))**2).OR.
	T .GE. SQRT ((X1(A)-X2(B))**2+(Y1(A)-Y2(B))**2)
A NEAR(T) B	T. GE. SQRT ((X1(A)-X1(B))**2+(Y1(A)-Y1(B))**2).OR.
	T. GE. SQRT ((X2(A)-X2(B))**2+(Y2(A)-Y2(B))**2).OR.
	T. GE. SQRT ((X2(A)-X1(B))**2+(Y2(A)-Y1(B))**2).OR.
	T. GE. SQRT ((X1(A)-X2(B))**2+(Y1(A)-Y2(B))**2).OR.

In scene analysis, trees are a common data structure. In order to facilitate the construction and use of trees, SPFOR will have an associative data type; an example of an associative declaration is ASSOCIATIVE TREE($,$,$). An example of an initial assignment of associative variables is: TREE(???)=TREE('A' 'FIS' 'ROOT'). Associative variables can be modified by a pattern matching statement like TREE($ $'LEAF')=TREE($ $ 'TERM'). Which changes all variables with 'LEAF' as the value of their 3rd term to variables with 'TERM' as the value of their third term; i.e.,

Before: TREE('B' "NTYPE' 'LEAF') AFTER: TREE('B' 'NTYPE' 'TERM')
        TREE('C' 'NTYPE' 'LEAF')   TREE('C' 'NTYPE' 'TERM')
        TREE('A' 'NTYPE' 'AND')   TREE('A' 'NTYPE' 'AND')

Finally, production rules are frequently used in scene analysis, thus the image processing enhancement to SPFOR will include a Production Rule Statement as shown below:

```
IF  (F1,F2,...)(W1,W2,...)(THRESHOLD)
   ACTION 1
       .
       .
       .

   END
```
It is patterned after the IF statement but allows for weights and
a threshold value. Thus, the actions are executed only if
Σ Fi*Wi $\geq$ THRESHOLD

IV. CONCLUSION

The MPP is a major advance in the attempt to process imagery
in real time. It will be capable of performing 6553 million 8-
bit additions and 1861 million multiplications per second.
Current estimates show that the MPP can perform a 3 x 3 pseudo
median filter on a 512 x 512 eight-bit per pixel image in 281.6
μsec. A 7 x 7 window convolution function requires about 18.32
milliseconds. Correlation of a 13 x 13 area takes about 128.3
μseconds. A 7 x 7 template match requires 147 μseconds. These
speeds and the ease of programming the MPP will make it invaluable
as an image processing computer.

REFERENCES

Batcher, K. E., (September 1980). "Design of a Massively
    Parallel Processor," IEEE Transactions on Computers.
Potter, J. L., (1978). "The STARAN Architecture and its
    Application to Image Processing and Pattern Recognition
    Algorithms," Proceedings National Computer Conference.
Rohrbacher, D. L., and Potter, J. L., (August 1977). "Image
    Processing with the STARAN Parallel Computer," Computer, P.
    54-59.
Potter, J. L., (1981). "Pattern Processing on STARAN,"
    Special Computer Architecture for Pattern Processing, CRC
    Press, Dr. K. S. Fu and T. Ichitawa (eds).

PIPELINE ARCHITECTURES FOR IMAGE PROCESSING

Stanley R. Sternberg

Environmental Research Institute of Michigan
and
University of Michigan
Ann Arbor, Michigan

Four computer architectures for implementing image processing as sequences of local neighborhood transformations are discussed. Each design is cellular, consisting entirely of identical modules or cells spatially configured in regular grids with regular interconnection patterns.

I. INTRODUCTION

Neighborhood processors are a class of devices that operate upon a first data array or matrix to generate a second matrix wherein each element has a value dependent upon the value of its equivalent element in the first matrix, and the values of its neighboring elements in the first matrix. These devices are useful for pattern recognition, image enhancement, area correlation and like image processing functions. One type of neighborhood processing device is constructed in a parallel array form with a single computing element for each matrix element or pixel. A parallel array neighborhood processor of this type is disclosed in U.S. Pat. No. 3,106,698 to Unger 1963. It comprises a matrix of identical processing cells, each cell including a memory register for storing the value of a single data element (pixel), and a neighborhood logic translator for computing the transformed value of that pixel as a function of the present value of the pixel and the neighborhood pixel values, and parallel connections between the translator and neighboring memory registers. The neighborhood logic may be fixed, in which case the same transformation is repeated indefinitely, or, may be

programmable, in which case the neighborhood transition function
may be modified at required times in the transition sequence of
the image processing scheme. A common clock causes a simultaneous
transition in the state of all the pixel value registers to
achieve a transformation of the entire matrix.

The principal advantage of such a parallel array processor is
speed. A neighborhood transformation of the entire image or
matrix requires only a single clock pulse interval so that
transformations may be performed at rates of millions per second.
The principal disadvantage of the parallel array processor
configuration is complexity since the neighborhood logic must be
replicated in every processor cell, making a processor for large
arrays, such as 1000 X 1000, which may be a reasonable size for a
digitized image, very large and costly.

II. SERIAL ARRAY PROCESSORS

A serial array processor represents an alternative approach to
neighborhood processing which greatly simplifies the processor
structure at the expense of speed when compared to the parallel
array. Such a system has been described by Sternberg[1] and has
been constructed at the Environmental Research Institute of
Michigan under the name Cytocomputer T.M.*.

That system employs a chain of serial neighborhood processing
stages, each stage capable of generating the transformed value of
a single pixel within a single clock pulse interval. The serial
neighborhood processing stage employs a neighborhood logic
translator identical to its counterpart in the parallel array
processor cell, and line delay memory for receiving a serial pixel
stream from a row by row raster scan of the input matrix and for
configuring the neighborhood window by providing the appropriate
matrix elements to the neighborhood logic translator. The
serialized input matrix is provided to the line delay memory and
the data bits are serially shifted through the line delays. When
the line delay memory has been filled with input data it contains
the neighborhood configuration for the first element to be
transformed. Taps at appropriate positions in the line delay
memory provide parallel neighborhood element values to the
neighborhood logic translator. These tapped memory elements in
the tapped line delay memory constitute the neighborhood window
registers.

*Cytocomputer is the trademark of Environmental Research
Institute of Michigan.

Referring now to FIGURE 1, the image being processed is represented by a matrix which is five elements wide but of unbounded height. Three successive time steps of the operation of the serial neighborhood processing stage are shown. The position of the input serial scan at each time step is represented by the darkened circle on the matrix positioned adjacent to the stage diagram. Previously scanned pixels are contained as indicated in the stage line delay memory.

The line delay memory of each stage is composed of window registers and delay lines. The number of memory elements in each delay line is equal to the number of pixels in a scan line of the data matrix less the three window register memory elements. As each pixel is successively scanned and fed into the stage, the stage's window register is effectively scanned over the input data matrix in a row-by-row fashion. In each clock pulse interval a new neighborhood configuration is entered into the window registers. The contents of the window registers are presented in parallel to the neighborhood logic translator which computes the transformed value of the center cell of the window register neighborhood in the same clock pulse interval.

The output of a serial neighborhood processing stage occurs at the same rate as its input and has the same format. This allows the output of one stage to be provided to the input of a subsequent stage, which may perform the same or a different neighborhood logic transformation. A chain of serial neighborhood processing stages constitute a serial array processor.

In a parallel array processor a neighborhood transformation is carried out as a two phase process of first identically programming each cell's neighborhood logic translator with the appropriate transition rule and secondly executing in parallel the neighborhood logic transformation in each cell. Thus an image processing algorithm is executed as a serial process of alternating programming and transforming steps. In contrast, all of the stages of a serial array processor are first programmed in a sequential manner by inserting appropriate but generally different transition rules into each stage's neighborhood logic translator. After the stages have been programmed the image is sequentially fed through the pipeline thus effecting the parallel execution of potentially as many different neighborhood transformations as there are stages in the pipeline.

The number of stages in a serial array processor equals the number of neighborhood logic transformations in the processing algorithm. For complex algorithms the length of any given pipeline may be insufficient, requiring external storage of a partially processed result. However, if the number of pipeline

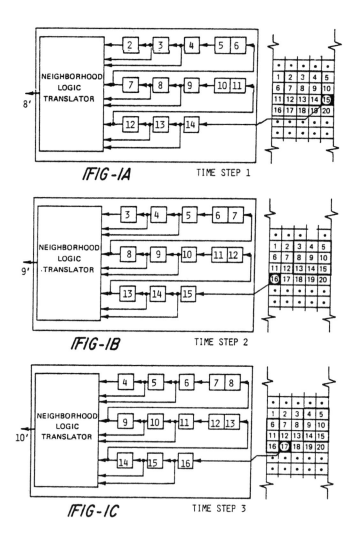

IFIG-IA   TIME STEP 1

IFIG-IB   TIME STEP 2

IFIG-IC   TIME STEP 3

stages is equal to the number of lines in the image, then the line delays in the stages are sufficient for storing the entire image and no external image storage is required if the output of the last stage is fed back into the input of the first stage, thus creating a cyclic serial array processor. The cyclic serial array processor is further discussed in section IV.

III.   PARALLEL PARTITIONED SERIAL ARRAY PROCESSORS

The most complex section of either a serial neighborhood processor stage or of a parallel array processor cell is the neighborhood logic translator. The serial array processor is conservative of neighborhood logic translator circuitry requiring only one translator circuit per stage while a parallel array requires one translator circuit for each matrix element.

In most practical design applications where the input matrix represents an image, the matrix size must be relatively large in order to achieve high resolution. For example, when the input matrix is generated by a state-of-the-art television pick-up tube it may be digitized into a matrix of about 1,000 X 1,000 pixels. The designer of a processor for this image is faced by the choice of a parallel array processor which can generate one transformation per clock time but will have 1,000,000 relatively complex cellular elements; or a serial array processor consisting of a chain of serial processing stages, one stage for each neighborhood transformation in the image processing algorithm.

The parallel array processor transforms images at the maximal rate of one neighborhood image transformation per clock pulse interval, while the serial array processor performs image transformations at the rate of K/P image transformations per descrete time step, where P is the total number of pixels in an image and K is the number of processing stages in the serial array. For large images, the ratio of serial array processor speed to parallel array processor speed can be very small. Since processing speed can only be increased by increasing the ratio of neighborhood logic translators to the total number of data elements in the matrix, the questions arises as to whether it is possible to incorporate more than one neighborhood logic translator per serial processor stage, or equivalently, whether it is possible to reduce the number of line delay memory elements associated with each neighborhood logic translator.

Another design problem arises from the desirability of forming the processor using integrated circuit techniques. The need to make all computations occur in a single clock pulse interval requires a high degree of internal parallelism in the circuitry of the neighborhood logic translator which may exceed the capability to produce chips economically. When efforts are made to divide the serial processor by function into a number of smaller chips the large number and variety of interconnections between the chips frustrates the modular design approach.

The Parallel Partitioned Serial Array Processor (PPSAP) disclosed by Sternberg[2] is a unique neighborhood processor which allows the achievement of transformation rates at a wide range of levels between the extremes presented by the parallel array processor and the serial array processor. The PPSAP is also well adapted to be realized using integrated circuit techniques and PPSAP systems can be formed of a number of identical integrated circuit modules requiring few and regular interconnections.

Broadly, the PPSAP takes the form of two or more serial array processors adapted to simultaneously process separate, contiguous segments of a data matrix in such a way as to allow each processor to make use of pertinent neighborhood information stored in the opposite processor.

In its simplest configuration, the PPSAP employs two serial array processors which equally divide the task of transforming a single data matrix. Assuming a data matrix N elements wide, the first N/2 columns of the matrix will be fed to the first serial array processor stage and the second N/2 columns will be simultaneously fed to the second serial array processor. In each case the data will be fed in a raster scan format on a half-row basis. These sequential data streams will be staggered so that one processor will always receive a pixel train delayed one row with respect to the pixel train feeding the other processor.

Each serial array processor in this pairwise configuration resembles a conventional serial processor but incorporates connections to certain of the window registers of the companion or "outboard" processor. These connections are provided by multiplexer elements, such as two-way gates, on the "inboard" processor. The multiplexer elements each have second inputs from one of the window registers of the inboard processor. The outputs of the multiplexer elements go to the inboard neighborhood logic translator. The multiplexer elements effectively switch either an inboard or outboard window register to the neighborhood logic translator depending upon the position within the matrix of the pixel for which a transformation is then being generated by the stage. If that pixel is separated from the partition of the two half matrix segments so that all of the neighbors of the pixel under consideration are in the inboard half-matrix, then the multiplexer switches the inboard window register value to the neighborhood logic translator. If, however, the pixel for which a transformation is generated is adjacent to the matrix partitioning line so that some of its pertinent neighbors are in the outboard half-matrix, then the multiplexing elements are controlled to

switch those outboard window register values to the neighborhood
logic translator in the inboard processor.

In this manner each serial array processor makes efficient use
of information stored in the other processor. The staggered
feeding of the two serialized pixel streams to the two serial
array processors and the multiplexing of matrix data between the
two processors allow the simultaneous utilization of both serial
array processors. Using this method the data matrix may be
laterally partitioned into any desired number of matrix segments
which are simultaneously processed by a like number of serial
array processor segments.

The serial neighborhood processing stages of the PPSAP may be
serially chained in the same manner as conventional serial
neighborhood processing stages. For example, the outputs of a
pair of serial neighborhood processor stage segments might be
provided to a second pair of serial neighborhood processor stage
segments in parallel, which perform a second transformation on the
output of the first pair.

All serial processor array segments are identical, and the
only interconnections required between lateral segments are the
relatively few window sharing connections. Accordingly, serial
neighborhood processing stage segments could be conveniently
formed as special integrated circuits and readily interconnected
with one another to form a parallel partitioned serial
neighborhood processing system.

Consider two serial neighborhood processing stages L and R,
illustrated in FIGURES 2A and 2B, operating independently on two
different serialized pixel trains derived from their respective
data matrices. The locations of the processing windows are shown
at two discrete sequential time steps. It is to be noted that the
pixel stream feeding the processor R has been delayed one scan
line from the pixel stream feeding the processor L. At time step
2, illustrated in FIGURE 2B, each processing stage window becomes
disjoint, the pixel locations comprising the window occupying
positions at the left and right-hand edges of the matrix. The
same condition exists at time step 3 (not illustrated). At time
step 4, (not illustrated) the individual window configurations are
once again connected.

At time step 2, FIGURE 2B, the pixels labeled 9, 10, 6', 14,
15, 11', 19, 20, 16' form a 3 X 3 virtual window which extends
across the boundary between the two serialized pixel arrays. The
virtual window is not processed by either neighborhood logic
translator, which, at time steps 2 and 3, are computing
neighborhood transformations on disjoint window configurations.
The pixels comprising the virtual window are, however, present in
the combined neighborhood window registers of the two serial
neighborhood processing stages.

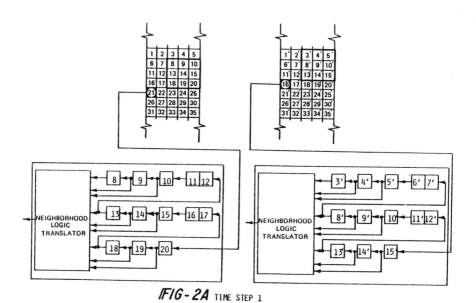

FIG-2A TIME STEP 1

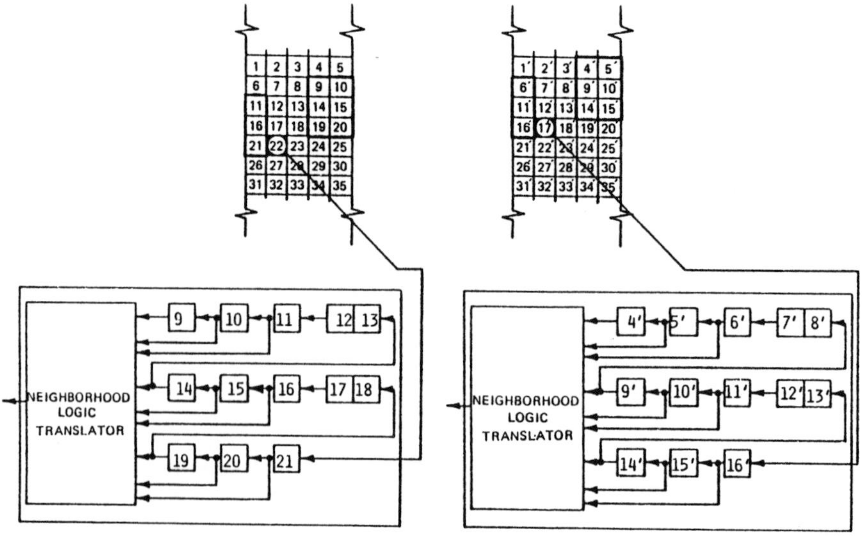

FIG-2B TIME STEP 2

Computing a neighborhood transformation on this virtual window configuration is illustrated in FIGURES 3A and 3B. The serial stages here differ from conventional serial neighborhood processors of the type illustrated in FIGURE 1 in that the output of the window registers are provided to three multiplexing elements rather than directly to the neighborhood logic translator. The outputs of these multiplexing elements are then each provided to the neighborhood logic translator. A second input to each of the multiplexing units is provided from the equivalent window registers on the neighboring processor. The multiplexer units each output one of their inputs, depending upon the nature of the signal received from a multiplexer control unit which provides its output signals to all of the multiplexers. When the control signal is high the multiplexers provide a first input to their output and when the control signal is low they provide the second input signal to their output. The multiplexer control simply consists of a recirculating shift register that is cycled once each clock time. The register is loaded so that one's or zero's occur at its output in order to achieve the desired control function. The input data matrix can be partitioned into any equal divisor of the number of columns in the matrix. When the division involves more than two processors, all of the processors except those operating on the edge of the matrix will require connections from both of their adjacent neighbor processors. FIGURE 4 illustrates three identical partitioned neighborhood processor modules all operating on sections of a single 15 pixel wide matrix.

IV. CYCLIC SERIAL ARRAY PROCESSOR

In the Cyclic Serial Array Processor (CSAP) feedback is provided for selectively coupling the output of the last stage in the pipeline to the input of the first stage. In a first version of the CSAP, enough stages are included in the pipeline to contain all of the pixels in the image matrix. The stages are first programmed to perform a first sequence of neighborhood transformations. The pixels are shifted through the pipeline with each of the stages performing the programmed transformation. Before the pixels are recirculated through the pipeline, the stages are progressively reprogrammed with new control instructions whereby the image pixels may be transformed a greater number of times than there are stages in the pipeline.

In a second version of the CSAP, the pipeline is used to transform two successive image frames of pixel data from a single framing data source. The transformation outputs of the pipeline are synchronously combined with the incoming frame. Accordingly, selective changes in the image contained in the two frames of

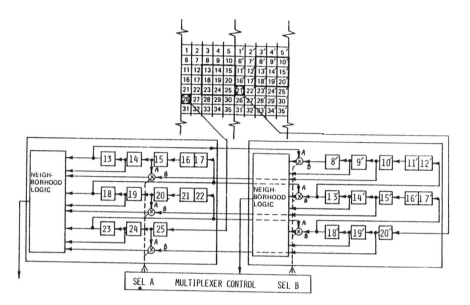

FIG-3A

TIME STEP 1

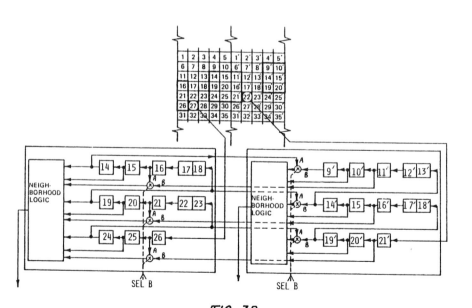

FIG-3B

TIME STEP 2

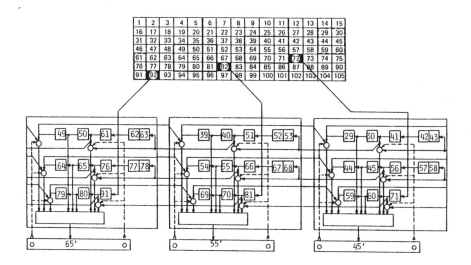

FIG-4

pixel data can be detected. Preferably, indicia of the transformation output of the pipeline for one of the frames is fed back and combined with pixel data from subsequent frames. The fed back indicia mark regions of previous change and constitutes apriori information in the processing of the subsequent frame. Accordingly, the steady state transformation output of the pipeline is a function of a time series of image frames.

Referring now to FIGURE 5, the CSAP takes the form of a pipeline of serially connected neighborhood transformation stages. In this example there are five stages, the image to be analyzed is represented by a matrix of thirty-five pixels arranged in a 5 X 7 array.

Assume that there are ten neighborhood transformation steps required in a particular image processing sequence. The neighborhood logic translator of the five stages 1 to 5 are first loaded with instructions from the controller for carrying out the first five steps of the sequence. The pixel data from the data source is then loaded into the pipeline. The position of a switching device S1 set by the controller represents that the data entering the pipeline is fresh data from the data source.

The pixel data enters the input of stage 1 and is shifted through the storage devices making up the window registers and line delay memories. The output of the first stage is coupled to the input of second stage 2 and so on up the pipeline. The numerals in the window registers and line delay memories in FIGURE 5 shows the position of pixels when all of the pixels in the data matrix have been loaded into the pipeline. The darkened lines in the data matrix surround those pixel values which are contained in the window registers for its adjacent stage. Each pixel value is transformed when it is in the center position of the window register. Thus, as the pixel data propagates up the pipeline, it is transformed by each of the stages. For example, pixel No. 1 which is in the center cell position of stage 5 has already been transformed by the previous stages by the time it reaches the position shown in FIGURE 5.

FIGURE 5 illustrates the condition of the pipeline after all of the pixels have been loaded from the data source. Note the condition of switch S1. The controller will now generate appropriate signals to cause switch S1 to couple the output of stage 5 to the input of the first stage. Consequently, the pixel data entering stage 1 is not fresh data from the source but

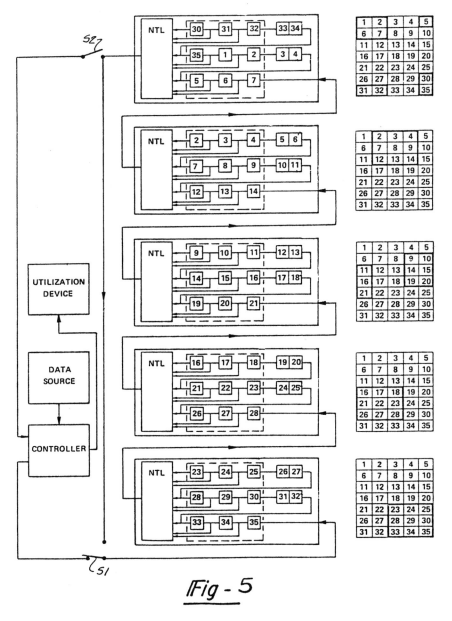

Fig - 5

instead represents transformed data which has been transformed by the previous stages which have carried out the first five of the ten step sequence in our example. In other words, pixel No. 1 in stage 1 has already been transformed five times, once by each of the stages 1 to 5.

When pixel number 35 has been shifted into the center position of the window register of stage 1, it will perform the last transformation pursuant to the previously programmed transformational control instructions. Before the next pixel shift, stage 1 is reprogrammed with new transformational control instructions from the controller for carrying out the sixth step in sequence. Thus, when pixel number 1 is shifted into the center position of the window register, the neighborhood logic translator will provide a transformation output for stage 1 pursuant to the new instructions.

Once the last pixel, here pixel number 35, has reached the center cell in the window registers of the second stage 2, the second stage neighborhood logic translator is likewise reprogrammed. The new transformation control instructions will condition the neighborhood logic translator of the second stage to carry out the seventh step in the ten step sequence. Thus, when pixel number 1 reaches the center cell, it will be transformed according to the new program instructions. This same operation occurs as the recirculated pixel data propagates up through the stages in the pipeline 10. In other words, stage 3 will be programmed with the eighth transformation step instructions, stage 4 with the ninth, and stage 5 will be reprogrammed for carrying out the tenth step in the operation. Thus, the pixel values eminating from the output of stage 5 will be transformed ten times by only five different neighborhood transformation stages. The output of stage 5 may be read by the controller by closing switch S2. The controller may then couple the transformed data to a utilization device.

The feedback approach just described has several advantages in addition to the decrease in number of stages required to perform a particular operation. Faster image processing is obtained because the pixel data does not have to be transferred through as many input/output devices which inherently slows down the processing time. Additionally, the amount of required external storage is substantially reduced.

A wide variety of alternative uses of the CSAP can be readily envisioned. For example, this feedback approach can be used for implementing noise filters for image data. It can also be used to detect movement of particular objects between successive frames of pixel data.

The parallel partitioned serial array processor concept may be combined with the cyclic serial array processor to form a cyclic parallel partitioned serial array processor, consisting of parallel cyclicly connected pipelines.

V. CONCLUSION

Several alternatives to the parallel array processor architecture for neighborhood processing have been presented. Each architecture provides unique advantages while at the same time preserving the cellular nature of the basic operations. Although the notion of parallel neighborhood operations translates directly into an obvious hardware architecture, less obvious but equally general kinds of architectures can be envisioned and may be preferable in particular applications.

VI. ACKNOWLEDGEMENT

This author is indebted to Greg Schivley of the firm Krass, Young and Schivley for his invaluable contributions to this paper.

VII. REFERENCES

1   Sternberg, S. R., <u>Automatic Image Processor,</u> U. S. Patent 4,167,728; September 11, 1979.

2   Sternberg, S. R., <u>Parallel Partitioned Serial Neighborhood Processors,</u> U. S. Patent 4,174,514; November 13, 1979.

MICRONET/MICROS - A Network Computer System
for Distributed Applications

Larry D. Wittie
Ronald S. Curtis
Ariel J. Frank

Department of Computer Science
State University of New York at Buffalo
Amherst, New York

MICROS is the distributed operating system for the
MICRONET network computer. MICRONET is a reconfigurable and
extensible network of sixteen loosely-coupled microcomputer
nodes connected by packet-switching interfaces to pairs of
high-speed shared communication busses. MICROS
simultaneously supports many users, each running
multicomputer parallel programs. MICROS is intended for
control of network computers of thousands of nodes.
Each network node is controlled by a private copy of
the MICROS kernel processes written in Concurrent Pascal.
Resource management tasks are distributed over the network in
a control hierarchy. System servers and user tasks are
Sequential Pascal programs dynamically loaded into the nodes.
Whether in the same or different nodes, tasks communicate via
a uniform message passing system. The MICROS command
language is a subset of UNIX which allows spawning of groups
of communicating tasks. Planned MICROS revisions will allow
large distributed application programs to run on MICRONET.

I. INTRODUCTION

MICROS is the distributed operating system that has been
developed to control MICRONET, a packet-switched network of
loosely-coupled microcomputer nodes linked together by 0.5
megabyte per second (MB/S) shared communication busses. Each
node contains a DEC LSI-11 microcomputer and a Zilog Z80 microcomputer, both controlled by a Signetics 8X300 microcontroller.
The prototype version of MICRONET consists of sixteen nodes

mounted in a single instrument cabinet. The network is easily
extended and reconfigured by adding nodes and changing external
cable connections.

Preliminary work on MICRONET began in 1975 when it became
clear that powerful microcomputers, complete with processor,
memory, and input/output (I/O) interfaces, would be available as
single chips within a decade. Supercomputers built as ensembles
of thousands of single-chip microcomputers will be economically
feasible, if inexpensive but effective ways to interconnect and
control them can be found. Single-chip computers with two bus
ports each can be connected so that maximum message delays grow
no faster than the logarithm of network size (Wittie, 1981).

Design of the MICRONET hardware began in 1976. MICRONET is
intended as a research test-bed for experimenting with network
computer architectures, with distributed operating systems such
as MICROS, and with distributed application programs to solve
problems in fields such as image processing, numerical analysis,
artificial intelligence, and network simulation. After 1977
when MICRONET development was funded, three years were needed to
design, construct, and test the high-speed frontends which
allow packet-switched transmission of messages among network
nodes. During 1978 and 1979 a two node version of MICRONET with
slow-speed interfaces was used to begin development of MICROS.
In early 1980, the memory in each frontend was trebled to
accommodate planned improvements to MICROS. A 169 megabyte (MB)
disk system was ordered in spring of 1981. A thirteen node
version of MICRONET with the new frontends is being tested
during the summer of 1981. The arrival of three more frontend
boards should result in the complete sixteen node version being
available by the end of the summer.

Many of the unsolved research problems in the development of
network computers lie not in the architecture and construction
of parallel hardware, but in learning how to organize and manage
distributed operating systems. Such systems must organize
networks of independent computer nodes to cooperate in sharing
resources and in running programs consisting of many small
cooperating tasks, called task forces (Jones, 1979).

Design and implementation of MICROS has progressed since
1978 as a series of projects for graduate seminars. Initial
testing of the 7000 lines of Concurrent Pascal source code for
the first version of MICROS was completed in June 1979. In
April 1980, 5000 lines of Signetics 8X300 microcode which pass
packets between nodes via direct memory access (DMA) and via
shared busses were completed and used to test the MICRONET
frontend hardware. The first version includes all of the code
for a single-node operating system and for the packet-switching
subsystem responsible for communication between nodes. During
1979 and 1980 debugging of MICROS was accompanied by extensions
to the Concurrent Pascal programming language and planning for a
new, more efficient implementation of MICROS.

II. DESIGN GOALS

The development of MICRONET/MICROS has already taken
eighteen man-years. Revisions of MICROS will eventually require
another ten. To harness all this effort, six major design
goals have been established for MICROS. They specify system
constraints which should ensure that MICROS will be applicable
to other network computers, including ones more powerful than
the first sixteen node version of MICRONET.

The main design goal is to develop methods of organization
for MICROS that can be extended to control network computers of
thousands of nodes. By the time distributed operating system
principles are understood, say by the mid 1980's, it will be
feasible to construct network computers of thousands of nodes.
Of course, MICROS should also be reasonably effective for
networks as small as a sixteen node version of MICRONET.

The other five design goals are consequences of the first.
MICROS must support a multiuser environment with a dynamically
varying mix of tasks. A network of thousands of computers must
be able to handle loads varying from a large number of users
performing small single-node tasks to a single task force
requiring most of the nodes in the network. In loosely-coupled
networks with no shared memory, closely-coupled computations
will not run as efficiently as collections of independent
tasks. However, MICROS should support task mixes anywhere
between the two extremes, regardless of hardware limitations.

Third, MICROS must adapt to changing communication loads,
i.e. flow and congestion control. Since a major cost component
of network computer hardware will be for internodal
communication as well as for memory and processing power, MICROS
must try to optimize utilization of communication facilities. In
particular, it must detect and eliminate dynamic communication
bottlenecks. To accomplish this goal, MICROS must contain
mechanisms to monitor message flow. It possibly will have to
migrate or replicate tasks or data within the network to relieve
overloaded communication links.

Fourth, MICROS must be tolerant of local faults. In a
network of thousands of nodes, some nodes and interconnection
links will be malfunctioning at any given time. Failures should
cause at most local loss of task and system information. Failed
components must be detected, removed, repaired, and replaced
while the rest of the network continues to serve user requests.
Of course, error tolerance must be built into the hardware. For
example, MICRONET hardware electrically disconnects failed nodes
which might otherwise block the shared communication busses. The
MICROS software must also recover gracefully from local failures
and prevent disruption of network operations.

Fifth, a distributed operating system designed for a large
network computer must not depend upon a fixed interconnection
configuration, or topology. MICROS should be designed to work
reasonably well on a network of any topology. There are two
reasons. First, even if the topology of a network is nominally
fixed, local hardware failures will eliminate nodes and
communication lines from the network; network extensions will
add components. MICROS must dynamically accommodate small local
changes in network topology. Second, there are many useful
topologies for large network computers (Wittie, 1981; Anderson,
1975): spanning busses (Wittie, 1976), trees (Despain, 1978),
binary hypercubes (Sullivan, 1977), and nearest neighbor meshes
(Thompson, 1977). Which is best depends on communication
technology and on the applications for which the network is
designed. Large networks will almost surely have different
topologies. Having to write different operating systems for
every conceivable network topology would be prohibitively
wasteful.

Sixth and finally, to simplify development and maintenance
MICROS should be designed in layers. The basic operating system
modules providing for message passing and task control should be
the same in all nodes. Specialized server tasks for more global
functions should be loaded as needed into selected nodes. The
resulting, structured MICROS operating system should provide
support for solutions to distributed task scheduling and
control, task communication, global resource management, and
abstracted operating system services for all layers.

III. MICRONET ARCHITECTURE

MICRONET is an extensible, reconfigurable network of
autonomous microcomputer modules, each sharing two of the many
communication busses over which packet-switched messages are
transmitted. The busses are passive connections whose control
logic is distributed onto the modules sharing each one in a
technique similar to that used in Ethernet (Metcalfe, 1976).

Figure 1 shows the organization of one node of MICRONET.
Each node consists of a host microcomputer and a communication
frontend microcomputer. The host and frontend memories are
linked by a 0.3 MB/S DMA channel controlled by the frontend.
The host computer is a DEC LSI-11 with a full complement of 28
kilowords of 16-bit random access memory (RAM) which executes
user tasks and higher level operating system servers. Each node
physically consists of a power supply and an LSI-11 backplane
containing circuit boards. About half of the nodes have a
serial interface board, connecting from one to four user
terminals to the LSI-11. A 169 MB Winchester disk will be
connected to a special LSI-11/23 node of MICRONET. Removable

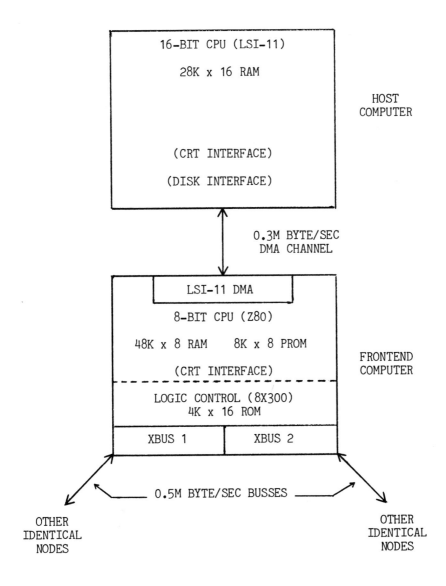

FIGURE 1. Each MICRONET node consists of separate host and frontend computers linked by a direct memory access channel. Each node has two ports to shared external busses. Some nodes have peripheral interfaces.

rack.modules house sets of four nodes.

The custom-built frontend in each node switches packets between its node and others in the communication subnet. The frontend has a Zilog Z80 microcomputer with 48 kilobytes (KB) of RAM and a serial interface. Initialization code resides in an 8 KB programmable read only memory (PROM) attached to the Z80. The frontend is managed by a Signetics 8X300 microcontroller capable of an I/O throughput rate of 4 MB/S. By executing instructions from a 4 kiloword 16-bit ROM, the 8X300 provides four DMA channels: one to the LSI-11 RAM, two externally to other nodes via 0.5 MB/S shared communication busses, and one internally to the Z80 RAM which contains the packet buffers for the frontend. The 8X300 code implements a "carrier sense multiple access" bus sharing protocol (Kleinrock, 1976). Physically, each frontend consists of two circuit boards inserted into its LSI-11 backplane: one full (22x26 cm) card for the 8X300 and port logic, and one half-card for the Z80 and memories.

IV. ORGANIZATION OF MICROS

The MICROS distributed operating system is designed to control network computers of thousands of nodes for users simultaneously running a varying mix of large and small task forces. MICROS forms a hierarchy of nested network resource pools, managed by processes running in nodes empirically selected to pass control messages efficiently. MICROS modules are written in Concurrent and Sequential Pascal for each LSI-11 and Z80. Concurrent Pascal was chosen because it allows operating systems to be written modularly, because compile-time syntax restrictions produce safe yet efficient code, and because it is well-documented (Brinch Hansen, 1977; Hartmann, 1977).

A major research problem in the design of operating systems for network computers is global resource management: the control of shared resources that may be requested by tasks throughout the network. Resources include: processors, communication links, memory space for user and system tasks, buffers for intertask communication, secondary storage files, peripheral devices, and system server modules. The management problem is especially acute for network computers such as MICRONET, whose nodes cannot directly access each others' memories. In particular, MICROS must not utilize centralized resource tables accessed by several management nodes.

In the first version of MICROS, each node has a resident, private copy of concurrent processes that handle I/O devices, execute sequential tasks, interpret task spawning commands, manage files and interprocess messages, and pass internodal messages using packet-switching protocols on the shared communication busses. Sequential Pascal program files for user tasks are dynamically loaded into hosts.

There is a common message interface for communication among resident processes and dynamically loaded tasks whether in the same or different nodes. Buffer pools, called channels, can be selected to manage high volume data transmissions, as during file accessing. Users and tasks can spawn linear task forces, called pipes (in UNIX), which are executed in parallel and linked by channels.

The global management structure proposed for MICROS to enable it to control networks of thousands of nodes is shown in Figure 2. It is a management hierarchy such as found in corporate, military, and other large social organizations. The circles represent nodes in the network. The arcs represent management paths between them, but need not directly correspond to physical communication links in the hardware.

The hierarchy is a truncated tree. The lowest, or leaf nodes in the tree perform user tasks and handle I/O devices connected to the network. The upper nodes manage resources and provide regional control for the nodes directly below them. Assuming that one computer node can manage resources associated with about ten other nodes of similar speed, the fanout at each level of the tree is about ten and there are about $\log_{10}(N)$ levels in a truncated tree containing N nodes.

Although globally accessible, network resources are managed in nested pools. Each management node controls, possibly indirectly, all resources within its subtree of the network hierarchy. Monitoring, scheduling, and allocation of each resource associated with a low-level node are performed by one or more of the management nodes in the chain between the resource node and the top of the hierarchy.

There is not a topmost node controlling the entire network. Having a single master node would make the network too vulnerable to a failure of that node. Instead, the highest management level consists of several nodes, forming a global control oligarchy. Each of the oligarchy members regularly passes summary information to other oligarchy members to protect against hardware failures. Similarly, lower level managers pass summaries to their immediate higher level managers including lists of their next lower level slaves. A distributed scheduling alogrithm, called Wave Scheduling (van Tilborg, 1981a; 1981b), utilizes the hierarchical structure for task force spawning.

An efficient control hierarchy (van Tilborg, 1980) can be selected regardless of the actual topology of physical communication links between nodes of a network. After major reconfigurations of busses or large changes in the number of nodes in MICRONET, MICROS must be initialized. During initialization, local groups of nodes can empirically determine which nodes are well enough connected via physical links to exchange messages easily with the others. Well-connected nodes

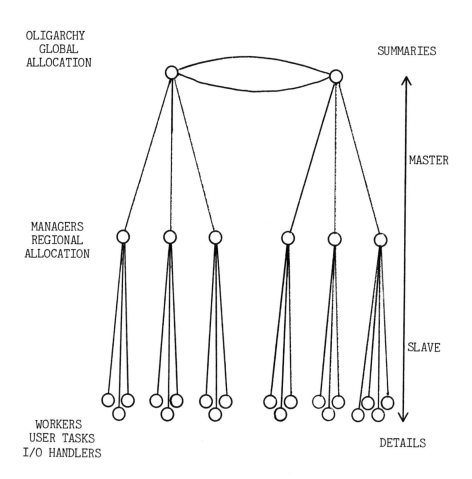

FIGURE 2. A logical hierarchy of global resource management nodes for MICROS. Paths between nodes may involve more than one hardware link. Nodes and paths are chosen for efficient message passing within each subtree.

are selected as managers of nearby nodes to form the hierarchy.
Management hierarchy paths are tested continually by the
communication subnet to detect and circumvent hardware
failures. After an isolated failure or the addition of nodes, a
new manager may be selected within a local area without fully
reinitializing the whole network. However, since only an idle
low-level node can easily be promoted to manager, control
messages may pass more slowly after a local reorganization.

V. PRESENT STATUS

A single-node version of MICROS has been running small
demonstration programs since June 1979. Production prototypes
for the MICRONET frontend cards were delivered in late 1979,
tested, and revised to treble memory size so that almost all
MICROS code resident in each node can run in the Z80's in the
planned second version of the operating system. As of June
1981, the nodes for a thirteen node version of MICRONET are
being tested. Testing has also begun on LSI-11/23 boards which
will allow a few nodes of MICRONET to have up to 240 kilobytes
of RAM to run very large programs such as compilers. Thus far,
integration of the parts has proceeded with no insurmountable
difficulties, although always at about half the speed and twice
the cost as planned.

The version of MICROS that has thus far been coded does
not contain routines to set up and maintain a distributed
management hierarchy. The first version of MICROS consists of a
set of Concurrent Pascal processes adequate to handle I/O
devices, interpret user commands, access diskette files, and run
user tasks in a network. Distributed global control routines
will be implemented in the second version of the system.

The current version of MICROS allows a user at an isolated
node attached to a terminal and a floppy disk to issue commands
to load and to control pipes of up to three tasks. Pipes of
linked simple programs, such as ones to "ECHO" or to "REVERSE"
whatever the user enters at a terminal, load and execute
successfully. Their execution demonstrates that at least 95% of
the basic first version of MICROS is working.

The frontend communication processes have been tested
separately both in Concurrent Pascal on a pair of LSI-11's and
in assembly code on two Z80's. Debugging of a Z80 interpreter
for Concurrent Pascal is nearly complete. It is fully
compatible with the output of the compiler for the LSI-11.

A new 6000 line version of the 8X300 microcode has been
coded and should be debugged by July 1981. It allows the LSI-11
and Z80 to send multiple I/O commands to the 8X300 or directly
to each other. It provides the necessary handshakes so that
both the LSI-11 and Z80 may asynchronously generate, execute and

acknowledge I/O commands. This new microcode is being
integrated with the formerly tested LSI-11 and Z80 sections of
MICROS to provide a working version of the system. After
debugging of the 8X300 interfaces to the LSI-11 and Z80, a
global resource management program will be added to complete the
first version of MICROS.

VI. FUTURE PLANS

    Besides the final integration of the current system, three
groups of extensions to MICROS are planned for the next two
years. The first is a major revision of MICROS both to reduce
its size in LSI-11 memory and to speed its responses. The part
of MICROS resident in each LSI-11 currently takes about 4000
lines of Concurrent Pascal source code. Along with the
Concurrent Pascal run-time system, it occupies about 47 KB of
memory, leaving only 9 KB for user tasks. A resident system
less than half this size is needed.
    Design of the revised version of MICROS is nearly
complete. The new design strives to separate user task
processing from communication overhead. The LSI-11 host runs
only dynamically loaded system servers and user tasks. The Z80
frontend is used for MICRONET control and communication.
    The host provides a number of envelopes which can load and
run a system server or user task. The envelope serves as an
encapsulating module, which provides for intertask communication
and abstracted system services. These services are provided by
standard prefix requests which are passed to the frontend for
forwarding to appropriate system servers. The use of the
frontend for almost all system services resident in each node
allows the host processor to be dedicated to execution of
dynamically loaded tasks.
    The portion of MICROS in the frontend of each node will
logically have three main parts: Packet Switching Manager (PSM),
Channel Manager (CM), and Node Controller (NC).
    The packet switching manager connects the node to the
communication subnet of MICRONET. The PSM implements the end-
to-end protocols for MICROS. The PSM shares control of the two
busses connected to its node and manages physical packets as
they enter and leave the node via the busses. Some entering
packets are relayed to other nodes. Packets received for the
node are assembled into complete messages which are then passed
to the channel manager. The channel manager supplies the
packets forming each message to the PSM for routing from the
node. The PSM controls packet flow to avoid overloading buffers
and congesting the communication subnet.
    The channel manager connects tasks in the LSI-11 host
processor to the frontend and consequently to the rest of

MICRONET. Its main goal is to transfer messages within
MICRONET. The CM provides this support by managing logical
circuits, called channels, which connect ports associated with
abstracted source and destination tasks. The logical messages
used for intertask communication are transferred on these
channels. The CM provides for buffering and transmission of
these messages. Exception messages can be sent as datagrams via
a virtual system channel.

The node controller manages resources local to its node
and interacts with the hierarchy of global resource managers.
The NC arbitrates the resource requests of the CM and PSM and
keeps track of the status of host envelopes. It interfaces with
global name servers to map symbolic resource names into unique
addresses. The NC keeps track of available system servers which
it may load locally or use remotely.

The second planned extension to MICROS is the elaboration
of global control and system initialization routines so that
MICROS can be used on large networks of arbitrary connection
topology. Global resource management will also require routines
to monitor communication traffic within MICRONET. At first,
MICRONET utilization statistics will be gathered via serial
interfaces controlled by an external minicomputer. Eventually,
network monitoring will be done internally by MICROS to increase
resource utilization.

The third planned extension to the MICROS system is to
provide an environment for the development of distributed
application programs. This environment must include compilers,
debugging tools, and processor allocation aids. MICRONET has
been developed not only to investigate the design of distributed
computer systems but also to provide a test-bed for implementing
distributed algorithms in areas such as artificial intelligence,
graphics, image processing, and numerical analysis.

A framework for debugging, testing and monitoring of
distributed systems is being developed. Although it will be
specific to the language used for MICROS, it is expected that
the framework will also be applicable to Ada and other
distributed programming languages.

For efficiency, the task force scheduler for a distributed
system must attempt to minimize communication costs between
cooperating processes and yet maintain parallelism. The
scheduler can obtain information from three sources: the user,
the compiler, and the runtime system.

First, the user may be able to specify, either at compile
or at load time, important interactions between modules. The
user should not have to assign tasks explicitly to physical
processors. A mechanism is needed that is independent of
network topology. It should allow the user to express logical
connections between modules. Second, a compiler may be able to
generate some heuristic information from the source code and
pass this information to the scheduler. Finally, the runtime

system should describe the actual observed utilization of
interconnections to both the user and the scheduler.

The fields of network computers architecture and
distributed operating systems design are still full of
interesting research topics. MICRONET and MICROS are intended
to provide a viable system for computer science research and
development of distributed applications programs.

REFERENCES

Anderson, G., and Jensen, E. (1975). "Computer Interconnection
    Structures: Taxonomy, Characteristics and Examples", ACM
    Computing Surveys, Vol. 7, No. 4, 197.
Brinch Hansen, P. (1977). "The Architecture of Concurrent
    Programs", Prentice-Hall, Englewood Cliffs.
Despain, A., and Patterson, D. (1978). "X-Tree: A Tree
    Structured Multi-Processor Computer Architecture", Proc. 5th
    Annual Symp. on Computer Architecture (IEEE), 144.
Hartmann, A. (1977), A Concurrent Pascal Compiler for
    Minicomputers, Springer-Verlag, New York.
Jones, A., et al. (1979). "StarOS, A Multiprocessor Operating
    System for the Support of Task Forces", Proc. 7th Symp.
    on Op. Sys. Principles (ACM), 117.
Kleinrock, L. (1976). Queueing Systems, Volume II: Computer
    Applications, Wiley, New York.
Metcalfe, R., and Boggs, D. (1976). "Ethernet: Distributed
    Packet Switching for Local Computer Networks", Comm. ACM,
    Vol. 19, No. 7, 395.
Sullivan, H., and Bashkow, T. (1977). "A Large Scale,
    Homogeneous Fully Distributed Parallel Machine, I", Proc. 4th
    Annual Symp. on Computer Architecture (IEEE), 105.
Thompson, C., and Kung, H. (1977). "Sorting on a Mesh-Connected
    Parallel Computer", Comm. ACM, Vol. 20, No. 4, 263.
van Tilborg, A., and Wittie, L. (1980). "High-Level Operating
    System Formation in Network Computers", 1980 Int. Conf.
    on Parallel Processing (IEEE), 131.
van Tilborg, A., and Wittie, L. (1981a). "Wave Scheduling:
    Distributed Allocation of Task Forces in Network Computers",
    2nd Int. Conf. on Distributed Comp. Sys. (IEEE), Paris, 337.
van Tilborg, A., and Wittie, L. (1981b). "Distributed Task force
    Scheduling in Multi-Microcomputer Networks", National
    Computer Conference, Chicago, 283.
Wittie, L. (1976). "Efficient Message Routing in Mega-Micro-
    Computer Networks", Proc. 3rd. Annual Symp. on Computer
    Architecture (IEEE), 136.
Wittie, L. (1981). "Communication Structures for Large Networks
    of Microcomputers", IEEE Trans. on Comp., Vol C-30, No. 4, 31.

PUMPS: A SHARED-RESOURCE MULTIPROCESSOR ARCHITECTURE FOR PATTERN
ANALYSIS AND IMAGE DATABASE MANAGEMENT[*]

Fáye A. Briggs, Kai Hwang, and K. S. Fu

School of Electrical Engineering
Purdue University
West Lafayette, Indiana

ABSTRACT

The PUMPS architecture consists of N Task Processing Units
(TPU) which share a pool of special peripheral processors and
VLSI functional units (PPVU) and a shared main memory (SM) via a
block transfer oriented interconnection network. A shared cache
is provided between the TPUs and SM for efficient MIMD interprocessor communication. The SM is also connected via a Backend Database Management Network (BDMN) with distributed Control to the
File Memories (FM), which are disk-based database storage devices.
A Front End Communication Processor (FECP) is used to switch
the control of I/O terminals. The FECP and the BDMN manage the
relationally-structured image databases in addition to providing
file manipulative primitives. The FECP and BDMN have unique
features such as a language interface for relational image database management using Query-by-Picture Example (QPE) and spatial
operators.

I. INTRODUCTION

Image analysis tasks require a wide variety of processing
techniques and mathematical tools. In most machine intelligence
systems, large computers are employed to process pictorial information. These computers are often designed for general applications and are not tailored for the special needs of pattern
analysis. Many image processing tasks require only repetitive
Boolean operations or simple arithmetic operations defined over
extremely large arrays of picture elements. Hence the use of a
large general purpose computer results in intolerable waste of

MULTICOMPUTERS AND IMAGE PROCESSING
ALGORITHMS AND PROGRAMS

319

resources. Moreover, a rigidly structured architecture does not lend itself to the flexibility required to process a wide spectrum of PRIP algorithms. Various image computer architectures have been proposed in the past (Fu, 1978). The existing image understanding system architectures consists of the array-structured machines ILLIAC IV, BSP and STARAN, (Hwang, et al. 1981) and the multi-mini/microprocessor systems, such as C.mmp and C.m* (Jones, et al. 1979). Other image processing systems are the real-time cellular-logic based CLIP-4 and the Toshiba TOSPICS which has a large image memory designed to reduce data transfer time. Most of the above machines are too rigid in their configurations. The PM^4 system proposed earlier by the authors (Briggs, et al. 1979) limits the reconfiguration to SIMD and MIMD.

These architectures do not satisfactorily address the bottleneck problems of manipulating large image databases especially in a real-time environment. Cost-effective solutions to many applications such as image analysis problems require some form of functional specialization in the computer architecture (Hon, 1977). A general-purpose multiprocessor system is not tailored to efficient execution of special classes of algorithms. Certain algorithms are more suited to SIMD array processing, some to MIMD multiprocessing. However, there are also some classes of image algorithms in which there is a need to update a common database (such as the same copy of a histogram), or the need to efficiently process a moving image, in which pipelining is most adequate. In order to meet the processing requirements of the explosive amount of pictorial information, concurrency must be exploited maximally at all levels of a computation. At the instruction preprocessing level, pipelining can be implemented within the processor. At the arithmetic and logic operation level, pipelining and parallelism can be implemented in VLSI or special attached processing units (Hwang and Cheng 1980, 1981). These units, when integrated with an efficient multiprocessing system enable configurations of parallelism (synchronized or asynchronous) (Kung, 1976) or of macropipelines (Händler, 1973). In addition, an efficient database system is harmoniously integrated into the reconfigurable multiprocessor system. The reconfigurability implies that for each given parallel algorithm, an optimum path may be selected through the structure to efficiently execute the set of task. Moreover, the system utilization and throughput may be increased by having several algorithms concurrently executed in different portions of the system.

The PUMPS architecture exploits the new emerging VLSI computing structures and special peripheral processors. The PUMPS configuration also resulted from the need to relax the processing modes defined in the PM^4 architecture. PUMPS organization introduces features such as distributed operating system and cost-effective resource pool configuration for a given applicative domain and workload. The PUMPS architecture can thus be easily reconfigured to suit a specially-chosen application area.

The emphasis of application of the PUMPS architecture is in pattern analysis and image database management. A general block diagram of an image analysis system, is shown in Figure 1. There are essentially five processing stages. In the preprocessing stage, image enhancement operations are performed to restore the noisy blurred images. The enhanced imagery data are then segmented according to feature regions. The classification stage recognizes the imagery pattern by indicating its membership in the pattern classes. Finally, structure analysis is performed to produce description of the pattern information.

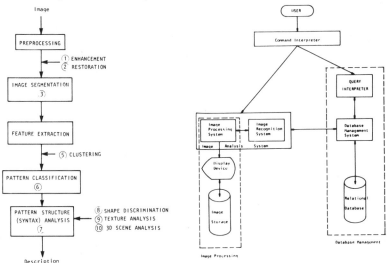

Fig. 1   A General Block Diagram
of Image Analysis System

Fig. 2   Integrated Image
Analysis and
Database Management
System

The image analysis system above is often applied to a continuous sequence of raw images. The processing time of such a repetitive task can be reduced by partitioning the task into a sequence of s processes which can be executed on a macropipeline configuration of the PUMPS. Each process can be executed in an autonomous segment, which operates concurrently with other segments. In the PUMPS, a segment can be an SISD, SIMD, MIMD, pipeline, or specially dedicated VLSI processors. Hence a macropipeline configuration of the PUMPS is composed of segments so that consecutive processes of a task can be assigned to distinct segments for processing. In the next section, we outline the architectural features of PUMPS. The design issues in the configuration of macropipelining are enumerated in Section III.

II.   SYSTEM ARCHITECTURE OF PUMPS

An integrated image analysis and image database processing

system is shown in Figure 2. It consists of three closely cooperating subsystems, namely, the man-machine interface using a command interpreter, the image processing and analysis subsystem and the database management subsystem. The PUMPS architecture integrates these subsystems into an efficient multiprocessor system.

The PUMPS is a high-performance multiple microprocessor computer operating with SISD/MIMD task processors and a set of shared PPVUs (Fig. 3). There are P Task Processing Units (TPUs) in the system, each of which is multiprogrammed. They also can operate in an interactive fashion through the shared PPVUs and shared memory system. The TPUs can communicate with each other via the Task Processor Communications (TPC) Bus. This intercommunication medium is very effective in passing interrupts, synchronization and other control signals. All the TPUs are connected to the shared-resource pool of special peripheral processors and VLSI units (PPVUs) via a Special Resource Arbitration Network (SRAN). This network provides connections between each TPU and the desired PPVU. The SRAN is a crossbar-like switch with increased connectivity. Besides connecting any TPU to any PPVU, it must provide for arbitrary inter-PPVU paths. As VLSI technology develops, modular crossbar switches (Franklin, 1981)

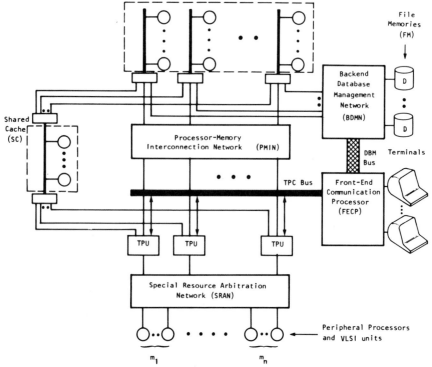

Fig. 3 The System Architecture of PUMPS

will become more cost-effective because of their regular and local connections. In case several TPUs reference the same functional unit, some priority must be established to resolve the
conflicts.

The allocation of the shared resources in the pool to the
TPU's is dynamic. The selection of the resource types in the
pool is tailored to special application requirements. For example, one may wish to include an FFT processor, a histogram
analyzer, and some VLSI array or pipeline processors in the pool
for image analysis applications. In this sense, the PUMPS has a
dynamically reconfigurable structure. Different applicative environment may be equipped with different functional units. The
remaining system resources such as TPU's and shared memories, are
designed for general-purpose MIMD computations.

The TPUs perform three basic types of functions: (i)
dispatching and initiating PPVU tasks, (ii) executing purely
sequential tasks, (iii) participating in MIMD processes and running the operating system. In order to perform the first type of
function, a local Task Memory (TM) is provided within each TPU as
depicted in Fig. 4. The TM is partitioned into several segments.
These consists of the unmapped memory, which is used for the
operating system Kernel and device drivers, the local image
buffers and the local scratch-pad. The local image buffers is
shared between the Task Processor (TP) and the PPVUs. The PPVUs
are generally passive and the TPU, acting as a controller, must

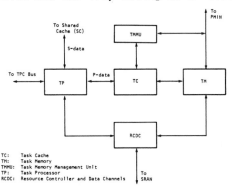

Fig. 4 Architecture of a Task Processing Unit (TPU)

provide the PPVU with a continuous flow of data when the PPVU is
processing a task. A Resource Controller and Data Channels
(RCDC) in the TPU is used to format and channel the data between
the TM and the PPVU.

To match the speed of the TP, a local Task Cache (TC), is
provided between the TP and TM. The TC stores the most recently
used private instructions and data of the active processes assigned to the TPU. The effectiveness of such caches in a multiprocessor system has been demonstrated (Briggs and Dubois,
1981a). Due to the locality property of programs, most of the
references can be made in the cache. A multiprocessor system

with private caches may encounter data coherence problem in which
several copies of the same block of shared data may exist in different caches at any given time. Basically, two methods have
been proposed to solve this coherence problem. In the first
method, whenever, a processor attempts to update a variable in a
TC, the possible copies of the variable in other caches must be
modified accordingly before the process execution proceeds. This
requires maintaining a central copy of the directories of each
cache or selectively tagging cache blocks with common variables
(Censier, 1978). This method although efficient may be cost-
prohibitive. The second method was proposed for the C.mmp in
which shared data is non-cacheable and reside in shared memory.
In the PUMPS architecture a compromise solution was sought to the
cache consistency problem by tagging the data as private (P-data)
or shared (S-data). The shared data are referenced in a unique
cache (SC) shared by all the processors. An analysis of the performance of the shared cache concept shows its effectiveness (Dubois and Briggs, 1981).

The PUMPS has a distributed memory organization using virtual
memory addressing based on paged segments. Within the TC, misses
are serviced by the TMMU which initiates the block transfer from
the TM to the TC, provided the block exists in the TM. If the
block does not reside in the TM but is known to the process, a TM
page fault occurs which is also serviced by the page-fault
handler that resides in the TMMU. The occurrence of a TM page
fault causes the current active process in the TP to be blocked,
whereupon a task switch is made to a runnable process also residing in the TPU. By distributing the memory management functions
over the entire system, the TPUs are relieved of performing
memory management functions and thereby increase their effectiveness in performing useful computations.

The interleaved TM in TPU serves as a high-spaced buffer
between the TPU and the Shared Memory (SMs). The SMs are semiconductor memories organized with multiported l lines and m
memory modules per each line as described in (Briggs, 1977) to
permit efficient block transfers of information. A block-
transfer oriented Processor-Memory Interconnection Network, such
as the delta network (Patel, 1979), is used between the TPUs and
the SMs. The shared memories are also connected to the File
Memories (FM) via a Backend Image Database Management Network
(BDMN), which is designed to handle the data transfers from multiple disks. The FM together with a backend computer and the
BDMN comprise the database machine for Image Processing (DMIP).

The architectural design of the database machine for image
processing is shown in Fig. 5. It consists of three parts: a set
of data modules each of which includes a disk with the associated
cellular logic for processing picture queries, a backend computer
and the BDMN. The design of an efficient image database system
of the PUMPS differs from the conventional database design for
alphanumeric information processing, because of the large amounts
of imagery data involved. The need to interface with image pro-

cessing packages and the necessity of providing a high-level
language interface to the users, complicates the design issues
(Chang and Fu, 1979). However, the design of the DMIP is based
on the identification of the properties of various image data
manipulation and retrieval operators. These operators are interpretable via a language interface which permits a logical
representation of the images. To permit a high degree of concurrency of accesses to the DMIP, its control is distributed.
The concurrent accesses are possible if the image database is
deadlock-free serializable and security controllable (Wah, 1979).
The control mechanism requires the partitioning and replication
of images and the placement of these partitions on the secondary
storage. The images on the file memory are also dynamically
reorganized for efficient retrievals.

A Front-End Communication Processor (FECP) is used to switch
the control of I/O terminals and performs preprocessing task such
as interactive file editing. The FECP together with the BDMN
provide efficient processing capabilities for classical file
manipulative primitives in addition to providing the users access
to the whole multiprocessor system.

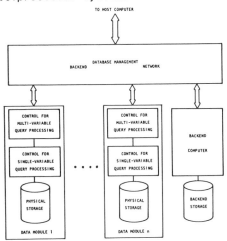

Fig. 5   Database Machine for Image Processing

III. CONFIGURATION OF MACROPIPELINES IN THE PUMPS

In general, the implementation of the possible configurations
of macropipelines in the PUMPS requires additional hardware support. Each processing segment or peripheral processor and VLSI
unit (PPVU) has one full duplex buffers for data operands. The
buffers are configured to enhance the access times of the
operands and to control the synchronization of data inputs to the
PPVU. The output of one segment is stored in the buffer of the
next segment in the macropipeline. This buffer is used as the

input to this next segment as shown in Fig. 6. Such structures
of macropipelines characterize most image processing algorithms.
Many examples of macropipelines are found in the literature:
analysis of motion (Agrawal, et al, 1980), image reconstruction
from projections (Farrell, 1978), image coding (Yasuda, et al.
1980), and radar signal processing (Armstrong, et al. 1979).

For the PUMPS architecture, the buffers, B_{ij} in Fig. 6 are
partitioned into several independent logical modules, whose sizes
are software selectable. Two successive segments cooperate in a
producer/consumer relationship. One segment stores its output as
successive batches of data in consecutive modules of the local
buffer of the next segment, where they are processed sequentially. Of course, the consumer or producer must be suspended when
the intermediate buffer is empty or full respectively. P and V
operations may be used to enforce the producer/consumer relationship. The flow of data through a PPVU is coordinated by a small
local control unit. In practice, a segment may receive input
data from the preceding segment and also from a TPU memory. The
communication between the two segments of a macropipeline might
be reduced to a few words, in which case, the local buffers are
sufficient. In some cases, the local buffers are used to limit
the congestion at the TPU's local memories when a picture is being streamed between two consecutive segments.

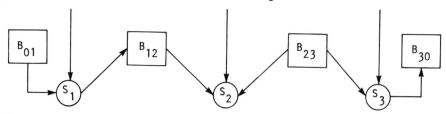

Fig. 6 Macropipeline in PUMPS Architecture with S_i a TPU or PPVU

For efficient macropipelining, the algorithms used for two
consecutive segments should be input/output compatible. Hence,
for example, if one segment produces an output image in row-major
format, then the succeeding segment should consume it also in
row-major format to avoid additional segments for data format
transformations (e.g. transposition). In this context, the
choice of the PPVUs will depend on the typical system workload.
The selection of the number of PPVUs of each type was discussed
in (Briggs, et al. 1981b).

If the execution times of the segments of the pipe are dissimilar, the overall performance of the task on the macropipeline
configuration may be poor. The performance is also affected by
the randomness of the processing times on the TPU (Kung, 1976).
The processing times on a PPVU are generally deterministic in nature. Some of the inputs to the pipeline may have deterministic
interarrival times and others are random. A general study which
takes into account these various aspects, would indicate decompo-

sition strategies for good performance of a macropipeline configuration.

Macropipelines are generally efficient when the same task is repeated on successive input data sets. This is typical of real-time processing applications. In such applications, the impact on the performance of the pipeline startup time is reduced. However, if the processing is performed on one input data set, a speed-up is still obtainable by applying the techniques discussed above. In this case, the intermediate data is not buffered between two segments but is forwarded directly to the next segment provided they are I/O compatible. The speed-up results by avoiding data manipulation which takes place in buffers. This technique is a generalization of chaining as implemented in the Cray-1 computer (Hwang, et al. 1981).

IV. CONCLUDING REMARKS

The PUMPS architecture generates a number of related research topics that need to be investigated. One basic topic is the development of an appropriate operating system and language that can be used to effectively control and express the processing tasks. Another operating system related task is the memory management method and its implementation. It could be said that Pictorial Image Analysis algorithms cannot be classified as having properties of the more general programs. For example, an algorithm for performing a specific function on a VLSI processor may be well behaved. In which case, parameters such as looping distance and page reference string will be well defined for a given task dimension. With such parameters we can develop an efficient image buffer management policy and determine appropriate buffer size for each VLSI unit. VLSI technology can be explored for fast manipulation and update of relational image data base and permit a modular growth.

In order to design special peripheral processors and VLSI units for analysis and management of imagery data, typical algorithms executed need to be developed. Some of the algorithms have been discussed in (Briggs, et. al. 1981c; Hwang and Su, 1981; Ni and Hwang, 1981; Chang and Fu, 1981; Fu, 1980; Hwang and Briggs, 1982; and Chiang and Fu, 1981).

Implementation of a processing task on the PUMPS architecture requires the efficient allocation of system resources. In particular, the effective utilization of the PPVUs is important. The match of the set of PPVU configurations with the specific application needs makes the PUMPS very attractive. The unification of the image database management subsystem with the user-oriented multiprocessor system in the PUMPS permits an effective on-line processing of pattern analysis problems and the interactive management of imagery data.

REFERENCES

Agrawal, D.P. and Jain, R. (1981). "Computer Analysis of Motion Using a Network of Processors," Proceedings of the 5th International Conference on Pattern Recognition, December.

Armstrong, C. V., et al. (1979). "An Adaptive Multimicroprocessors Array Computing Structure for Radar Signal Processing Applications," Proceedings of the 6th Annual Symposium on Computer Architecture.

Briggs, F. A. (1977). "Organization of Semiconductor Memories for Parallel-Pipelined Processors," IEEE Trans. on Comput., C-26, 2, February, 162-169.

Briggs, F. A., Fu, K. S., Hwang, K. and Patel, J. H. (1979). "PM^4 - A Reconfigurable Multiprocessors System for Pattern Recognition and Image Processing," Proc. of NCC, Vol. 48, AFIPS, June, 255-265.

Briggs, F. A. and Dubois, M. (1981a). "Performance of Cache-based Multiprocessors," Proc. of ACM/Sigmetrics Conference on Measurement and Modeling of Computer Systems, September.

Briggs, F. A., Dubois, M., and Hwang, K. (1981b). "Throughput Analysis and Configuration Design of a Shared-Resource Multiprocessor System: PUMPS," Proceedings of the 8th Annual Symposium on Computer Architecture, May, 67-80.

Briggs, F. A., Hwang, K., Fu, K. S. and Dubois, M. (1981c). "PUMPS Architecture for Pattern Analysis and Image Database Management," Proceedings of IEEE Pattern Recognition and Image Processing Conference, August.

Censier, L. M. and Feautrier, P. (1978). "Solution to Coherence Problems in Multicache Systems," IEEE Trans. on Computers, C-27, 12, December.

Chang, J. M. and Fu, K. S. (1981). "Extended K-d Tree Database Organization: A Dynamic Multiattribute Clustering Method," IEEE Trans. Software Engr., SE-7, 3, May, 284-290.

Chang, N. S. and Fu, K. S. (1979). "Parallel Parsing of Tree Languages for Syntactic Pattern Recognition," Pattern Recognition, 11, 213-222.

Chang, N. S. and Fu, K. S. (1979). "A Relational Data Base System for Images," Technical Report TR-EE 79-28, School of Electrical Engineering, Purdue University, May.

Chang, N. S. and Fu, K. S. (1981). "Query-By-Pictorial Example," IEEE Trans. Software Engineering, SE-6, November, 519-524.

Cheng, J. K. and Huang, T. S. (1980). "Algorithms for Matching Relational Structures and Their Applications TR-EE 80-53, TR-EE 80-53, School of Electrical Engineering, Purdue University, December.

Chiang, Y. T. and Fu, K. S. (1981). "Parallel Distance Computations in Syntactic Pattern Recognition," Proc. IEEE Computer Society CAPAIDM Workshop, November 11-13, 1981, Hot Springs, VA.

Dubois, M. and Briggs, F. A. (1981). "Efficient Interprocessor Communication for MIMD Multiprocessor Systems," Proceedings of the 8th Annual Symposium on Computer Architecture, May, pp. 187-196.

Farrell, E. J. (1978). "Processing Limitations of Ultrasonic Image Reconstruction," Proceedings of the Conference on Pattern Recognition and Image Processing, June.

Franklin, M. A. (1981). "VLSI Performance Comparison of Banyan and Crossbar Communications Networks," IEEE Trans. on Computers, C-30, 4, April.

Fu, K. S. and Rosenfeld, A. (1976). "Pattern Recognition and Image Processing," IEEE Trans. on Computers, C-25, 12, December, 1336-1346.

Fu, K. S. (1976). Digital Pattern Recognition. Springer-Verlag.

Fu, K. S. (1978). "Special Computer Architectures for Pattern Recognition and Image Processing," Proc. 1978 National Comp. Conf., 1003-1013.

Fu, K. S. (1980). "Panel on Special Computer Architecture for Pattern Recognition and Image Processing," Proc. 5th International Conference on Pattern Recognition, December, Miami Beach, FL.

Händler, W. (1973). "The Concept of Macropipelining with High Availability," Elektronische Rechenanhagan, 15, 269-274.

Hon, R. and Reddy, D. R. (1977). "The Effect of Computer Architecture on Algorithm Decomposition and Performance," in High-Speed Computers and Algorithm Organization, (Kuck, et al. editors), pp. 411-421, Academic Press.

Hwang, K. (1979). Computer Arithmetic: Principles, Architecture and Design, John Wiley, New York.

Hwang, K. and Su, S. P. (1981). "VLSI Pattern Classification With Partitioned Matrix Algorithms," IEEE Workshop on Computer Architecture For Pattern Analysis and Image Database Management, Hot Springs, VA, November.

Hwang, K. and Cheng, Y. H. (1980). "VLSI Computer Structures for Solving Large-Scale Linear System of Equations," Proceedings of the 1980 International Conference on Parallel Processing, 217-230.

Hwang, K., Su, S. P. and Ni, L. M. (1981). "Vector Processing Computer Architecture," in Advances in Computers, 20, Academic Press, New York, 115-197.

Hwang, K. and Cheng, Y. H. (1981). "Partitioned Algorithms and VLSI Structures for Large-Scale Matrix Computations," Fifth Symp. Computer Arithmetic, Ann Arbor, Michigan, May 18-19, pp. 222-232.

Hwang, K. and Ni, L. M. (1980). "Resource Optimization of A Parallel Computer for Multiple Vector Processing," IEEE Trans. Computers, C-29, No. 9, September, 831-836.

Hwang, K. and Briggs, F. A. (1982). Computer Architecture and Parallel Processing, McGraw-Hill Book Co., New York (In press).

Jones, A. K. and Schwartz, P. (1979). "Experience Using Multiprocessor Systems: A Status Report," Carnegie-Mellon University, Technical Report CMU-CS-79-146, October.

Kung, H. T. (1976). "Synchronized and Asynchronous Parallel Algorithms for Multiprocessors," in Algorithms and Complexity: New Directions and Recent Results, (J. F. Traub, ed.), Academic Press, New York.

Mead, C. A. and Conway, L. A. (1980). Introduction to VLSI Systems, Addison Welsey Publishing Company, Reading, Mass.

Ni, L. M. and Hwang, K. (1981a). "Performance Modeling of Shared Resource Array Processors," IEEE Trans. Software Engr., Vol. SE-7, July.

Ni, L. M. and Hwang, K. (1981b). "A Multiprocessor System for Office Image Processing," IEEE Workshop on CAPAIDM, Hot Springs, VA, November.

Patel, J. H. (1979). "Processor-Memory Interconnection for Multiprocessors," 6th International Symp. Comp. Arch., April, 168-177.

Wah, B. (1979). "A Systematic Approach to the Management of Data on Distributed Data Bases," Ph.D. Dissertation, University of California, Berkeley.

Yasuda, Y., Dubois, M., and Huang, T. S. (1980). "Data Compression for Check Processing Machines," Proceedings of the IEEE, 68, 7, July.

---

*This research was supported by NSF-Grant ECS 80-16580

REMOTE SENSING ON PASM AND CDC FLEXIBLE PROCESSORS

Howard Jay Siegel
Philip H. Swain
Bradley W. Smith

Laboratory for Applications of Remote Sensing
and School of Electrical Engineering
Purdue University
West Lafayette, Indiana 47907, U.S.A.

I. INTRODUCTION

Multispectral image data collected by remote sensing devices aboard aircraft and spacecraft are relatively complex data entities. Both the spatial attributes and spectral attributes of these data are known to be information bearing (Swain et al., 1978), but to reduce the computation involved, most analysis efforts have focused on one or the other. Characteristic spatial features include, for example, shape, texture, and structural relationships. Some useful research has been accomplished in recent years in the direction of incorporating spatial information into the data analysis process (Haralick et al., 1973; Kettig et al., 1976; Weszka et al., 1976).

One way to approach spatial information in image data is to recognize that the ground cover associated with a given pixel, i.e., its "class" is not independent of the classes of its neighboring pixels. Recent investigations have demonstrated the effectiveness of a contextual classifier that combines spatial and spectral information by exploiting the tendency of certain ground-cover classes to occur more frequently in some spatial contexts than in others (Swain et al., 1980a,b; Tilton et al., 1981; Welch et al., 1971). The practical utilization of this contextual classsfier in remote sensing has awaited the solution of two key problems: (1) lack of an effective method for characterizing and extracting contextual information in multispectral remote sensing imagery, and (2) the need to reduce the execution time of the very computation-intensive contextual classification algorithm. The

This work was sponsored by the National Aeronautics and Space Administration under Contract No. NAS9-15466 and the Air Force Office of Scientific Research, Air Force Systems Command, USAF, under Grant No. AFOSR-78-3581. The United States Government is authorized to reproduce and distribute reprints for Government purposes nothwithstanding any copyright notation hereon.

first of these problems has been solved by development of an unbiased estimation procedure (Tilton et al., 1981) which provides a
good characterization of the contextual information without
requiring exhorbitant amounts of classifier training data ("ground
truth"). But although the resulting improvement in classification
accuracy, compared to conventional no-context statistical classification methods, is significant, the cost-effectiveness of the
contextual classifier depends on solution of the second problem,
which is the subject of this paper.

A method to reduce the execution time of classification
algorithms such as the contextual classifier (and even much simpler algorithms used for remote sensing data analysis) is through
the use of parallelism. There are several types of parallel processing systems. An SIMD (single instruction stream -- multiple
data stream) machine typically consists of a control unit, N processors, N memory modules, and an interconnection network. The
control unit broadcasts instructions to all of the processors, and
all active (enabled) processors execute the same instruction at
the same time. Each active processor executes the instruction on
data in its own associated memory module. The interconnection
network provides a communications facility for the processors and
memory modules. An MIMD (multiple instruction stream -- multiple
data stream) machine typically consists of N processors and N
memories, where each processor can follow an independent instruction stream. As with SIMD architecture, there is a multiple data
stream and an interconnection network. CDC Flexible Processor
(FP) systems are MIMD architectures that have been built (CDC,
1977a,b). PASM is a proposed PArtitionable SIMD/MIMD multimicroprocessor system for image processing and pattern recognition
(Siegel, 1981).

Maximum likelihood classification (Swain et al., 1978), often
used in remote sensing, classifies each pixel independently of all
others. Using either the SIMD or MIMD mode of parallelism, the
image can be subdivided among the processors, each processor classifying its subimage. However, parallel implementations of contextual classifiers are, in general, not so straightforward, due
to the use of neighborhood information. The way in which parallel
machines such as the CDC FP system and PASM perform contextual
classifications is examined in the following sections.

II. CONTEXTUAL CLASSIFIERS

The image data to be classified are a two-dimensional I-by-J
array of multivariate pixels. Associated with the pixel at "row
i" and "column j" is the multivariate measurement n-vector $X_{ij} \in$
R^n and the true class of the pixel $\theta_{ij} \in \Omega = \{\omega_1, \ldots, \omega_C\}$. The
measurements have class-conditional densities $f(X|\omega_k)$, k =
1,2,...,C, and are assumed to be class-conditionally independent.
The objective is to classify the pixels in the array.

In order to incorporate contextual information into the classification process, when each pixel is to be classified, p-1 of its neighbors are also examined. This neighborhood, including the pixel to be classified, will be referred to as the p-array. For each pixel, for each class in Ω, a discriminant function g is calculated. The pixel is assigned to the class for which g is greatest. Each value of g is computed as a weighted sum of the product of probabilities based on the pixels in the neighborhood. This is described below mathematically for pixel (i,j) being in class ω_k. (The description is followed by an example to clarify the notation used. Further details can be found in (Swain et al., 1980a,b).)

$$g_{\omega_k}(\underline{X}_{ij}) = \sum_{\substack{\underline{\theta}_{ij} \in \Omega^p, \\ \theta_{ij} = \omega_k}} \left[\prod_{\ell=1}^{p} f(X_\ell | \theta_\ell) \, G^p(\underline{\theta}_{ij}) \right]$$

where

$X_\ell \in \underline{X}_{ij}$ is the measurement vector from the ℓth pixel in the p-array (for pixel (i,j))

$\theta_\ell \in \underline{\theta}_{ij}$ is the class of the ℓth pixel in the p-array (for pixel (i,j))

$f(X_\ell | \theta_\ell)$ is the class-conditional density of X_ℓ given that the ℓth pixel is from class θ_ℓ

$G^p(\underline{\theta}_{ij})$ = $G(\theta_1, \theta_2, \ldots, \theta_p)$ is the a priori probability of observing the p-array (neighborhood) $\theta_1, \theta_2, \ldots, \theta_p$.

Within the p-array, the pixel locations may be numbered in any convenient but fixed order. The class-conditional density of pixel measurement vector X given that the pixel is from class k is:

$$f(X|\omega_k) = e^{-[\log(2\pi)^n |\Sigma_k| + (X-m_k)^T \Sigma_k^{-1} (X-m_k)]/2}$$

where the measurement vector for each pixel is of size four, Σ_k^{-1} is the inverse covariance matrix for class k (four-by-four matrix), m_k is the mean vector for class k (size four vector), "T" indicates the transpose, and "log" is the natural logarithm. This same function is computed for maximum likelihood classification.

Consider as an example the horizontally linear neighborhood shown in Fig. 1, and assume there are two possible classes: $\Omega = \{a, b\}$. Then the discriminant function for class b is explicitly:

1	2	3
(i,j-1)	(i,j)	(i,j+1)

i-1,j-1	i-1,j	i-1,j+1
i,j-1	i,j	i,j+1
i+1,j-1	i+1,j	i+1,j+1

Fig. 1. Size three horizontally linear and size nine square (non-linear) neighborhoods.

$$g_b(\underline{X}_{ij}) = f(X_1|a)f(X_2|b)f(X_3|a)G(a,b,a)$$
$$+(f(X_1|a)f(X_2|b)f(X_3|b)G(a,b,b)$$
$$+ f(X_1|b)f(X_2|b)f(X_3|a)G(b,b,a)$$
$$+ f(X_1|b)f(X_2|b)f(X_3|b)G(b,b,b)$$

Based on the discriminant functions g_a and g_b, pixel (i,j) is assigned to the class having the larger discriminant value.

A non-linear three-by-three context array (neighborhood) is shown in Fig. 1. In general, for each g there are C^{p-1} product terms, each term having p+1 factors, where C is the number of classes and p is the neighborhood size. All of the calculations are done using floating point data.

The algorithm, shown in Fig. 2, implements the size three contextual classifier. Let "hold(m,k)" be a two-dimensional array of size three-by-C, i.e., $0 \leq m \leq 2$ and $1 \leq k \leq C$. For m=cr, hold(cr,k) is a vector of length C containing the class-conditional density values ("compf"s) for the pixel (i,j) ("cr" is an abbreviation for center). hold(lt,k) and hold(rt,k) are the analogous vectors for the pixel (i,j-1) (the left ("lt") neighbor) and pixel (i,j+1) (the right ("rt") neighbor), respectively. By using this array to save the class-conditional densities, each density (for a given pixel and class) is calculated only once.

In the Landsat data used in the test described in (Swain et al., 1980b) the percentage of a priori probabilities (G^Ps) that were non-zero was about 1% (based on a size nine neighborhood and 14 classes). The memory requirements of the classifier can be reduced greatly if the zero values are ignored and only the non-zero values stored. Assume that each non-zero G value is a floating point number requiring 32 bits. In memory, alternate each non-zero G value with a 16-bit word that specifies the three classes associated with that class, e.g., if G(3,3,2) is non-zero, the word preceding it is a representation (concatenation) of 3, 3, and 2. This would allow $\lfloor 16/3 \rfloor$ = 5 bits per pixel for specifying the class, i.e., up to 32 classes. Variations on this method can be employed for larger neighborhhoods and greater numbers of classes. This technique is used in the following sections.

The complexity of the algorithm is proportional to $I*J*C^3$ assignments, multiplications, and additions, and $I*J*C$ "compf" calculations. Typically, $10 \leq C \leq 60$ for the analysis of Landsat data.

The algorithm can be extended for a non-linear contextual classifier with a neighborhood of size nine (as shown in Fig. 1). The complexity of the algorithm would have growth proportional to $I*J*C^9$ assignments, multiplications, and additions. The number of "compf" calculations would still be $I*J*C$. In this case, "hold" would be a (2*J+3)-by-C array (assuming the neighborhood window moves along rows). The 2*J+3 pixels whose "compf" values are stored in "hold" are chosen to make it unnecessary to perform

Main Loop

```
for i = 0 to I-1 do /* row index */
    for k = 1 to C do /* for each class */
        for m = 0 to 2 do hold(m,k) = compf(i,m,k) /*cols.0-2*/
    lt = 0 /* hold(lt,k) is left neighbor */
    cr = 1 /* hold(cr,k) is pixel being classified */
    rt = 2 /* hold(rt,k) is right neighbor */
    for j = 1 to J-2 do /* column index */
        value = -1; class = -1 /* max "g" and class */
        for k = 1 to C do /* for each class */
            current = g(lt,cr,rt,k)
            if current > value /* compare with max */
                then value = current; class = k
        print pixel (i,j) is classified as "class"
        if j ≠ J-2 then /* update hold pointers */
            tp = lt; lt = cr; cr = rt; rt = tp
            for k = 1 to C do /* compf's for next col */
                hold(rt,k) = compf(i,j+2,k)
```

Discriminant Function Calculation

```
function g(lt,cr,rt,k)  /* for pixel cr, class k  */
sum = 0  /* initialize sum, used to accumulate g  */
for r = 1 to C do  /* all classes for pixel (i,j-1)  */
    for q = 1 to C do  /* all classes for pixel (i,j+1)  */
        if G(r,k,q) ≠ 0  /* do not multiply if G = 0  */
            then sum = hold(lt,r) * hold(cr,k)
                        * hold(rt,q) * G(r,k,q) + sum
return (sum) /* sum contains value of g(lt,cr,rt,k)  */
```

Class-Conditional Density Calculation

```
function compf(a,b,k)  /* for pixel (a,b), class k  */
x = A(a,b)  /* x is the pixel (a,b) measurement vector  */
```
$$\text{expo} = -[\log|\Sigma_k| + (x-m_k)^T \Sigma_k^{-1}(x-m_k)]/2$$
```
return (e^expo)  /* return value of f(A(a,b)|k)  */
```

Fig. 2. Uniprocessor contextual classification algorithm
for a size three horizontally linear neighborhood.

redundant "compf" calculations. In general, when classifying
pixel (i,j), "hold" has the "compf" values for pixels j-1 to J-1
of row i-1, pixels 0 to J-1 (all) of row i, and pixels 0 to j+1 of
row i+1. After the classification of pixel (i,j), the values for
pixel (i-1,j-1) are removed from "hold" and values for (i+1,j+2)
are added. When the pixels on a new row are to be classified,
call it i', then the values for pixels (i'-2,J-3), (i'-2,J-2), and
(i'-2,J-1) are removed and the values for (i'+1, 0), (i'+1, 1) and
(i'+1, 2) are added. (This assumes row i' is classified after
i'-1.) Given this, the rest of transforming the algorithm for the
size nine square neighborhood case is straightforward.

III. MIMD IMPLEMENTATION

The CDC Flexible Processor (FP) system consists of up to 16
FPs linked together, providing much parallelism at the processor
level. The FPs can communicate among themselves through a high-
speed ring or shared bulk memory. An FP is programmed in micro-
assembly language, allowing parallelism at the instruction level.
The basic components of an FP are shown in Fig. 3. The following
list summarizes its important features:

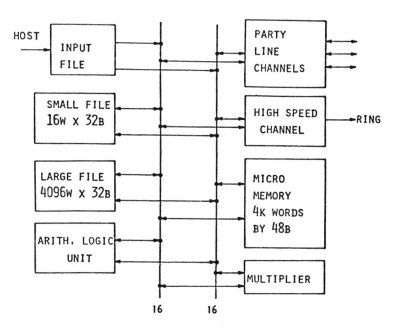

Fig. 3. Data paths in a Flexible Processor

User microprogrammable.
Dual 16-bit internal bus system.
Able to operate with either 16- or 32-bit words.
125 nsec. clock cycle.
125 nsec. time to add two 32-bit integers.
250 nsec. time to multiply two 8-bit integers.
Register file of over 8000 16-bit words.
Up to 16 banks of 250 nsec. bulk memory (each holds 64k words).
In order to debug, verify, and time FP algorithms, we developed a
micro-assembly language assembler and an FP simulator (Smith et
al., 1980).

Consider using an FP system to implement the contextual classifier based on a horizontally linear neighborhood of size three
(Fig. 1). Divide the I-by-J image into subimages of I/N rows J
pixels long. Assign each subimage to a different FP. The entire
neighborhood of each pixel is included in its subimage. Each FP
can therefore execute the uniprocessor algorithm in Fig. 2 on its
own subimage. No interaction between FPs is needed, i.e., each FP
can process its subimage independently.

The pixel measurement vectors, covariance matrices, logar-
ithms of the determinants of the covariance matrices, a priori
probabilities, and hold array are all stored in the Large File
(see Fig. 3). Execution times per pixel vary because all floating
point operations are done in the software. For the purpose of
testing the FP contextual classifier program, 30 rows of 16 pixels
were classified. Each measurement vector consisted of four 32-bit
floating point representations of 8-bit integers. The data set
consisted of a four-class subset of the Landsat data used in
(Swain et al., 1980b). To provide a basis for comparison, a similar contextual classifier was run on a PDP-11/70 over the same
test data (Tilton, 1981). It was found that lack of exponent
range in the 11/70 floating point hardware required extra handling. FP floating point algorithms are implemented in the software, so a 14-bit exponent was used to overcome this problem.
Twenty non-zero G^Ps were chosen for the benchmark tests. Running
under the above constraints, the single FP classifier took .035
sec./pixel, while the PDP-11/70 required .050 sec./pixel. If the
image data are too large to fit in the register files, bulk memory
can be used, adding, at most, 1 microsec./pixel to the classification time, assuming one 16-bit bus between an FP and its associated banks of bulk memory. A 16 FP configuration, where each FP
had its own bulk memory, would perform classifications at a rate
of 457 pixels per sec., as opposed to 20 pixels per sec. for a
single PDP-11/70. There are, of course, cost differences between
these two systems; however, the purpose here is to show the gains
made possible by a multiprocessor system.

Consider the non-linear neighborhood as shown in Fig. 1. An
FP is capable of addressing up to three channels of 16 banks of
memory. The sharing of bulk memory is a scheme that can be used
for shared data. One possible implementation is shown in Fig. 4.
Assume each FP will classify the pixels in I/N rows, using the

Howard Jay Siegel *et al.*

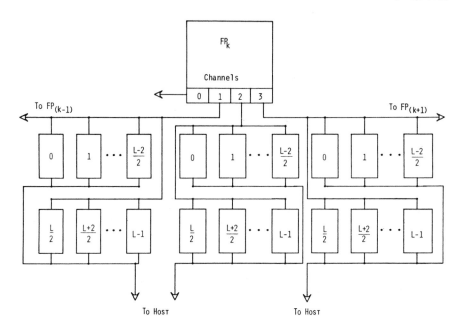

Fig. 4. FP system configuration for performing non-linear contextual classification.

algorithm in Section II. FP k will store its first $(I/N)/4$ rows in the banks shared with FP k-1, its middle $(I/N)/2$ rows in its local banks, and its last $(I/N)/4$ rows in in the banks shared with FP k+1, $0 \leq k \leq 15$. An FP processes the rows in order, moving from the banks shared with FP k-1, to its local banks, to those shared with FP k+1. This scheme gives each FP access to the data for adjacent subimages when processing its edge pixels. There will be no memory contention, i.e., at no time will two or more FPs attempt to access the same set of memory banks simultaneously. For processor k to "catch up" with processor k+1, processor k will have to process more than 75% of its data in the time it takes processor k+1 to process less than 25% of its data. Furthermore, when FP k computes the "compf" values for the pixels in the first row of its subimage, it stores these values in the memory banks it shares with FP k-1 for that FP's later use. Thus, no "compf" values are computed more than once. This scheme will allow N FPs to classify an image N times faster than one FP. An FP will be allowed to address only half of its memory banks at one time. This is done to facilitate double buffering. The other half will be accessible by the host. This permits the FP to be classifying the current image while the host unloads and stores the results of the previous classification and then loads the next image to be processed.

Timings run with Landsat data from (Swain et al., 1980b) show
that, on the average, the FP implementation of the four class,
size nine square neighborhood contextual classifier requires .137
sec./pixel. A PDP-11/70 implementation of the same algorithm
requires .154 sec./pixel (Tilton, 1981). Tests for the 11/70 were
run with 50 non-zero Gs and four spectral classes on 52 lines of
16 pixels. A 30-line by 16-pixel subset of the above image was
used to derive the FP timings for a 52-line image. Based on the
above timings, a 16-FP array can classify 116 pixels/sec., while a
PDP-11/70 can classify 6 pixels/sec.

Several points should be noted. First, the FP was programmed
in micro-assembly language while the PDP-11 was programmed in "C."
Second, the Section II algorithm could be further optimized.
Third, the efficiency of the FP micro-assembly code is a function
of the experience of the programmer.

IV. SIMD IMPLEMENTATIONS ON PASM

PASM is a dynamically reconfigurable multimicrocomputer system whose design will support as many as 1024 processors. SIMD
implementations of contextual classifiers based on PASM are discussed in this section. First, a brief overview of PASM, limited
to those aspects of PASM that are needed to understand the SIMD
algorithms in the following sections, is presented.
Fig. 5 is a block diagram of PASM. The heart of the system
is the Parallel Computation Unit (PCU), which contains N processors, N memory modules, and the interconnection network. The PCU
processors are microprocessors that perform the actual computations. The PCU memory modules are used by the PCU processors for

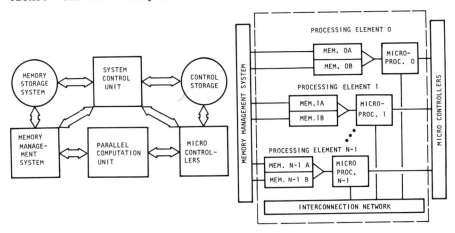

Fig. 5. Overview of PASM and structure of Parallel
Computation Unit.

data storage in SIMD mode. The interconnection network provides a
means of communication among the PCU processors and memory
modules. The Micro Controllers (MCs) are a set of microprocessors
which act as the control unit for the PCU processors in SIMD mode.
Control Storage contains the programs for the Micro Controllers.
The Memory Management System controls the loading and unloading of
the PCU memory modules. It employs a set of cooperating dedicated
microprocessors. The Memory Storage System stores these files.
Multiple devices are used to allow parallel data transfers. The
System Control Unit is a conventional machine, such as a PDP-11,
and is responsible for the overall coordination of the activities
of the other components of PASM.

The processors, memory modules, and interconnection network
of the PCU are organized as shown in Fig. 5. A pair of memory
units is used for each PCU memory module so that data can be moved
between one memory unit and the Memory Storage System while the
PCU processor operates on data in the other memory unit. This is
controlled by the Memory Management System. Each PCU memory unit
may be as large as 64K 16-bit words. First consider the SIMD
implementation of a contextual classifier based upon a horizontally linear neighborhood of size three. The approach to decom-
posing the task will be similar to that used for the FP system,
allowing N microprocessors to achieve a factor of N improvement
over a single microprocessor. There are three main differences
between the FP and SIMD implementations.

First, it is technologically feasible to construct a multimicroprocessor SIMD machine with many more than 16 processors. Second, there are differences in computational capabilities, i.e., 16
FPs may be faster than 32 microprocessors. Third, in SIMD mode,
the program is stored in the control unit (MCs), which broadcasts
it to the PCU microprocessors. The control unit also stores the G
array, decoding and broadcasting each element as needed. In the
FP system, each FP stores a copy of the program and G array.

Consider the SIMD implementation of a size nine non-linear
square neighborhood contextual classifier on PASM. The approach
taken is different from that for the FP system since the processors are synchronized and there is no directly-wired shared
memory.

The I-by-J image is divided into N subimages, each an
$(I/\sqrt{N})$-by-$(J/\sqrt{N})$ array as shown in Fig. 6. Each PE stores one
such subimage. All of the PEs execute the algorithm discussed in
Section II. Each PE can classify all the pixels in its subimage
which are not on the subimage edges. All PEs can do this simultaneously. To classify subimage edge pixels, the PEs must share
data by passing information through the interconnection network.
For example, in order for PE 0 to classify pixel $(0, (J/\sqrt{N})-1)$ it
needs to get the "compf" values for pixel $(0, J/\sqrt{N})$ from PE 1.

One way to do this is to have each PE first compute and store
the "compf" values for their edge pixels in a vector called EDGE.
(Later, when a PE needs the "compf" values for these pixels in

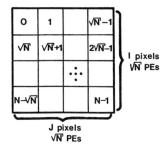

Fig. 6. Dividing an image using a "checkerboard" pattern. Each
square represents one PE with a (I/√N)-by-(J/√N) subimage. The PE number is in the square.

order to classify pixels in its own subimage, they are fetched
from EDGE, not recomputed.) Each PE sends copies of these values
to the appropriate "adjacent" PE. A PE saves the value it
receives in a vector OUTEREDGE. Each PE accesses its own
OUTEREDGE vector when it is ready to classify its edge pixels.
This method requires only $(2(I+J)/\sqrt{N})+4$ parallel data transfers.
For each of the required transfers, the networks being considered
for PASM will allow all PEs to perform the transfer simultaneously. A checkerboard division of the image was used since, in
general, it requires fewer inter-PE transfers than dividing the
image by rows or columns. For arithmetic operations and "compf"
calculations, a perfect factor of N speedup is attained. This is
done at the "cost" of $(2(I+J)/\sqrt{N})+4$ inter-PE transfers. These
data transfers are negligible when compared with the $I*J*C/N$
"compf" computations.

The FP and PASM approaches could be combined. A multimicroprocessor SIMD machine with shared memories (as in the FP
approach) and no interconnection network would be an efficient
special-purpose system for performing maximum likelihood classification and contextual classifications with various size and shape
neighborhoods.

V. CONCLUSIONS

Through the use of parallel computer systems, such as PASM
and CDC FPs, the types of computations required for contextual
classifiers and other computationally demanding remote sensing
processes can be implemented efficiently. This will not only
reduce the computation time required to do contextual classification but will also allow the investigation of techniques which may
otherwise be considered infeasible.

REFERENCES

CDC (1977a). "Cyber-Ikon image processing system concepts," Digital Systems Division, Control Data Corp., Minneapolis, MN.
CDC (1977b). "Cyber-Ikon Flexible Processor programming textbook," Digital Systems Division, Control Data Corp., Minneapolis, MN.
Haralick, R., Shanmugan, K., and Dinstein, I. (1973). "Textural features for image classification," IEEE Trans. Syst., Man, Cybern., SMC-3, 610.
Kettig, R. and Landgrebe, D. (1976). "Classification of multispectral image data by extraction and classification of homogeneous objects," IEEE Trans. Geosci. Electron., GE-14, 19.
Siegel, H.J. (1981). "PASM: a reconfigurable multimicrocomputer system for image processing," in "Languages and Architectures for Image Processing" (M. Duff and S. Levialdi, ed.), Academic Press, London.
Smith, B., Siegel, H.J. and Swain, P. (1980). "A multiprocessor implementation of a contextual image processing algorithm," LARS Tech. Report 070180, LARS, Purdue Univ., W. Lafayette, IN.
Swain, P. and Davis, S. (1978). "Remote Sensing: The Quantitative Approach," McGraw-Hill, Inc., New York.
Swain, P., Siegel, H.J., and Smith, B. (1980a). "Contextual classification of multispectral remote sensing data using a multiprocessor system," IEEE Trans. Geosci. and Remote Sensing, GE-18, 197.
Swain, P., Vardeman, S., and Tilton, J. (1980b). "Contextual classification of multispectral image data," LARS Contr. Report 011080, LARS, Purdue Univ., W. Lafayette, IN.
Tilton, J. (1981). "PDP-11 contextual classifier timings," unpublished report.
Tilton, J., Swain, P., and Vardeman, S. (1981). "Contextual classification of multispectral image data: an unbiased estimator for context distribution," 1981 Symp. Machine Processing of Remotely Sensed Data.
Welch, J. and Salter, K. (1971). "A context algorithm for pattern recognition and image interpretation," IEEE Trans. Syst., Man, Cybern., SMC-1, p.24.
Weszka, J., Dyer, C., and Rosenfeld, A. (1976). "A comparative study of texture measures and terrain classification," IEEE Trans. Syst., Man, Cybern., SMC-6, 259.

TEMPLATE-CONTROLLED IMAGE PROCESSOR (TIP) PROJECT

Shin-ichi Hanaki
Tsutomu Temma

C&C Systems Research Laboratories
Nippon Electric Co., Ltd.
Kawasaki-city, Japan

I. INTRODUCTION

One of the digital image processing applications, remote sensing, requires more and more data processing capabilities both in data amount to be processed and in processing speed (Hanaki, 1980). Data amount for one scene in remote sensing is large in comparison with that of imagery in other application fields. For example, data amount for one scene reaches up to 30 Mbytes in four band LANDSAT multispectral scanner (MSS) imagery.

Also, data amount daily obtained is rapidly increasing. Several new earth observation satellites are planned in a few years, which will carry imaging sensors, for example, LANDSAT-D, SPOT, MOS-1, ERS-1(Japan), ERS-1(European Space Agency), SEO-2 (India) etc. Total data amount obtained from satellites in one day is estimated to increase up to 10^{13} bits/day during the 1980s (Strong et al., 1979).

On the other hand, processing speed up requirements have, so far, emerged from, primarily, man machine interactive environment for digital image analysis, in which shorter systems response time for processing image is the main concern for a smoother man machine communication (Hanaki, 1979).

Recent development in synthetic aperture radar (SAR) digital image processing has brought about a new field where a tremendous amount of calculations are necessary to obtain the final earth surface radar reflection image, which is similar to an aerial photograph (Bennet and Cumming, 1979; Nohmi et al., 1980). For example, data processing amount for generating a 100x100 Km SEASAT/ SAR image (4000x4000 pixels) is estimated to be about $4x10^{10}$ floating operations. It implies that, at least, 40,000 seconds (about 11 hours) are required to generate a SAR image with a processor having a processing capability of 1 MFLOPS (million

MULTICOMPUTERS AND IMAGE PROCESSING
ALGORITHMS AND PROGRAMS

folating operations per second) (Hanaki et al., 1980).

Processing speeding up efforts have been devoted to several aims.   One of them, processor speeding up for image display with simple image processing capabilities, has succeeded in realizing video speed processing which completes such operations as density transformation, pseudo color conversion etc. in one TV frame time (about 33 ms) for a 512x512 pixel image (Sorimachi et al.,1981). Most of these video speed processing are achieved by hardwired processors.   Incessant requirements, however, for new kinds of image processing functions to these image display processors have led to programmable high speed image processor development.

Similar needs have emerged from satellite image data processing.   It is not an easy job to process even LANDSAT/MSS data with a conventional general purpose computer.   Moreover, new image sensors are expected in a few years, which require more data processing than present LANDSAT/ MSS, for example, thematic mapper (TM) for the future LANDSAT-D will have 85Mbps data transmission rate, compared with 15Mbps for the present LANDSAT 1 to 3 MSS.

A harder task is to digitally produce such a radar image as SEASAT synthetic aperture radar (SAR) from raw data.   SEASAT/ SAR raw data are equivalent to a microwave hologram.   Reconstructing a radar image digitally from SAR raw data has been achieved by several organizations independently (Bennet and Cumming, 1979; Nohmi et al., 1980).   The total amount of necessary operations for single SAR image production from SEASAT raw data is about 4×10^{18} floating operations for 4000x4000 pixel output image.   Digital image processing for these new imaging sensors also requires a high speed programmable processor.

II. IMAGE PROCESSING SYSTEMS

Several LANDSAT image processing and analysis systems are now commercially available.   For example, a blockdiagram of a stand-alone image processing system NEC N7835 is shown in Fig.1. The system has "T" configuration, in which image data are stored in the image memory (2 Mbytes MOS-IC memory) and are processed by special hardware image processors.   The host minicomputer controls each system component, handles man-machine-communication control and manages data.   An image memory and image processor combination handles image data under the host computer control. The computer avoids processing image data, either in its own memory or in its arithmetic-logic-unit, so that the entire system should maintain a high efficiency for image data processing. The processed results obtained in the image memory can be displayed on a monitor in order to aid human comprehension of image data and judgement on processing effectiveness.   In display mode, the image memory is used as a display refresh memory (Hanaki, 1979; NEC, 1980).

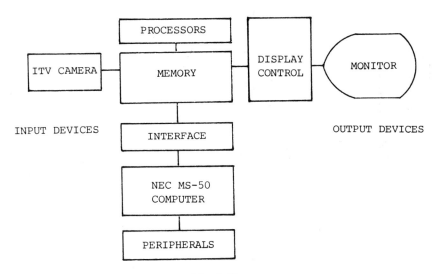

FIGURE 1. *N7835 system blockdiagram.*

III. DATA DRIVEN IMAGE PROCESSOR

In order to improve hardware processing capabilities, the
pipeline technique and parallel implementation of operation
modules are usually employed.   The pipeline processor has disadvantages, in that it is not flexible to match processing change
and must be synchronized at all pipeline stages.

Template-controlled or data-driven architecture resolves these
problems (Dennis and Misunas, 1975; Plas et al., 1976; Davis,
1979).   Also, operation module parallel implementation takes
place easily to increase the processing speed.

As shown in Fig.2, a template-controlled image processor (TIP)
is composed of control unit, addressor unit and operational unit
(Temma et al., 1980; Temma et al., 1981).   The control unit
contains a bit sliced microprocessor, which manages both the
operational unit initialization and the addressor unit initialization and starts the addressor unit.

The addressor unit contains generator, reader, writer, bit-
operator, distributor, concatenator and ring interface, all of
which are individually connected to the address ring through bus
interfaces.   The addressor unit accesses the memory (it is planned to use N7835 system MOS-IC image memory for this purpose),
picks up the data and supplies them via main ring to the operational unit.

The operational unit receives the data, performs indicated
calculations and sends back the resultant data to the addressor
unit.   The resultant data are written back into the memory via
writer according to the parameter given during system initialization.

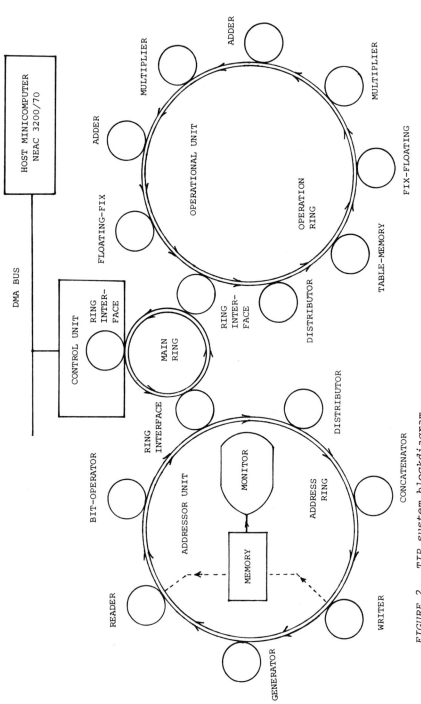

FIGURE 2. TIP system blockdiagram.

A datum consisting of destination flags, identifier and token
flows through address ring, main ring and operation ring.
Destination flags are used in the destination module selections.
Each module decides by destination flags whether to accept the
datum or not.   If the datum is accepted, then a template is found
which corresponds to the datum by identifier.   Then, the module
executes the computation using the token and outputs a datum which
consists of new destination flags, a new identifier from the
identifier-table and result token.   Each datum is transferred
from one module to the next module in a clock cycle through the
ring bus and disappears when all destination flags are off.

      There are two operation modes; system initialization mode and
data processing mode.   In the system initialization mode, a set
of templates to control data flow and image data definitions
declared by the user are transferred from the host computer to the
control unit, where they are interpreted and sent to each module.
After initialization is completed, a start command is given from
the host computer to the control unit and the data processing mode
starts.

      In the addressor unit, modules are connected to directional
address ring bus.   The generator generates a data stream sequence
according to the parameters given in the initialization mode.
The reader reads data from the memory by using an input token as a
memory address.   The concatenator generates a new token from two
old tokens, for example, it generates a memory address from a
known (x,y) coordinate pair.   The bit operator carries out bit
shift, bit reversing and bit data masking operations.   In order
to control the data stream, the distributor modifies arriving data
identifiers according to the rule.

      The operational ring contains such modules as distributor,
table memory for table look up operation, two adders, two multipliers, a pair of modules achieving transformation between floating point data and the fixed point data and ring interface to the
main ring.

      A ring interface is composed of two bus interfaces.   Each
module, except the ring interface, has a bus interface and a
pipeline operation element.

      There are two kinds of bus interfaces.   One is used for one-
operand-operation module.   Another is used for two-operand-
operation module.   The latter bus interface is the same as the
former, except for a queue control and a data queue, as shown in
Fig.3.   When a one-operand-operation module accepts a datum, its
execution starts immediately.

      If two operand tokens are provided, the queue control loads
the token from the data queue and sends it to the pipeline operation element with the input token.   Otherwise, it stores the
input token in the data queue.

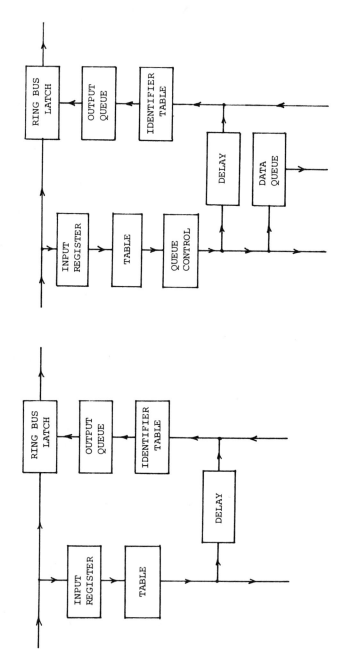

FIGURE 3. Bus interfaces; one-operand operation module interface
(left) and two-operand operation module interface (right).

IV. DATA FLOW EVALUATION (Temma et al., 1981)

TIP's processing speed capabilities can be discussed based upon the following items:
(1)   Time required to initially set each processing module.
(2)   Number of data inputs to or outputs from each processing module.
(3)   Amount of data passing through the ring bus between processing modules.
(4)   Pipeline cycle time for each processing module.
(5)   Time required to obtain processed result data response after the last data were input.
(6)   Number of busy ring buses, which may occur when bus transfer cycle time is shorter than module pipeline cycle time.

Effective pipeline route is determined at the end of the initialization mode for TIP.   Times described in above items (1) and (5) are negligible, in comparison with total processing time. Factors (2) and (3) can be analyzed by the data flow graph technique described in Section V.   Busy ring bus, described in (6), is difficult to analyze.   It might be examined through computer simulation.

V. DATA FLOW GRAPH (Temma et al., 1981)

Let us consider an example of Fast Fourier Transform (FFT) and the corresponding flow graph representation.
Fundamental butterfly operation for the FFT algorithm is given by the following expressions.

$$X_R = A_R + B_R \cdot W_R - B_I \cdot W_I$$

$$X_I = A_I + B_R \cdot W_I + B_I \cdot W_R$$

$$Z_R = A_R - B_R \cdot W_R + B_I \cdot W_I$$

$$Z_I = A_I - B_R \cdot W_I - B_I \cdot W_R$$

(1)

A flow graph corresponding to Eqs.(1) is shown in Fig.4.   For explanation clarity, only the operational unit is discussed.
In Fig.4, pentagonal terminals represent input/output tokens to/from the operational unit.   The circular node represents processing module.
In an ideal situation, a complete pipeline could be built, provided that 6 adders and 4 multipliers were avialable with buses which have sufficient transfer capacities.
In the real TIP shown in Fig.2, however, all calculations

should be executed with 2 adders and 2 multipliers.   Thus, it is
necessary to allocate a limited number of resource modules to
graph nodes in Fig.4.

Assume that bus transfer time is t, pipeline cycle time in
multiplier is 3t and pipeline cycle time in other modules than
multipliers is 2t.   Figure 5 shows data flow with the operational
ring, which can be built from Fig.4, as follows.

In Fig.4, the number on the right beneath each node indicates
the template number.   For example, in Fig.4 the pentagonal terminal for B_R has template number 0.   This token B_R should be transferred to module x1 and x2, respectively.   In Fig.5, this data
flow is represented by the arrows in the top, which show the token
is transferred from the ring interface to both x1 module and x2
module with template number 0.   In the same manner, other arrows
in Fig.5 can be drawn from the flow graph in Fig.4.

From Fig.5, it is possible to count the number of data transfers between modules, which is indicated by numbers with the mark
*.   For example, there are 10 data transfers from module x1 to
module +1.   Also, the number of data inputs to / outputs from
each module can be counted, which is indicated by symbols △ and o,
respectively.

Figure 5 shows that the largest number of data transfers is
10, which requires time 10t/BOP, where BOP represents butterfly
operation.   The longest data input time is 12t/BOP, which occurs
at x1, x2, +1 and +2 input.   The longest data output time is

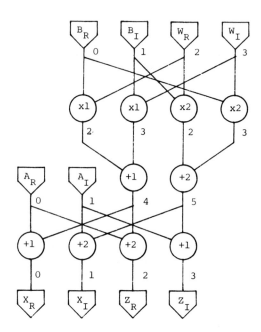

FIGURE 4.   Flow graph for butterfly operation.

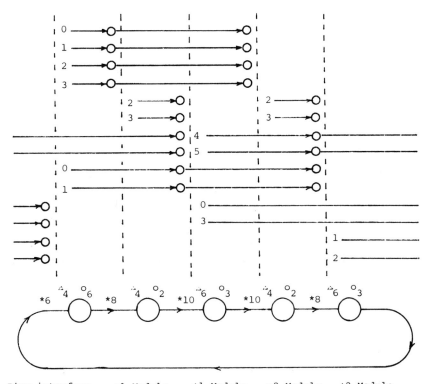

Ring interface   x1 Module   +1 Module   x2 Module   +2 Module

FIGURE 5.   Data flow amount.

12t/BOP, which occurs at ring interface output.   Thus, both data
input time to modules x1, x2, +1 and +2 and data output time from
the ring interface limit the total processing speed.

If t is assumed as 50ns, then 600ns are required to complete
one butterfly operation.

VI. PROJECT STATUS

Computer simulation is undertaken to get a better understanding of data flow at processing modules and on the ring bus.

Hardware implementation is going on, in which data width is 24
bits (8 bits for exponent and 16 bits for mantissa).   At the
moment, project resources are concentrated to build TIP hardware.

A very small amount of software has been developed, including
control unit cross assembler and coding support software for
operational and addressor unit programming.   Programming is
carried out with machine level language, using a similar technique
to that described in Section V.

VII. CONCLUSION

Requirements for image processor development were discussed, both from data amount increase and processing speeding up viewpoints.

As an improvement in pipeline technique that is conventionally employed for data series processing e.g. image data processing, a template-controlled image processor, named TIP, was described.

The TIP project is going on, concentrating resources on hardware implementation.

After basic hardware is completed, software development will be undertaken, which will require tremendous effort and include a high level language to control this data driven image processor. This would be one of the major challenge targets in the TIP project.

The authors would like to express their thanks to TIP project members and Mr. M. Ogiwara for his continuous encouragement in the project.

REFERENCES

Bennet, J. R., and Cumming, I. G., (1979). *Proc. 13th Intn'l Sympo. on Remote Sensing of Environment,* p.337. ERIM, Ann Arbor.

Davis, A. L. (1979). *Proc.NCC,* p.1079.

Dennis, J. B. and Misunas, D.P., (1975). *Proc. 2nd Ann. Sympo. on Computer Architectures.* p.126

Hanaki, S. (1979),(J). IP-23-3. IPSJ.

Hanaki, S. et al. (1980). *Proc. of the 1st Asian Conference on Remote Sensing,* JARS, Tokyo.

NEC, (1980). Cat. No. J55351.

Nohmi, H. et al. (1980), (J). SANE80-24. IECEJ.

Plas, A. et al. (1976). *Proc. 1976 Intn'l Conf. on Parallel Proc.*

Sorimachi, Y. et al. (1981). NEC R&D *60,* 26.

Strong, J. P. et al. (1979). *Proc. 13th Intn'l Sympo. on Remote Sensing of Environment,*

Temma, T. et al. (1980). *Proc. ISMM intn'l Sympo. MIMI'80 Asilom,* Vol.5,3. ACTA PRESS, Anaheim.

Temma, T. et al. (1981), (J). IE81-6, IECEJ.

MEMORY STRUCTURES FOR AN IMAGE PROCESSING SYSTEM

Mitsuo Ishii
Yasushi Inamoto

Fujitsu Laboratories Ltd.
Kawasaki, Japan

1. INTRODUCTION

Image processing systems are being used effectively in a variety of fields. Such systems include computer graphic systems for the layout and routing design of LSIs, three-dimensional graphics for mechanical engineering, image processing systems for feature extraction and pattern recognition, and simulation systems for coding of images. For research which deals with images, convetional computer systems are inadequte which results in demand for a system specially designed for image processing. Such image processing systems must be capable of realtime input and output of large quantities of image data, and interactive processing by visual checking images on monitors. The quality of these functions significantly influences the efficiency of research and its outcome.

We have developed a general-purpose image processing system as a tool for research which requires various kinds of image data processing. The development goal was to produce a system capable of a variety of functions, which can handle various images ranging from binary graphics to continuous-level natural images, and from a single large drawing to scores of TV images. The following sections describe the design goals, the hardware configuration, and the structure of the image memory which is the core of the system.

2. DESIGN GOALS

One special characteristic essential in research which requires processing of images is that, in making decisions, the operator must be able to visually check images displayed on the CRT screen. The results of feature extraction and image coding as

MULTICOMPUTERS AND IMAGE PROCESSING
ALGORITHMS AND PROGRAMS

well as graphics, must be evaluated subjectivly. Hence the
human-computer interface is especially important. Processing image
data is normally time-consuming due to its large volume: if
ongoing processing can be visually checked, unnecessary subsequent
processings can be eliminated.

With these considerations in mind, we concentrated our efforts
on developing a versatile system which would achieve the following
goals:

    (1) Graphics functions with high quality image display
    (2) Sophisticated interactive processing
    (3) Input and output of television signals in realtime
    (4) Motion picture display
    (5) Easy expansion to combine with other special-purpose devi-
ces such as array processors.

While conventional systems are dedicated to a particular purpose, graphics, images, or video signals, our system is capable of
processing all of them. A versatile system which can accept TV
images and graphics has two advantages. First, it allows efficient development of research which requires image processing since it widens the scope of the research. Secondly, the system is
cost-effective in comparison with dedicated systems.

To suit each application, the image memory configuration can
be varied by mode signals. Thus, the large-capacity image memory
is the core of the system.

3. HARDWARE

The hardware configuration of the system is shown in Fig.1.
The conversational devices are the keyboard, tablet, and trackball. The system is also equipped with two monitors, one for NTSC
color television signals, the other for RGB non-interlaced signals. There are three converters each for A/D and D/A used for
input and output of video signals, a large-capacity image memory,
lookup tables, function memories, and a built-in processor
(i-8086). The processor controls the entire system, generates
images such as vectors and circles, and fills in areas. In addition, there is a character display with a lightpen .

There are three switch-selectable interfaces between system
and host computer: the bus connection with the host mini-computer,
modem interface used under the TSS environment of a large scale
computer, and a current loop interface which uses optical fiber
cables.

The memory also has an interface for input and output of 8-bit
parallel data so it can easily be connected with conventional
digital equipment such as an array processor.

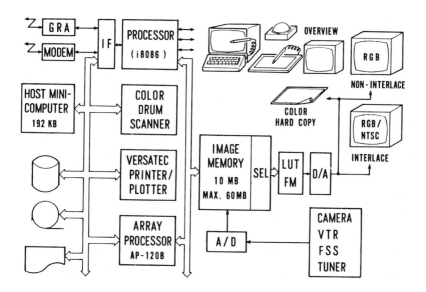

FIGURE 1.   Hardware configuration.

4. IMAGE MEMORY

The image memory consists of 64 cards of 16K RAM giving a capacity at present of 10M bytes: capacity can be expanded to 60M bytes.   The image memory is also used as a refresh memory for CRT display.   It functions in parallel with accesses from the processor.   One of the special characteristics is two addressing modes to suit different applications and image sizes.   The Graphics or G-mode is used for 5120 x 4096 picture elements (pixels), the Image or I-mode is used for 640 x 512 pixels.   Fig.2 shows the concept of memory structures.   A one-bit plane is known as a channel, and 8 channels make up a bank.   The system handles channel by channel, bank by bank, or, for color images, in units of 3 banks.

Additional memory increases the number of channels: for the maximum capacity of 60M bytes, channels can be expanded from 4 to 24 for G-mode, and from 256 to 1536 for I-mode.

5. GRAPHICS MODE

G-mode is suitable for interactive processing of large patterns.   640 x 512 pixels out of the large image size, i.e. 5120 x 4096 pixels, are displayed at a rate of 57 frames/sec on the non-

interlaced monitor. The panning function allows the operator to
roam freely throughout the display area in realtime by means of
the trackball.

Each pixel consists of 4 bits, and the lookup table (LUT) and
function memory (FM) permit any 16 desired colors out of 2**24 to
be selected. A typical image obtained by our scanner consists of
1600 lines of 2240 samples per line, and 8 bits of luminance information for each pixel. If a pattern is too large to be completely displayed at one time, it is normally divided up and
displayed in sections. This panning feature permits realtime
display of the divided image sections thus significantly improving
operability and response time in interactive processing.

To indicate which (640 x 512) part of the (5120 x 4096) entire
image is being displayed, the overview monitor shown in Fig.2 is
under development. This monitor is a graphic terminal with a
resolution of 1024 x 1024 which displays a rectangle indicating
the entire image and which contains a small rectangle indicating
the displayed area. It can also compress 5 x 4 pixels into one
pixel and display the outline of an entire pattern with an over-
layed rectangle to indicate the displayed area. The entire image
is updated approximately every ten seconds .

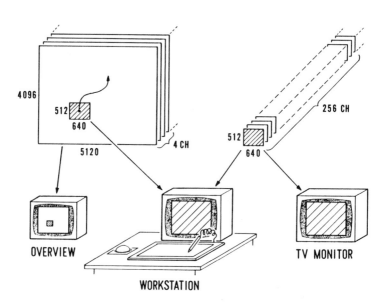

FIGURE 2. Memory structures.

6. IMAGE MODE

The I-mode shown on the right in Fig.2 is suitable for display of picture manipulation. In this mode, an image occupies an area of 640 x 512 pixels. Depending on the requirement, 1, 8, or 24 bits per pixel can be selected. A special characteristic of this mode is the way in which channels are used. 256 channels can handle 256 1-bit images, 32 8-bit images, and 10 red, green, and blue color component images. The channel and bank selector selects individual sequential images so that pictures can be manipulated. Standard frame rates of 57 (non-interlaced) and 30 (interlaced) are available. In addition, for forward or backward slow motion display, frame repetitions of 2, 4, 8, 16, 32 or 64 are selectable.

As shown in fig.3, the system provides with advanced display features such as zooming, panning, pseudo and false coloring, and linearity compensation by function memories. There are also three min-max registers, and overlay gates for three overlay channels. All these functions can be controlled in realtime by changing the contents of RAM in the directly addressable memory space of the i-8086.

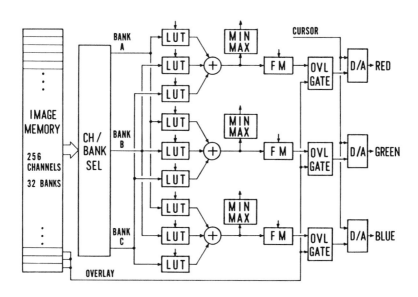

FIGURE 3.   Video output schematics.

7. TV SIGNAL INPUT/OUTPUT

Fig.4 is schematic of television signal input and output. Television signals are sampled at 8 bits per pixel which means composite and 3-component signals can be handled. Output signals from memory go to the D/A converters either directly or through LUTs and FMs. The memory can store up to 32 consecutive frames at a rate of one frame per bank for NTSC composite signals. For color component signals, it can store up to 10 consecutive frames. The sampling frequency can be varied up to 14.3 MHz (4 fsc). Fig.5 shows the sampling range for composite signals. The signals are sampled from the beginning of each horizontal line until the

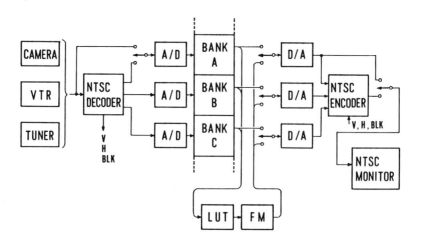

FIGURE 4.   Input/output of TV signals.

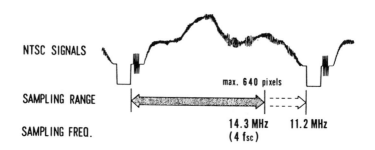

FIGURE 5.   Sampling range for composite signals.

maximum of 640 pixels, which includes color bursts, is reached, thus enabling color reproduction. The variable sampling frequency is a significant feature because sampling rate is an important design parameter in coding simulations of bandwidth compression schemes. Ability to handle consecutive frame signals also permits statistics on differential signals between frames to be taken and various data required for reseach on interframe coding to be collected. In addition, subjective evaluation of motion pictures, which was not easy in the past, can be done for one second, or six seconds if the maximum memory is installed.

8. MEMORY ACCESS TIME SHARING

The system handles four different kinds of signals, non-interlaced refresh data, interlaced refresh data, random access data from the processor, and realtime television signals. Fig.6 shows time sharing for memory access for these signals. The top is the normal clock operation and the bottom is a variable clock operation.

Time slots for displaying and processing are assigned so as not to influence each other, therefore the images on the monitors are always clearly visible and completely stable. Looking at the screen enables checking of intermediate results without having to wait for the final results over an entire image.

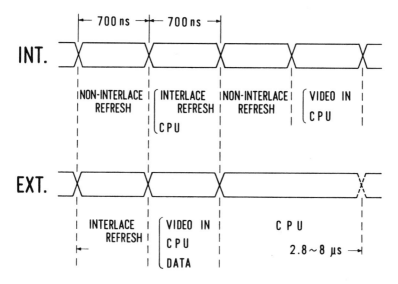

FIGURE 6. Memory access time sharing.

9. CONCLUSION

The system is intended as an interactive CAD system for the design of electronic circuits and as a simulation system for image processing. Our efforts were concentrated on developing hardware with a large image memory as its core and the basic structures of the memory have been confirmed. As for software, the development of device interfaces and system service routines is underway. System evaluation is scheduled using several applications programs.

ACKNOWLEDGMENTS

The authors would like to express thier gratitude to Dr. K.Kurokawa for his encouragement and Mr. S.Sasaki for his valuable advice.

REFERENCES

Adams, J., and Wallis, R. (1977). Computer, Vol.10, No.8, pp.61-69.
Latta, J. (1978). Image Processing Application Note, COMTAL, Pasadena.
Troxel, D. E. (1981). IEEE Trans. Pattern Analysis and Machine Intelligence, Vol.PAMI-3, No.1, pp.95-101.

MULTIMICROPROCESSOR SYSTEM PX-1
FOR PATTERN INFORMATION PROCESSING

Makoto Sato[1]
Hiroyuki Matsuura[2]
Hidemitsu Ogawa
Taizo Iijima

Department of Computer Science
Tokyo Institute of Technology
Tokyo, Japan

I. INTRODUCTION

In pattern information processing, such as digital image processing and pattern recognition, the amount of handled data is ex-
ceedingly large. So it is very difficult for conventional computers to perform real-time processing of pattern information.

Recently there have been done a lot of researches and developments of special hardwares for pattern information processing,
especially for digital image processing. Most of these hardwares
aim at parallel processing of local process of digital images,
which may be effective in particular for preprocessings of images.
But for more sophisticated processing such as labelling process by
relaxation and image understanding,they are not necessarily effective. More general purpose processor, which is possible to treat
various numerical analyses and optimization problems, is required.

Parallel processors, such as Illiac IV, aimed at fast processing of numerical analysis, have also been studied. Most of these
processors are very large in scale, since numerical operations
with high accuracy are required. In pattern information processing,
high-powered processing ability of numerical operations is necessary. But the accuracy of the operations need not be so high. Hence

[1]*This work was supported by the Grand-in-Aid for Scientific
Research, No.342030, of the Ministry of Education, Japan.*
[2]*Present address: Yokokawa Electric Works, Ltd., Tokyo, Japan.*

361

it is desirable to develop a not only powerful but also compact
processor suited to the demands of pattern information processing.

PX-1 is a multimicroprocessor system aimed at real-time processing of pattern information. The performance of microprocessors
is not so high. Therefore, in order to achieve real-time processing ability, it is necessary to construct an architecture suited
to the principal style of pattern information processing. In section II, we show the outlines of the basic design. The architecture of PX-1 is shown in section III. In section IV, the performance of PX-1 is estimated.

II. DESIGN OUTLINE

In this section, we will show the outline of the basic design
of PX-1.

A. *Local Process and Global Process*

Pattern information processing can be divided, for the most
part, into a couple of processing types, local process and global
process. For instance, preprocessing of digital images and computation of similarities in pattern recognition belong to the local
process. Global feature extraction of images and discrimination in
pattern recognition belong to the global process. In general, the
local process is not difficult to handle in parallel. As the dimension of pattern information become larger, the amount of the local
process increases rapidly and therefore the effect of parallel
processing grows remarkably.

The system PX-1 is a multimicroprocessor composed of one parent-system and dozens of child-systems. The local process is executed by the child-systems in parallel. The parent-system controls
the child-systems and execute the global process.

B. *Flow of Information*

Generally, the problem of data access is one of the primary
causes of lowering the efficiency of parallel processing. Therefore, it is required to connect the parent-system and the child=
systems in order to smooth the flow of data.

In pattern information processing, there often arise problems
of the form that the local process and the global process are
repeated by turns as shown in FIGURE 1. At the change of the local
process and the global process, a large amount of common data must
be transfered among the parent-system and the child-systems. Therefore a data bus, which transfer a lot of data among the systems,
is required.

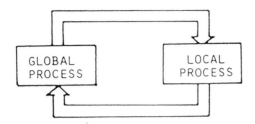

FIGURE 1. Local process and global process.

The child-systems perform the local process in parallel independently. However, it sometimes happens to use information of the adjacent child-system or to exchange information between a couple of child-systems. In order to give flexibility to the parallel processing by the child-systems, it is also desirable to have a data bus which make it possible to transfer local data between child-systems in various forms.

C. Flow of Control

Repetition of global processing by the parent-system and local processing by the child-systems is the basic flow of pattern information processing. The parent-system controls the flow of processing. After transfering necessary data to each child-system, the parent-system activates all the child-systems. Each child-system performs a given task independently. When the child-tasks have been finished, the control is returned to the parent-system. Here a couple of modes are considered (see FIGURE 2):

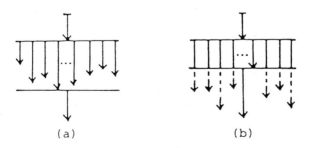

(a)   (b)

FIGURE 2. Two modes of control flow.

(a) When all the child-systems have finished their tasks, the control is returned to the parent-system. After gathering the results from all the child-systems, the parent-system proceeds to the next global processing.

(b) When one of the child-systems has finished its task, the control is returned to the parent-system. The parent-system breaks off the tasks of the other child-systems and proceeds to the next processing.

It is required to have a control structure, which is possible to get fast the necessary information for the control of the above modes.

D. *Basic Arithmetic Operations*

To make real-time processing of pattern information possible, basic arithmetic operations, such as addition, multiplication and product-accumulation, must be executed sufficiently fast.

As an example, let us take a filtering problem of 512x512 digital images by two-dimensional Fourie transform. To perform this problem in real-time (e.g., in a second), it is required to execute about $2x10^7$ times of multiplications in a second.

Generally it is necessary to perform multiplications of more than 10^6 times in a second. In the system PX-1, ordinary microprocessors are used as the central processors of the child-systems. The performance of microprocessors is not so high. It takes about 100 μs to execute a single multiplication of 8 bits data. Therefore in order to achieve a sufficient performance of arithmetic operations, it is necessary to equip a child-system with a special hardware for arithmetic operations.

III. ARCHITECTURE OF PX-1

FIGURE 3 shows the block diagram of the system PX-1. This system consists of one parent-system and 32 child-systems.

The parent-system is a computer system composed of CPU(Z-80) unit, memory unit(64kbyte), floppy disk drive, CRT terminal and so on. In addition, the parent-system has special function units to ˎcontrol child-systems and to execute data transmission. The parent-system is eqipped with fast product-accumulator and data search module for global processing. Data search module is able to find out, for instance, the maximum value in a certain data area quickly.

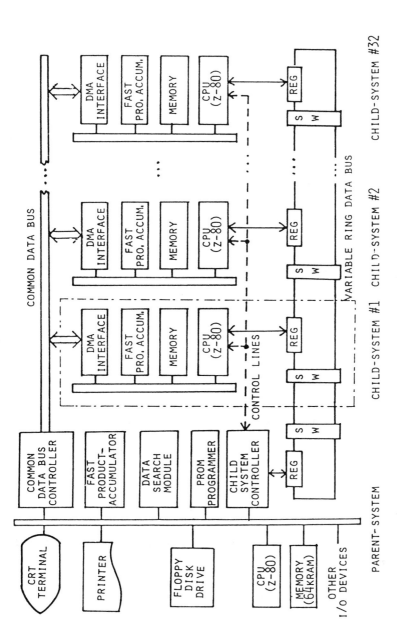

FIGURE 3. Block diagram of PX-1.

A child-system is composed of CPU unit, memory unit and interface module for data transmission. In addition, a child-system is equipped with fast product-accumulator making use of special LSI (TDC 1008J,TRW) for signal processing.

Fast product-accumulator has the ability to execute product-accumulation of 8 bits data very fast. The processing time $T(k)$ of product-accumulation of k-dimensional vectors is given as

$$T(k) = 1.5 \ k \ + 17.0 \ \ \mu s \ .$$

Let $k = 256$. Then the processing time is about 400 μs.

PX-1 has two types of data bus, that is, common data bus and variable ring data bus.

Common data bus joins the parent-system and the child-systems together in order to transmit global data among them. Making use of DMA method by special controller, high speed data transmission is possible. As transmission of global data, various modes can be considered. In the system PX-1, nine kinds of transmission modes are adopted taking accout of the usefulness of transmission and price performance. These modes are illustrated in FIGURE 4. Most of pattern information processing can be performed by combining these modes.

Variable ring data bus for transmission of local information is a ring bus connecting all the registers of the child-systems into ring form as shown in FIGURE 5, (1). At transmission time, data in a register is shifted to the next register under the control of the parent-system. The ring data bus has ring switches between registers, which make it possible to take various ring structures as shown in FIGURE 5, (2). Some examples of ring structure are shown in FIGURE 6. The variable ring data bus is suitable to transmission of local information among child-systems in various modes.

Finally, we show the control lines of PX-1.
The parent-system activates all the child-systems by interrupt signals. There are two kinds of control signals from the child-systems to the parent-system as shown in FIGURE 7. One is the AND signal of normal terminations of all the child-systems and the other is the OR signal of abnormal terminations. The former informs the parent-system that all the child-systems have finished their own tasks and the latter informs that an abnormal termination arose in one of the child-systems. By these signals, it is possible to control the two modes of child-system control shown in FIGURE 2.

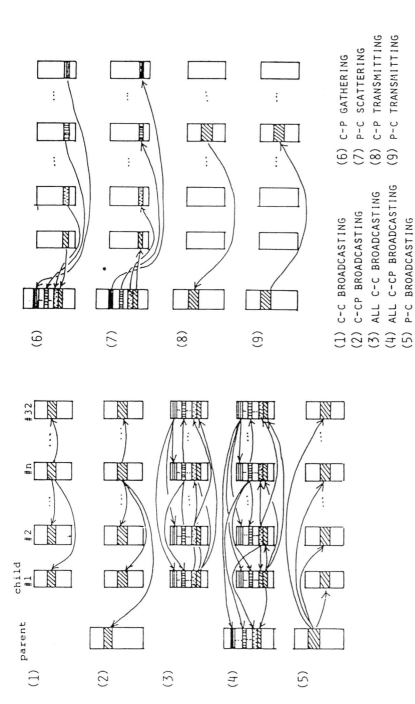

FIGURE 4. Transmission modes of common data bus.

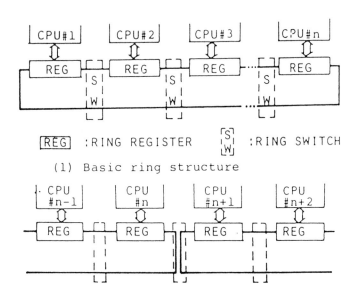

[REG] : RING REGISTER   ⌈S⌉ : RING SWITCH
                         ⌊W⌋

(1) Basic ring structure

(2) Ring cutting by ring switch

FIGURE 5. Variable ring data bus.

IV. PERFORMANCE OF PX-1

As an example of repetitive processings, we consider a numerical solution of linear equation by Gauss' iteration method.
The iterative solution of a linear equation $Ax = b$ is given
by

$$x_{n+1} = x_n + (b - A x_n) \quad , \qquad n = 0,1,2,\dots ,$$

when all the diagonal elements $\{ a_{ii} \}$ of matrix A are normalized
to the unity. Starting from any initial value x_0, the approximate
solution x_n is given by iterating the above equation.
Let the dimension of the equation be 128. In each iteration,
a child-system computes the values of 4 (= 128/32) components of
the vector x_{n+1} from the vector x_n. Gathering these values from
all the child-systems, the parent-system makes the vector x_{n+1} and
returns it again to all the child-systems. The operations required
in one iteration are: (i) four product-accumulations of 128-dimen-
sional vectors by fast product-accumulator and some additions and
substractions, (ii) one C-P gathering and one P-C broadcasting by
common data bus and (iii) some control operations for iteration.
Table I illustrates the actual times of the operations and their

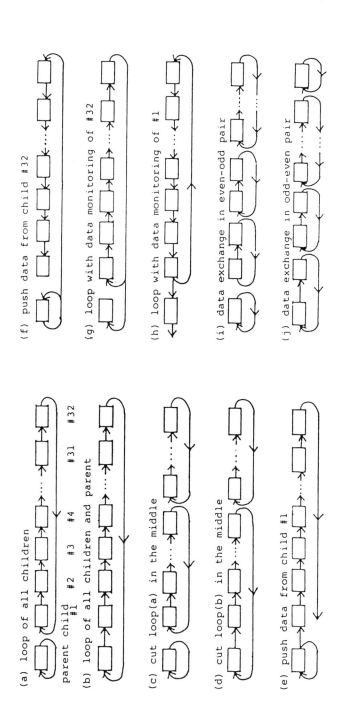

(a) loop of all children

parent child #1 #2 #3 #4 #31 #32

(b) loop of all children and parent

(c) cut loop(a) in the middle

(d) cut loop(b) in the middle

(e) push data from child #1

(f) push data from child #32

(g) loop with data monitoring of #32

(h) loop with data monitoring of #1

(i) data exchange in even-odd pair

(j) data exchange in odd-even pair

FIGURE 6. Examples of connection of variable ring data bus.

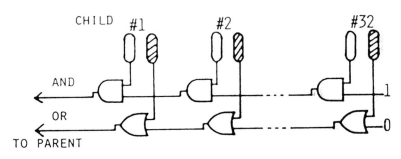

FIGURE 7.   *Control signals from child-systems to parent-system.*

ratios.

It takes about 1 ms for one iteration. Suppose that the matrix A is diagonally dominant and the solution converges within 50 times of iteration. Then it is possible to solve 128-dimensional linear equation 20 times a second.

Table I. *Comparison of Times in Iteratioe Method*

Operation	Time(μs)	Ratio(%)
Product-Accumulation	822	82
Addition and Substraction	20	2
Gathering and Broadcasting	153	15
Iteration Control	7	1
Total	1002	100

V. CONCLUSIONS

Multimicroprocessor system PX-1 is designed and developed for real-time processing of pattern information. Though PX-1 is a MIMD machine based on ordinary microprocessors, this system has the sufficiently fast processing ability. The principal features of the system PX-1 are: (i) two types of data bus, common data bus and variable ring data bus, that make it possible to transfer data fast between the parent-system and the child-systems in various modes, (ii) fast product-accumulator, which reinforces the processing ability of arithmetic operations of child-systems, and (iii) simple and efficient control structure for repetitive processing of local process and global process.

ACKNOWLEDGMENT

The authors would like to thank M.Akagi, T.Sakai and M.Kawamoto for the contribution to the development of PX-1.

A SYSTOLIC 2-D CONVOLUTION CHIP[1]

H. T. Kung[2]

S. W. Song[3]

Department of Computer Science
Carnegie-Mellon University
Pittsburgh, Pa. 15213

I. INTRODUCTION

With recent technological advances in the VLSI circuitry, the chip capacity (or component count on a chip) is increasing at an astonishing rate (7). Both the opportunities and challenges regarding effective use of VLSI are tremendous.

[1]This research was supported in part by the Office of Naval Research under Contracts N00014-76-C-0370, NR 044-422, and N00014-80-C-0236, NR 048-659, in part by the National Science Foundation under Grant MCS 78-236-76, and in part by the Defense Advanced Research Projects Agency under Contract F33615-78-C-1551 (monitored by the Air Force Office of Scientific Research).

[2]Currently on leave from Carnegie-Mellon University at ESL's Advanced Processor Technology Group in San Jose, California. (ESL is a subsidiary of TRW.)

[3]Supported in part by CNPq, Conselho Nacional de Desenvolvimento Cientifico e Tecnologico, Brazil, under Contract 200.402-79-CC, and is on leave from the Institute of Mathematics and Statistics of the University of Sao Paulo, Brazil.

Systolic design is an architectural concept proposed for VLSI
(3). Chip designs based on systolic architectures tend to be
simple, modular, and of high performance. In (6) a general
discussion on the attractiveness of systolic architectures is
given. Systolic architectures are particularly suited to chip
implementation of operations in signal and image processing such
as filtering, correlation, and discrete Fourier transform (5).
In this paper we describe a chip, based on a novel systolic
design, for performing the 2-D convolution operator. The chip
consists of essentially only one type of simple cells, which are
mesh-interconnected in a regular and modular way, and achieves
high performance through extensive concurrent and pipelined use
of these cells. Denoting by u the cycle time of the basic cell,
the chip allows convolving a kxk window with an nxn image in
$O(n^2u/k)$ time, using a total of k^3 basic cells. The total
number of cells is optimal in the sense that the usual
sequential algorithm takes $O(n^2k^2u)$ time. Furthermore, because
of the modularity of the design, the number of cells used by the
chip can be easily adjusted to achieve any desirable balance
between I/O and computation speeds.

Examples of application of the 2-D convolution operator
include noise smoothing, linear edge enhancement, edge
crispening, etc.. Designs similar to the one described in this
paper can be used for digital filtering and numerical
relaxation.

II. DESCRIPTION OF THE SYSTOLIC DESIGN

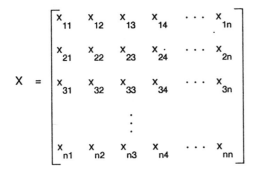

Figure 1. nxn matrix and kxk window with k=3.

To define the 2-D convolution problem, consider a matrix (or image) $X = \{x_{ij}\}$ and a kxk window with weighting coefficients w_{ij} as shown in Figure 1. For the purpose of illustration, we use k=3. Now slide the 3x3 window along the matrix. For each position of the window, the weighted sum of the entries (or pixels) in the submatrix covered by the window is to be computed. More precisely, if the entry at the center of the submatrix is x_{rs}, then we wish to compute

$$\sum_{i=1}^{3} \sum_{j=1}^{3} w_{ij} \ x_{r+i-2, \ s+j-2}.$$

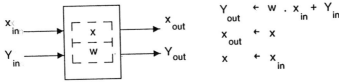

[EACH X STAYS INSIDE THE CELL FOR ONE CYCLE]

[VALUE W IS A PRELOADED WEIGHTING COEFFICIENT]

Figure 2. Basic cell.

The 2-D convolution chip is based on a variant of the systolic FIR filtering array proposed in (4) and (5). The chip uses essentially only one type of basic cells (see Figure 2), which are interconnected in a regular and modular way to form a two-dimensional systolic array. The function of a basic cell is to update a result Y as indicated in the figure. Note that in each cell, w is a preloaded weighting coefficient and each x stays inside the cell for a cycle.

The three kernel cells that form the systolic array are shown in Figure 3. Each kernel cell is composed of nine basic cells and one row-interface cell whose function will become clear later. Five rows of the matrix X advance synchronously from left to right. The small dots in the figure denote appropriate delays needed to make each column advance as a whole. (This does not mean that these delays have to be implemented on-chip; the inputs can be spaced accordingly when fed into the device.) During the cycle subsequent to that an entry x_{ij} enters the upper right basic cell of a kernel cell, the weighted sum corresponding to a submatrix with that entry as

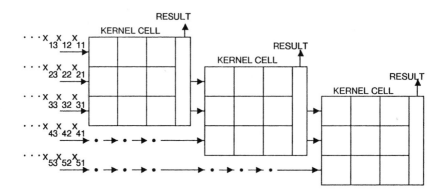

Figure 3. Systolic array consisting of three kernel cells.

the upper left entry will be output from that kernel cell. Therefore by sweeping the first five rows, a systolic array consisting of three kernel cells can compute the first three output rows on-the-fly. In general, by sweeping rows $3i+1$ to $3i+5$ the array computes rows $3i+1$ to $3i+3$. Thus an entry in the nxn matrix is input to the cell array at most twice. In the rest of the section, we describe the underlying idea of the kernel cell design.

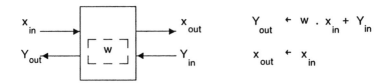

[VALUE W IS A PRELOADED WEIGHTING COEFFICIENT]

Figure 4. A two-way flow basic cell.

Suppose we want to compute the following weighted sum in the first row of a kernel cell:

$$Y = w_{11}\, x_{i\ j-2} + w_{12}\, x_{i\ j-1} + w_{13}\, x_{ij}.$$

One way to do this is to use a special case of the linear systolic FIR filtering array discussed in (5). We use a two-way flow basic cell (to distinguish it from the one-way flow basic cell defined earlier) as in Figure 4, and the desired value for Y can be obtained as illustrated in Figure 5 (a) through (c). In Figure 5 (a), we have denoted by Y the particular Y_{in} to the rightmost basic cell, with its value initialized to zero.

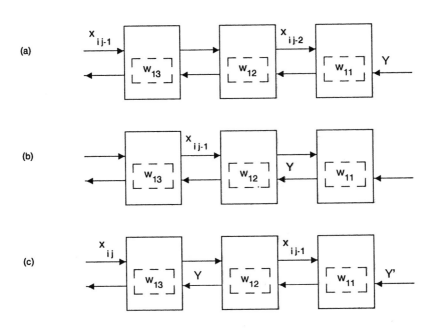

Figure 5. Computing Y with two-way flow.

An improved way is to use the one-way flow basic cell as defined in Figure 2. Consider Figure 6 (a) where three basic cells in the first row of a kernel cell are shown. Consider the moment the entry x_{ij} enters the leftmost basic cell. Denote by Y the particular Y_{in} to this cell, whose value is initialized as zero. During the cycle x_{ij} enters the leftmost cell, Y becomes $w_{13} x_{ij}$. In the next cycle $w_{12} x_{i\,j-1}$ is accumulated to Y, as shown in Figure 6 (b). Finally, as in Figure 6 (c), $w_{11} x_{i\,j-2}$ is further accumulated to Y. As a result, when Y emerges from the rightmost cell, it has already accumulated the weighted sum of the first row of a submatrix. In the meanwhile, the same has

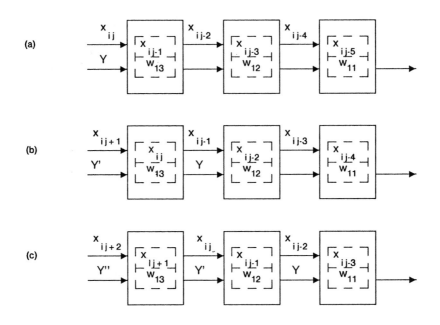

Figure 6. Computing Y with one-way flow.

been happening at the two bottom rows of the kernel cell. Therefore, by summing the partial results at the row-interface cell, the desired output, which is the weighted sum of all the nine entries in a 3x3 submatrix, is obtained (see Figure 3).

The filtering array of Figure 6 differs from the previous one in that both the x-stream and Y-stream now move from left to right and the x-stream travels at half the speed of the Y-stream. This variation results in the active use of every cell at any given time. As a result each kernel cell can produce a pixel every cycle rather than every other cycle, though each basic cell now requires the additional storage to provide the needed delay for the x-stream.

III. BALANCING I/O AND COMPUTATION

Because of the modularity of the basic design of the

systolic convolution array, its size can easily be adjusted to
match the bandwidth between the array and the host to which it
is attached. Let u denote the cycle time of the basic cell.
Suppose that the image is nxn, and that in time u, 2k-1 pixel
values can be transferred from the host to the array, and k
output pixels can be sent back to the host. For processing the
whole image, the I/O time alone is thus $O(n^2u/k)$. To balance
this, the convolution array allows convolving a kxk window with
the whole image in $O(n^2u/k)$ time, using k kxk kernel cells. We
thus see how I/O complexity can be used to guide the design of
special-purpose devices. The reader is referred to (2) for
several lower bound results on the I/O complexity for a number
of problems including the fast Fourier transform. The total
number of cells, k^2, used in the proposed convolution array is
optimal in the sense that the usual sequential algorithm takes
$O(n^2k^2u)$ time.

IV. IMPLEMENTATION

Using part of a chip we have implemented a 3x3 kernel cell,
consisting of nine basic cells plus one row-interface cell. The
implementation uses a bit-serial word-parallel organization.
Pixels are input as 8-bit samples and output with 16 bits. The
weighting coefficients can be changed during the loading phase
prior to the convolution computation. Their values are
restricted to be powers of two in $[-16, 16]$, to reduce the area
of multiplier circuits in each basic cell. (See the remark in
Section IV.D for a technique of using the chip to handle cases
where weighting coefficients are not powers of two.)

We expect that based on this prototype design a full chip
can contain three 3x3 kernel cells and output a pixel in less
than 175 ns. The anticipated high computation rate is mostly
due to the high degree of concurrency inherent in the overall
design.

A. I/O Description

A total of twelve I/O pins are used. Eleven of these are
for input; only one is for output. Weighting coefficients are
loaded into the convolution chip prior to the computation proper
and therefore the inputs of these coefficients and the pixel
values share the same pins. A cycle is composed of 16 minor
cycles each of which includes phases ϕ_1 and ϕ_2. During each

cycle one pixel value is input and one result is output. Since
an input pixel value has only 8 bits which are input bit
serially, input will occur every other half-cycle. In the
following we give a list of names of each pin accompanied by a
brief description. Some of the control signals can in fact be
generated on-chip.

Name	Description
Vdd	Power
Gnd	Ground
ϕ_1	Clock phase 1
ϕ_2	Clock phase 2
$xorw_1$	Pixel value or weighting coefficient for row 1
$xorw_2$	Pixel value or weighting coefficient for row 2
$xorw_3$	Pixel value or weighting coefficient for row 3
Load	Control signal indicating weighting coefficients are being loaded
LSB	Indicates the least significant pixel bits are being input
Circulate	Control signal indicating the 2nd half cycle (absence of input)
Y_{aux}	Optional input to be accumulated to the computed result (to allow repeated computation for non-power-of-2 weights)
Y	Result of the weighted sum

B. A General Layout

In Figure 7 we illustrate a general layout of a kernel cell.
Power and ground, as well as the clock signals, have been
omitted in the illustration.

C. Implementation of the Basic Cell

The basic cell is represented schematically as in Figure 8.
Each pixel value remains in a basic cell for two cycles while
each result stays there for one cycle. Two 8-bit shift
registers are provided for the storage of two 8-bit pixel
values. During the first half cycle, pixel bits entering a
basic cell advance from left to right into the first 8-bit shift

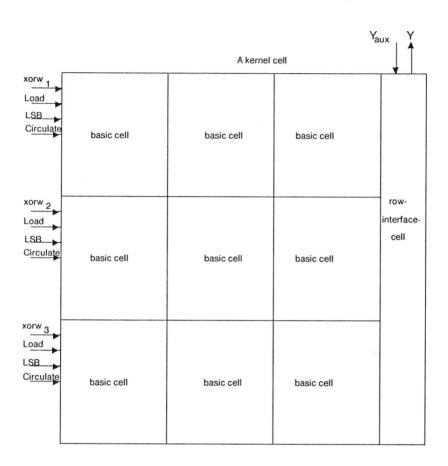

Figure 7. General layout of a kernel cell.

register, and at the same time enter the multiplier, as
indicated by the solid lines. During this same time, pixel bits
already inside the first 8-bit shift register are moved into the
second 8-bit shift register, while pixel bits originally inside
the second 8-bit shift register are moved to the next basic
cell. During the second half cycle all pixel bits circulate
inside the 8-bit shift registers. Flow during the second half
cycle is indicated by the dash lines.

Multiplication in this particular case is merely a shift

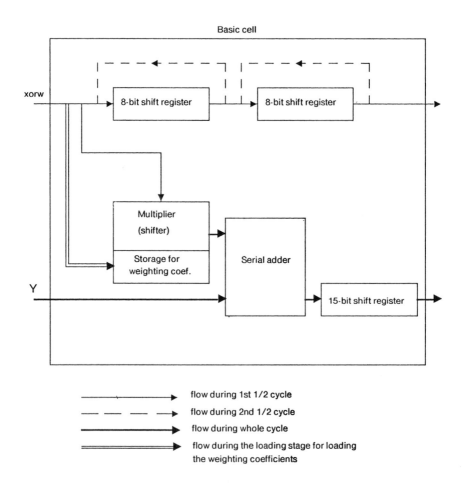

Figure 8. Implementation of the basic cell.

operation. It takes one minor cycle for an input bit to enter
the shifter and the result bit to exit the serial adder. A
15-bit shift register therefore is provided to hold the result
bits so that they will arrive at the next basic cell at the
appropriate time.

D. The Row-Interface Cell

This cell computes the sum of the three partial results from the three rows of basic cells. It also takes an optional input Y_{aux} which will be accumulated to the computed result. This will allow the handling of cases in which weighting coefficients are not powers of two by repeating the computation several times. In most image processing applications it is sufficient to have weighting coefficients as powers of two and, in such cases, Y_{aux} will be zero.

V. CONCLUDING REMARKS

We have developed a chip design for the 2–D convolution operator, which uses essentially only one type of basic cells interconnected in a regular and modular way. While this experiment is about a particular design, we wish to make the following general remark concerning the balance of I/O and computation. Since a special-purpose device is typically attached to a host, from which it gets the data to be processed and to which it outputs results, I/O considerations play an important role on the overall performance. The ultimate goal of a special-purpose hardware design is (and should be no more than) that of achieving a computation rate that balances the available I/O bandwidth. Thus it is important that the design be modular so that its size can easily be adjusted to match the bandwidth between the device and the host. In the design presented basic cells are implemented in a bit-serial manner. As a result, the time for a basic cell to process a pixel is likely to be longer than that for the memory to input and output a pixel. This calls for parallel inputs and outputs of multiple pixels at each cycle of the basic cell. This is the reason why we made the systolic array of this paper input five pixels and output three pixels every cycle. If, on the other hand, basic cells are implemented by bit-parallel schemes (which imply a much shorter cycle time for the basic cell), then the chip is likely able to input and output only one pixel during each cycle. A one-dimensional, rather than two-dimensional, systolic array would be an appropriate design for this case.

ACKNOWLEDGMENTS

We would like to thank all those people who made the MPC79

project (1) possible, through which a prototype design of our
chip could be implemented. We would also like to thank the
following people at CMU who helped on this research. Dave
McKeown and John Kender provided valuable information regarding
typical requirements in image processing applications. Dan Hoey
provided the design of the serial adder. Mike Foster and Dan
Hoey checked for design rule violations.

REFERENCES

1. Conway, L., Bell, A., and Newell, M. E., MPC79: The
    Large-Scale Demonstration of a New Way to Create Systems in
    Silicon. LAMBDA 1(2), pages 10-19, 1980.

2. Hong, J.-W., and Kung, H. T., I/O Complexity: The Red-Blue
    Pebble Game. In Proceedings of the Thirteenth Annual ACM
    Symposium on Theory of Computing, pages 326-333, May, 1981.
    Also available as a CMU Computer Science Department
    technical report CMU-CS-81-111, March, 1981.

3. Kung, H. T., and Leiserson, C. E., Systolic Arrays (for
    VLSI). In Duff, I. S., and Stewart, G. W. (editors), Sparse
    Matrix Proceedings 1978, pages 256-282. Society for
    Industrial and Applied Mathematics, 1979. A slightly
    different version appears in Introduction to VLSI Systems by
    C. A. Mead and L. A. Conway, Addison-Wesley, 1980, Section
    8.3.

4. Kung, H. T., Let's Design Algorithms for VLSI Systems, In
    Proceedings of Conference on Very Large Scale of
    Integration: Architecture, Design, Fabrication, pages 65-90.
    California Institute of Technology, January, 1979. Also
    available as a CMU Computer Science Department technical
    report, September, 1979.

5. Kung, H. T., Special-Purpose Devices for Signal and Image
    Processing: An Opportunity in VLSI. In Proceedings of the
    SPIE, Vol. 241, Real-Time Signal Processing III, pages
    76-84. The Society of Photo-Optical Instrumentation
    Engineers, July, 1980.

6. Kung, H. T., Why Systolic Architecture. To appear in
    Computer Magazine, 1981.

7. Mead, C. A., and Conway, L. A., Introduction to VLSI
    Systems, Addison-Wesley, Reading, Massachusetts, 1980.

ADVANCES IN PROCESSOR ARCHITECTURE, DEVICE TECHNOLOGY, AND COMPUTER-AIDED DESIGN FOR BIOMEDICAL IMAGE PROCESSING

Barry K. Gilbert[1]
Thomas M. Kinter
Loren M. Krueger

Biodynamics Research Unit
Department of Physiology and Biophysics
Mayo Foundation
Rochester, Minnesota

Many new biomedical imaging modalities have arisen during the past decade which exploit the creation of images via computed tomography, or the processing of images generated by conventional radiographic techniques. These improved imaging capabilities are motivating advancements both in the architectures and capabilities of the computers which carry out the processing steps, and in the hardware and software tools by which the computers themselves are designed and fabricated. Several important trends in the technology of advanced processors relevent to the biomedical environment will be described.

I. INTRODUCTION

The past decade has witnessed a rapid growth in a variety of imaging techniques in biomedical research and clinical diagnostic environments, the most widely publicized of which are based upon the creation of images by means of x-ray, ultrasound, nuclear magnetic resonance, and emission computer-assisted tomography (CT) (Kak, 1979). Simultaneously with the advent of CT techniques, a variety of image processing applications have also arisen, either in support of CT imaging or as independent techniques for the enhancement of projection radiographic images and ultrasound images generated by, e.g., ultrasound sector scanners. The new imaging modalities have themselves become increasingly powerful, in part because of substantial improvements in the technology of energy sensitive detectors, and also as a result of advances in electronic scanning and storage of image data in video format (Gilbert, 1976; Robb, 1979). Electronic scanning

[1]This work was supported in part by U.S. Public Health Service Grants HL-04664, RR-00007 from the National Institutes of Health, U.S. Air Force Contract F-33615-79-C-1875, and a grant from the Fannie E. Rippel Foundation.

and video formatting of x-ray generated fluorescent images, of
optical transmission and phase contrast microscopy images, and
most recently of ultrasonic microscopy, has created the possibility of large image data bases residing in an easily manipul-
able format. These electronically formatted images are readily
convertible into digitized image arrays for computerized post
processing to enhance features present therein but not readily
discernible, or to extract alternate forms of information therefrom. This paper will discuss the current techniques by which
digital computer-like devices are configured to handle a wide
range of image generation and processing problems, and will describe the probable evolution of such techniques during the next
few years.

Initial feasibility studies in the development of a new
method for computerized enhancement of biomedical images are
usually performed with the aid of general-purpose mainframe computers or minicomputers using higher level language computer
programs (e.g., FORTRAN or PASCAL), a technique which may be
suitable indefinitely if the image processing is in a research
environment and if the amount of data to be processed is small.
However, when the image processing task is eventually refined
to the level of an independent tool or a portion of a larger
technique, operational considerations frequently preclude the
execution of these algorithms on all but the most cost-effective
hardware configurations dedicated to a narrow range of functions.
Three fundamentally different approaches to the development of
special-purpose processors for a wide range of image generation
and processing tasks will be described, which are categorized
according to the number of computational and bookkeeping operations which must be executed per second to assure an acceptable
turnaround time with respect to the human observer: 1) computational tasks requiring from 10^6 to 10^7 operations per second
(assigned at present to mainframe computers and minicomputers);
2) tasks requiring from 10^7 to 10^8 operations per second (such
tasks presently can be executed only on large mainframe computers),
and 3) tasks requiring computational rates much greater than 10^8
operations per second (these tasks are not presently executable
on any available general-purpose or special-purpose computers).
These three categories of computational requirements will be
referred to respectively as low, medium, and high computational
demand tasks.

II. LOW COMPUTATIONAL DEMAND TASKS

During the past few years, tasks requiring up to 10^7 logical
and arithmetic operations/second have been assigned either to
mainframe computers, or where possible, to the largest minicomputers, frequently augmented by commercially available "programmable array processors". Available in a variety of designs,

the array processors are specifically intended to execute up to 10^7 arithmetic and bookkeeping operations per second in fixed precision or floating point arithmetic under the supervision of a general-purpose host computer. Although most current array processors execute all primitive arithmetic and numerical functions except division, they are generally optimized for a small subclass of special operations, e.g., the FFT "butterfly" used to compute the Fast Fourier Transform. Costs of these special-purpose array processors in 1980 vary from \$10,000-\$150,000. The combined minicomputer-array processor combination is capable of executing five to ten million operations/second, with primary emphasis on arithmetic functions.

As a result of advances in digital device technology, the "single component" microprocessor, which has been described extensively (Morse, 1978; Stritter, 1979; Ziegler, 1981), will play an increasing role for computation of fixed precision operands and eventually floating point operands up to rates of approximately 2×10^6 operations per second. The most recent generations of these single-component microprocessors operate directly on 32-bit operands, contain on-board read only memory (ROM) and random access memory (RAM), and can directly access the address space of large blocks of solid-state digital memory. These devices execute a large, sophisticated set of assembly language instructions predetermined by the component vendor, including single-instruction arithmetic and inter-register operations on full width operands (16 or 32 bits). Single component microprocessors, of the type well exemplified by the new Intel IAPX 432 processor, are generally designed to facilitate communication with external arithmetic components which augment the processor's own intrinsic arithmetic capability, and if desired, with special-purpose single component communications controllers which support all input/output functions for the microprocessor. Hence, though not really single component computers in the most restricted sense, these microprocessors can provide considerable computational and communications capability in a total of less than 10 components.

Each succeeding generation of these devices has exhibited increasingly more powerful arithmetic capability; the very newest employs 32-bit operands in its internal structure, though the input/output pathways retain a sixteen bit width as a result of pin count limitations on the integrated circuit packages. At present, the slow gate propagation delays of these devices limit their minor cycle clock rates to 5-8 MHz (every major instruction is executed in several consecutive small segments, each denoted as a minor cycle). This restriction may be somewhat relaxed with the advent of low power high-speed device technologies (e.g., H-MOS, V-MOS) in combination with improved photolithographic and electron beam lithographic fabrication techniques which create the fine structure of each microcircuit. The low per-unit costs and the availability of support aids for the development of hardware systems and computer programs based around these components allows them to execute rather complex arithmetic and logic functions in a cost-effective and straightforward manner. These features, combined with a rapidly increasing availability of microprocessor-oriented higher level language compilers (e.g., PASCAL, FORTH, PL/M, and MP/C), will permit them gradually to supplant the use of minicomputers for a wide variety of image processing algorithms in environments where absolute minimum processing time is not required.

The second generic type of microprocessor, which is being used to implement the next generation of minicomputers, is known as a "bit-slice" processor. These devices are not a single component, but are instead a family of related part types; each part type performs a given set of functions on a few adjacent bits of a large operand. Each part type can be interconnected with others of the same type to form a functional operator for an operand of any desired width. Such part types include arithmetic and logic units, input/output controllers, and so on. For example, if each basic component is capable of processing a 4-bit portion, or "slice", of an operand of any width, a computer capable of direct operation on 32-bit operands can be created by interconnecting in parallel eight components of each required part type (Alexandridis, 1978). Advances in the fabrication technology of these devices will permit the production of bit-slice components capable of handling even larger portions of a full operand; a family of bit-slice devices is already available in which each component processes a byte-wide (i.e., 8-bit) portion of a full operand. Technological advances will also permit these building block elements to support an increasingly large set of internal functions, as well as continually increasing device speeds and system clock rates. The most popular family of bit-slice devices is for all practical purposes not capable of clock rates in excess of 7 MHz; a new generation of such devices currently in development will support clock rates of 30–40 MHz; 100 MHz rates will be feasible by the mid-1980s. However, unlike the single component microprocessors, instruction sets of the bit-slice microprocessors are determined by the system designer, not by the component vendor. Thus, a substantially larger commitment both to hardware and to software is mandatory for implementations employing the bit-slice devices.

Although several recent papers (Morse, 1978; Stritter, 1979) have commented about the functional competition and overlap between the single component and bit-slice microprocessors, each device type possesses unique strengths and weaknesses; their capabilities are only superficially interchangeable. The strengths of the single component microprocessors include their pre-established instruction set, their physical compactness and low cost, and the substantial software packages with which they are supported; conversely, the operand widths of the single component processors are still somewhat restricted. The variable operand width of the bit-slice devices makes them an ideal set of building blocks for the design of full scale processors of 32 and 64 bits

capability; however, as noted earlier, the requisite hardware
commitment and software support are usually very substantial.
Both generic types can employ microcycle clock rates in the 4-8
MHz range; however, the single component units require an average
of 5-8 microcycles to perform a complete instruction, while bit-
slice processors can frequently be designed by the user to execute
an "average" complete instruction within 1-3 microcycles.

It is of interest that the bit-slice families include specialized components, called microprogram sequencers, which are
ideal for the execution of control functions for large multiprocessor hardware arrays which do not otherwise exploit the bit-
slice design concept. Although usable in a similar manner,
single component microprocessors are not as suitable for such
control functions in parallel processor designs. The microprogram sequencers are ideal for the control of special-purpose processors whose designs deviate substantially from classical single
instruction single data operand computer architectures. Bit-
slice components have been used successfully as the basis for at
least one family of very powerful "maxi" minicomputers, and have
been exploited by the vendors of commercial x-ray computed tomography units for the design of special-purpose reconstruction
processors. With increasing enhancement of the capabilities of
the bit-sliced elements, it is probable that they will virtually
dominate the implementation of processors in the $2-10 \times 10^6$ opera-
tion/sec speed range during the coming years.

As alluded to in the prior discussion of microprocessors such
as the IAPX 432, the performance of both the single component and
bit-slice microprocessors can be further enhanced by recently
developed specialized components which function as peripheral
arithmetic units. These arithmetic components, which are intended
to extend the capabilities of the single component multipliers and
multiplier/accumulators available for several years, carry out a
variety of fixed precision and floating point arithmetic functions
at very high speed under the operational control of a small internal instruction set.

III. MEDIUM COMPUTATIONAL DEMAND TASKS

A. Programmable Array Processors

A new generation of programmable array processors is currently
evolving which will fulfill requirements for computation in the
range of 10^7 to 10^9 operations/second. The new array processors
will either be user-programmable, or will be supplied with a vast
armamentarium of vendor-prepared subroutines which can be employed
as desired by the user. Advanced design array processors will be
particularly useful for processing of moderate numbers of high
resolution (e.g., 2000 x 2000) images, such as medical radio-
graphs, or very large numbers of lower resolution (e.g., 512 x

512) images. The pressures for the development of such devices
are arising principally from the earth resources management programs for the processing of high resolution multispectral imaging
satellite data, from oil and minerals exploration groups for the
reduction of seismic data, and from the military for the real-time
analysis of moderate bandwidth microwave and radar signals. The
biomedical community will recognize the benefits of such technological advances without having to support their development
costs.

Unlike present array processors which are optimized for a
specific computation such as the FFT "butterfly" operation, the
next generation of such machines will possess more general and
hence more flexible architectures. With the availability of
appropriate software design support aids, and when operated in
conjunction with general-purpose host computers, these devices
will be capable of executing a wide variety of image and array
processing tasks with extremely high efficiency.

The architectures of the new array processors are fundamental
to their computational and programming flexibility. An example of
one promising design (Control Data Corporation, 1979) is depicted
in Figure 1. Sixteen distinct arithmetic, control, external communications, and memory modules are connected in parallel to one
another through a large digital crossbar network. The crossbar
simultaneously allows the outputs of any one or more of the
modules to be connected to the inputs of any one or more of the
other modules for a duration as short as a single cycle of the
system clock, or for as long as desired. The inputs and outputs
of every module are equipped with data holding registers, creating in effect a parallel multiprocessor which is completely reconfigurable under software control during every clock cycle. It is
of interest that the input and output data registers contribute a
so-called "pipeline" capability to the design; a complicated
numerical operation is said to be "pipelined" when it is subdivided into a number of sequential hardware subunits, with each
executing only a portion of the total computation and separated
from the others by intermediate data registers. The effective
throughput of a process divided into "n" pipelined stages is
generally n times greater than if pipelining is not employed.

If implemented in the appropriate high-speed logic (see
below), overall machine clock frequencies of 50 MHz are feasible,
with each processor subunit executing an entire operation in one
to three clock cycles; in addition, individual processor subunits
are themselves amenable to pipeline design, allowing a completed
result to emerge from each subunit during every machine cycle.
The processor of Figure 1 will execute operations on 8, 16, or
32-bit integer operands, or, with some degradation in throughput,
on 64-bit floating point operands. In the theoretical limit,
which is never achieved in practice, the processor should support
a maximum of 3.5×10^7 64-bit floating point arithmetic operations per second, a maximum of 1.1×10^8 32-bit fixed precision

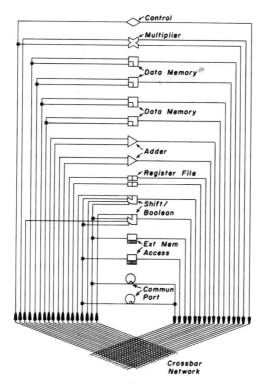

Fig. 1.   Example of next-generation programmable array processor.
Note the sixteen computational and communications subprocessors
connected in a parallel structure through a large crossbar switch
network.   A single microcoded control unit provides operating
instructions during each clock cycle for all computational, communications, and crossbar substructures.   (Reproduced with permission from Control Data Corporation, 1979).

arithmetic operations per second, or a maximum of 5×10^8 8-bit
fixed precision arithmetic operations per second.

The designs of present and future generation array processors
emphasize arithmetic operations and input/output functions rather
than data dependent branch operations.   Limited capability for
branch execution is not critical for a processor intended to execute array-oriented arithmetic manipulations on large blocks of
data, since most image and signal processing algorithms are
heavily computation dependent, are easily pipelined, and rarely
deviate from a predetermined set of numerical substeps.   The more
arithmetically-intensive image processing algorithms generally
contain a very low percentage of data dependent branch operations.
Nonetheless, a minimal provision for data dependent decisions can
be incorporated in the design of the control unit of the array

processor. In general, the single component microprocessor implementations described earlier, and the network processors described
below, possess the most well developed branching capability;
general-purpose array processors (e.g., Figure 1) have significantly less decision making capability but considerably more
arithmetic capacity. As described below, specially designed processors which execute a small set of algorithms on extremely large
data bases tolerate branching operations least well in order to
maximize computational throughput.

B. Computer Networks

Another fundamentally powerful architectural approach which
is a candidate for high performance processing applications is
the computer network, in which a number of "processing elements"
and "resources" are interconnected by a grid of communications
channels through which are transmitted control and synchronization instructions, as well as partially processed data. The
individual processing elements in the network may be assigned to
the execution of portions of a large algorithm, followed by a
merging of their partial results. The processing elements and
resources are referred to as nodes; the operational characteristics of different nodes may be identical or vary widely, the
only requirement being that the interface between the node and
the network obey a pre-established protocol. Such an approach
allows nodes to be general-purpose processors, large solid state
memories, rotating discs, or dedicated arithmetic units capable
of executing one or a few algorithms with extremely high efficiency.

The above description might appear to apply equally well to
the array processors described earlier, and in fact there are
often many similarities between the two types of architectures.
However, a general distinguishing feature of the network structure is the autonomous ability of its processing nodes to execute
program streams stored in their own local program memories,
including decision and branch operations. Hence, network processing nodes are much more general purpose in their range of capabilities than are the slaved arithmetic units comprising the elements
of an array processor. This latter class of machines usually contain one master control unit which issues step by step instructions
to each of the interconnected arithmetic elements, which possess
no autonomous control or decision making capacity of their own.

A large number of topologically different network schemes have
been proposed, all of which possess unique strengths and weaknesses; however, it has not yet been demonstrated through actual
hardware development efforts which of the network structures are
the most flexible. Two examples of the many such networks devised
during the past two decades will thus be presented to illustrate
the wide variation in function and capabilities available among
the various structures.

The crossbar network, for many years employed in early elec-
tromechanical telephone exchange networks, is depicted in the
lower portion of Figure 1. This approach is conceptually the
most straightforward and also the most flexible of all network
designs. Processing elements and resources are interconnected
through the ends of the orthogonally intersecting buses; externally controllable switches at each of the bus intersections allow
any processor or resource to be connected to any other in the
network. In an N x N crossbar, up to N pairs of processors or
resources can be interconnected at a given time. Since the complexity of this architecture grows as N^2, the crossbar arrangement has been used only in those instances in which maximum
interconnection flexibility is required for a minimum number of
nodes, and when large scale integrated circuit technology can be
exploited in the fabrication of the intersection point switches
to reduce overall network costs (Control Data Corporation, 1979).
The cube network is another form of interconnection array
currently undergoing intensive theoretical analysis (Siegel,
1979; Siegel, 1980); many variations of this basic design have
been proposed. In this approach, a group of nodes are interconnected through several ranks of special switching elements called
exchange boxes. As depicted in Figure 2, each exchange box
accepts a pair of N-bit bus inputs from exchange boxes in the

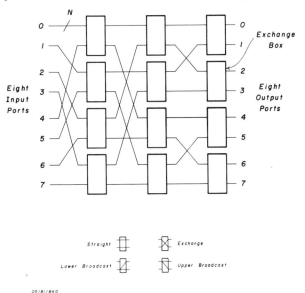

Fig. 2. Topology of a cube interconnection network, with a
detail view of the four possible data transmission modes of each
"exchange box". (Reproduced with permission from B. K. Gilbert,
"Architectures, Tools, and Technologies for the Development of
Special Purpose Processors" in Modular Computer Systems Series,
Volume 2 (In Press)).

previous rank, and is capable of connecting either of the two
inputs to either or both of the two outputs. Communications pathways between nodes on the left and right of the network in Figure
2 can be established by either of two mechanisms. First, the
switch settings may be established by an independent routing and
control unit, e.g., a minicomputer which monitors and acts upon
all traffic requests. Alternately, the data can be converted into
"packets" by the source devices (e.g., minicomputers connected to
the network) and transmitted from the source nodes to the destination nodes. As they are constructed by the source device, all
packets are prefixed by a destination address which can be decoded
in real-time by each successive switching node to determine the
proper setting to transmit a given packet to the next rank of
switches. Transmission of data through a cube network in packets
requires that each node contain a small amount of memory to support the temporary storage and forwarding of each packet, and also
sufficient logic to decode the packet destination address. Packet
transmission through a cube network is not entirely free of problems, however, since network blocking can occur if a packet is
transmitted to an exchange box whose paths are already engaged by
other packets. The blocking problem can be reduced or eliminated
at the expense of real-time network flexibility by controlling
the individual exchange boxes via an external controller. Both
methods have been studied intensively and both have advantages
and weaknesses (Siegel, 1979; Siegel, 1980).

Regardless of the network architecture selected, an algorithm
must be developed which can partition a complex problem among
many processing elements. One well-known and internally consistent terminology refers to these partitioning approaches as
"SIMD", or "single instruction stream, multiple data stream";
"MIMD", or "multiple instruction stream, multiple data stream";
and "MSIMD", or "multiple, single instruction stream, single data
stream". In this nomenclature, the conventional mainframe or
minicomputer architecture is usually referred to as "SISD", or
"single instruction stream, single data stream". In the SIMD
approach, all processor nodes in a network simultaneously execute
the same algorithm or subsection of an algorithm on a lockstep
instruction-by-instruction basis, but on different data blocks;
for example, ten processing elements in a network may each perform an inner product operation on ten different pairs of vectors
of the same length. In the MIMD architecture, each processor is
capable of executing a unique stream of instructions on different
blocks of data; nowhere else in the processor network need a
similar instruction stream be undergoing execution simultaneously.
With an MIMD architecture, a complicated problem may thus be subdivided into many different sections, each of which may be
assigned to one of the processor nodes for execution. If the
algorithm is such that certain stages cannot be executed prior to
the completion of other stages, several of the processing nodes
may not be initiated until the appropriate data block becomes
available from a prior processing step executed by one of the

other nodes. Conversely, a processing node may execute one portion of the algorithm, and then be reloaded with a new stream of
instructions to execute yet another portion of the same algorithm
representing a later stage of computation. In a comparison of the
SIMD and MIMD approaches, SIMD machines are easier to program
because the operation of all processing elements is simultaneous
and in lockstep; however, SIMD machines cannot achieve the flexibility of the MIMD architectures. Conversely, the MIMD approach
is very much more difficult to program and, particularly, to
verify for correct operation.

The operational advantages of a network processor over a
dedicated architecture are apparent. These structures can be
extremely flexible since the task definition for individual processing elements and for the entire network is via software rather
than through a fixed hardware configuration. With appropriate
design of the network communications protocol, the total number
of nodes in the network can be increased or decreased even after
initial fabrication. Similarly, a sufficiently versatile network
communications protocol allows the inclusion of processing and
resource nodes with a wide variety of speeds, computational capabilities, physical configurations, and so on, since the only
constraint on node design is that the node interface to the communications network obey a predetermined protocol. The network
processor concept is so powerful that numerous projects are currently underway to exploit the shared resource features of such
systems (Dias, 1981; LeLann, 1981; Wu, 1981).

During the 1970s, many attempts were made to develop computationally powerful processors by interconnecting large numbers
of inexpensive single component microprocessors into a network
of one type or another. Several examples (Wittie, 1980; Tasto,
1977) have demonstrated that the synchronization and communication requirements of a large number of semi-independent but low
performance processors, frequently increase rapidly with the number of processors in the network. As a result, the individual
processors consume much of their capability satisfying the network communications protocol. Networks of microprocessors have
demonstrated performance improvements over that of a single microprocessor far less than their numbers would imply; in many cases
the performance improvement has approached the square root of the
number of processors employed. This finding has been taken into
account in several recent network processor projects in which the
number of network nodes has been minimized, while the power of
each node has been increased to the maximum feasible level. In
one well known ongoing project, the optimum number of processing
nodes has been set at 16, and the maximum number of memory module
resources has been set at 8, with the individual processors and
memory resources interconnected by a high-speed crossbar switch;
conversely, the power of each processing node has been increased
to 400 million floating point operations per second (S-1 Project
Staff, 1979). The foregoing comments should not be interpreted
as a statement that large processor networks are not potentially

powerful and should not be attempted, but rather as a suggestion
that the total number of nodes be increased in conjunction with a
maximization of the power of each node as well. Computing power
may be maintained at high levels by making each subprocessor as
powerful as possible (Ornstein, 1975; S-1 Project Staff, 1979),
either by an augmentation of its parts count or by the use of a
faster digital technology; problems of interprocessor communications are thereby circumvented to the maximum possible degree
(S-1 Project Staff, 1979).

IV. HIGH COMPUTATIONAL DEMAND TASKS

For a small class of tasks which require processors capable
of throughput in the range of 2×10^8 to 5×10^9 operations/
second, none of the approaches described earlier, with the possible exception of next-generation network processors, are sufficient; special measures are always required to achieve such high
computational rates. The demand for such processors is already
becoming substantial in the aerospace, military, seismic, and
earth resources disciplines. Although such large computational
tasks are less likely to be encountered in the biomedical world,
biomedical applications requiring very high capacity computers do
occasionally occur.

Achievement of computational rates above several hundred
million arithmetic operations per second, in a sufficiently cost
conservative manner to allow more widespread usage than can be
expected for the general-purpose giant computers, will require
the development of special-purpose computers optimized to execute
a very small class of well understood algorithms. Though limited
in flexibility, highly specialized processors can be very efficient in comparison to general-purpose processors since all computational elements not required to execute a pre-selected class of
algorithms are carefully eliminated from their design.

A recent biomedical research project requiring computational
rates approaching several billion arithmetic operations per second
is an advanced generation truly three-dimensional real-time x-ray
computed tomography scanner, the Dynamic Spatial Reconstructor
(DSR). Unlike commercially available x-ray computed tomography
machines (Nahamoo, 1981), which collect projection data in 5-20
seconds to reconstruct only a single cross-sectional image, the
DSR collects within an 11 msec duration sufficient projection data
to reconstruct up to 240 adjacent 1 mm thick cross sections encompassing a cylindrical volume of tissue (e.g., the thoracic or
abdominal contents) 22 cm in axial extent and 24 cm in diameter
(Robb, 1981). Since the entire data collection procedure can be
repeated 60 times each second, a scan lasting only ten seconds
produces enough projection data to reconstruct up to 150,000 cross
sections (Gilbert, 1979). Reconstruction and appropriate display

of these cross sections results in a time varying, truly three-
dimensional x-ray image of the structure under study, with potentially powerful biomedical research and clinical diagnostic
possibilities (Gilbert, 1980).

   If such an x-ray scanning capability is to prove useful in
biomedical research and clinical diagnostic environments, the
duration necessary to completely process and display the volu-
metric images must be reduced to at most a few minutes for a few
seconds of actual data collection. Since the best throughput
times achieved to date using large minicomputers combined with
presently available programmable array processors is approximately
2-5 seconds per cross section, several orders of magnitude speed
improvement will be necessary to achieve the required throughput,
equivalent to a computational rate of at least three billion
arithmetic operations per second. Since such a large throughput
is not supportable by next-generation programmable array processors, and possibly not by next generation network processors, it
was recognized that a special-purpose processor executing a
limited subclass of algorithms would have to be developed.

   Figure 3 is a schematic diagram of a special-purpose parallel
hardware processor designed to execute several closely related
reconstruction algorithms (Gilbert, 1979). The design employs
multiple interlinked, cooperating arithmetic sections, each executing a portion of the entire image reconstruction algorithm.
The arithmetic unit depicted in the upper portion of the figure
executes a linear filtration operation on individual projection
data vectors each containing up to 512 elements, and then transfers its intermediate results to the multiple processors in the
lower portion of the figure, which complete the remaining portions
of the algorithm. In its full embodiment, the processor will consist of 29 separate arithmetic sections, each executing a portion
of the procedure in parallel. None of the arithmetic units is
capable of controlling its own step-by-step operation; all rely
on a pair of microprogram sequencers for execution and coordination of every computational step.

   A small scale engineering prototype model of the parallel
processor design of Figure 3 has been developed to verify basic
architectural concepts and the software subroutines for the microprogram control sequencers which will control the processor. The
engineering model processor contains only five arithmetic units,
but is otherwise architecturally complete. Although the engineering prototype executes a complete set of parallel operations every
160-200 nsec, the next-generation full scale processor will execute a much larger number of parallel operations every 35-40 nsec,
resulting in an eventual capability in the range of 3-4 billion
fixed precision arithmetic operations per sec.

   The design and development of a special-purpose processor
with very narrowly defined capabilities generally requires several
years of engineering effort, from the original assessment of the
numerical analytic characteristics of the algorithms to be executed, through the final operational verification of the full

Generation of Projection Addresses (Ray Sum Indices)
& Back Projection Weights via Table Lookup

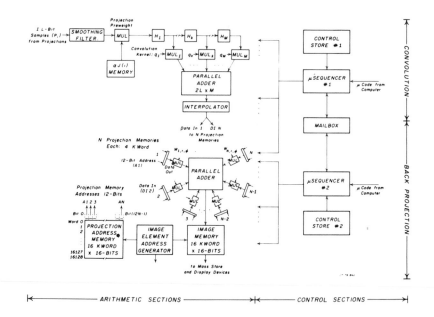

Fig. 3. Design for a special-purpose parallel pipeline processor
to execute high-speed image reconstructions for x-ray computed
tomography. Note pair of interlocked microprogram control units
which control filter subprocessor and back projection subprocessor,
respectively. (Reproduced with permission from Gilbert and Harris,
IEEE Transactions on Nuclear Science   NS-27(3)1197-1206 (June)
1980).

scale processor. Hence, such a special design can only be justi-
fied in those instances in which neither microcomputers, programmable array processors, nor large general-purpose mainframe
computers will yield sufficient cost-effective computational
throughput to satisfy the operational constraints of the complete
system.

V. INCORPORATION OF IMPROVED DEVICE TECHNOLOGY IN PROCESSOR DESIGN

A. Very Large Scale Integrated Circuitry

Incorporation of advanced digital device technologies into
new processor architectures can often simultaneously result in
a simplification in their design and an increase in processor
capacity. Exploitation of high density device technology also

frequently permits the newest versions of processors using long
established architectures to operate at considerably higher
throughput levels. For example, the array processor of Figure 1
also exists in an earlier version which employs standard digital
device technology. In part, as a result of improved large scale
integrated (LSI) device technology, the newer design operates with
30-50 times the throughput of the prior design but without a con-
comitant increase in acquisition or operating costs. The impact
of ever increasing levels of circuit density was also illustrated
earlier in the description of the newest families of "single
component" 32-bit microprocessors, which in effect will allow
algorithms previously executable only on large minicomputers to
be supported with considerably less hardware and at considerably
lower cost.

B. Ultra High-Speed Integrated Circuitry

An additional advance in the state of the integrated circuit
art which will become increasingly important for the fabrication
of ultra high-speed processors is the development of extremely
fast digital devices with both high circuit density and subnanosecond gate propagation delays (Gilbert, 1980). For purposes of
comparison, gate delays in the newest single component very large
scale integrated circuit (VLSI) microprocessors are in the 2-5
nsec range. The new families of ultra high-speed components will
allow the redesign of several traditional architectures to achieve
increases in throughput and/or cost/performance ratios by factors
of 3-5 while executing the same instruction set. These new subnanosecond devices are being incorporated into the next generations of giant commercial computers (Kozdrowicki, 1980) and programmable array processors (Control Data Corporation, 1979) of
the type depicted in Figure 1, into highly specialized processors
such as that depicted in Figure 3 (Gilbert, 1980), and also into
the designs of very powerful signal processing computers under
development for the military (S-1 Project Staff, 1979). Since
the system design rules for these new subnanosecond devices are
different from and considerably more stringent than those for more
familiar families of digital integrated circuits, their use will
be restricted to a small number of cases in which extremely high
computational throughput is required.

These new subnanosecond device technologies are under investigation at the Mayo Foundation for possible use in the fabrication of the next generations of ultra high-speed special-purpose
image processing and reconstruction computers (Gilbert, 1979;
Gilbert, 1980). Using conventional transistor-transistor logic
(TTL) device technology with 3-6 nsec gate delays, system clock
rates much in excess of 25 MHz have in the past resulted in processors with poor noise margins which often experienced transient,
unrepeatable errors. Using an earlier family of emitter coupled
logic (ECL) exhibiting 2 nsec gate delays, it was possible to

fabricate large special-purpose processors containing in excess
of four thousand integrated circuits and employing a 40 MHz system clock (Gilbert, 1976). Several applications now envisioned
for special-purpose high-speed computers will routinely require
clock rates in the range of 75–150 MHz.

To verify the performance improvement which might be achieved
with the newest subnanosecond devices over that achievable with
less advanced technologies, several prototype processors have been
designed and are undergoing rigorous operational testing here.
For example, we have recently fabricated an engineering prototype
nine element linear filter processor, the design of which exploits
the even function symmetry required of all filter kernels employed
in computed tomographic reconstruction algorithms (Gilbert, 1981).
The prototype filter processor, which contains more than one
hundred integrated circuits, was designed to allow operation at
very high system clock rates. Figure 4 depicts measurements from
selected test points while the processor was operating at a clock
rate of 125 MHz. The processor can accept a new operand and produce a processed operand every 8 nsec, filtering an entire vector
containing five hundred elements in four microseconds (this processing rate exceeds that of present generation programmable array
processors by a factor of roughly 10^3). The high-speed oscillo-
scope traces depict the 125 MHz clock waveform, as well as single

(6-Bit Input & 18-Bit Output Data; 9 Elements; Wire Wrap Interconnects)

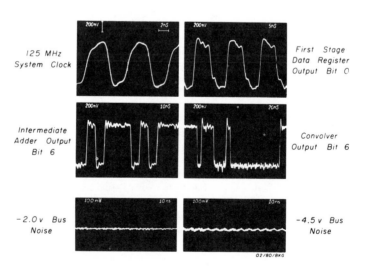

Fig. 4. Operation of prototype direct convolution filter processor fabricated with subnanosecond emitter coupled logic. Note
particularly the high system clock rate, short rise/fall times,
and low system noise levels. (Reproduced with permission from
Gilbert and Harris, IEEE Transactions on Nuclear Science
NS-27(3):1197–1206 (June) 1980).

single bit-lines at various locations within the functioning
processor. The high performance of this circuit is similar to
that of several larger systems designed in our laboratories.

C. Configurable Gate Arrays

Improvements in the performance of the special-purpose and
programmable array processors will be accompanied by a decrease
in their costs, through the recent introduction of a technique
for producing LSI and even VLSI circuits containing a nearly
custom designed section of an entire computational algorithm.
Computer designers are well aware that the cost of a processor
which is to be replicated many times is proportional to the total
number of integrated circuits used in its fabrication; hence, the
cost of such a processor can be markedly reduced by employing
high density integrated circuits incorporating a large portion
of the entire processor. Until very recently, designs for large
scale integrated circuits were feasible only for those devices,
e.g., fixed precision multipliers, which were in such high demand
that production runs in the tens of thousands of components could
be virtually guaranteed.

In a new approach to the fabrication of "custom" components,
a single large scale integrated circuit structure containing
several thousand individual unconnected logic gates is mass produced, with the final interconnection of the individual gate
elements carried out with highly automated equipment during the
final stage of fabrication at a very low per-design cost. These
special LSI devices, termed "configurable gate arrays" (CGAs),
presently contain more than 2000 unconnected gates; similar
designs presently under development will contain four thousand
equivalent gates; the maximum size limit imposed upon CGAs by
technological contraints is likely to be in the range of 10,000-
15,000 gates, which will probably not be attained until the mid
to late 1980's. At least one family of CGAs will also exhibit
propagation delays in the range of .4-2.0 nsec with moderate
levels of power dissipation. Hence, both the performance and the
device density of individual integrated circuit CGAs is already
high, and can be expected to improve continually.

As an example of the improvements in processor performance to
be gained by exploiting CGAs, the prototype convolution filter
processor of one hundred integrated circuits described earlier
could be placed on a single 2000 gate CGA with a 20% unused gate
reserve. A fully augmented direct convolution filter processor
of a similar design would require approximately 800 of the currently available subnanosecond ECL integrated circuits; the same
design could be reduced to a parts count of approximately 25-30
CGA components of only three distinct device types, which would
execute an entire direct filtration of a 500 element fixed precision vector in less than 2 microseconds (Gilbert, 1981).

VI.   COMPUTER AIDED DESIGN SOFTWARE FOR RAPID DEVELOPMENT OF NEW PROCESSOR ARCHITECTURES

The logical and physical complexity of modern general-purpose computers and special-purpose processors has been increasing rapidly since the late 1960's; as a result, large machines containing from one hundred thousand to one quarter million logic gates are now the norm. Whether these gates are physically distributed among thousands of small scale and medium scale integrated circuits, as is still the case for many mainframe computers (Kozdrowiski, 1980), or are, in the future, all placed on a small number of VLSI integrated circuits (Ziegler, 1981), the designs of these systems are becoming too complex for engineers to develop within a reasonable duration using traditional paper and pencil methods.

In an attempt to mitigate the design complexity problem, there is an increasing effort to develop specialized operator interactive software systems called computer aided design, manufacturing, and test (CAD/CAM/CAT) packages, which assist the computer architect at every stage of the design, layout, and operational verification of a new processor. By means of a high resolution interactive computer graphics terminal, the engineer can construct the logical design of a single subsystem or of the entire computer. Using appropriate keyboard commands and a light pen or X-Y cursor, individual logical elements or components are retrieved from a pre-established file and presented on the computer terminal in the desired topological relationships; as depicted in Figure 5, interconnections are then specified by the engineer, and are incorporated by the CAD program into a growing software data base which is a symbolic representation of the design. As the design process proceeds, the CAD program provides the engineering staff with intermediate information and partial results to allow the correctness of the design to be verified continually, including inadvertent specification conflicts which violate pre-established design rules, and worst-case timing limits on all subsections of the processor.

Following the initial design stage, integrated CAD/CAM/CAT programs then "compile" the symbolic representation of the final system design; the output of the compilation step is a set of instructions for the physical fabrication of the processor. This "recipe" includes extremely detailed information regarding component by component layout on the individual logic boards, interboard cabling, power supply and cooling requirements, and so on. In addition, a variety of detailed documentation aids are prepared for use by the engineering staff during the postfabrication checkout phase to verify that the machine is operating properly.

Modern CAD programs employ an integrated data base concept which allows additional software modules in the CAD/CAM/CAT package to carry out the next phase of the design procedure, i.e., direct support of the actual manufacturing process for the first and all subsequent machines. This portion of the integrated

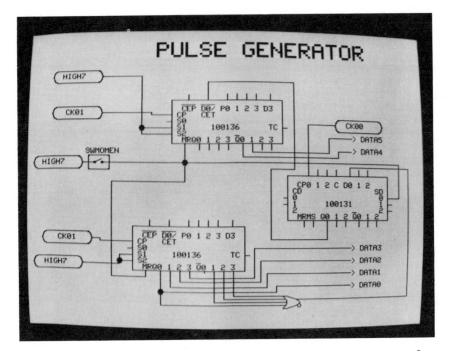

Fig. 5. A view of the operator interactive graphics screen of a modern computer aided design terminal. Graphical representations of various components are retrieved from a "library" of such parts stored on computer disc, drawn on the screen, and then interconnected by the designer using a light pen. A data base of components and their interconnects is built up in this manner.

software package is referred to as computer aided manufacturing (CAM). Unlike the design modules described earlier, the operational characteristics of the CAM module are closely related to the details of the fabrication process to be used. For example, if multilayer printed circuit boards are to be employed, the CAM module must perform the layout of individual component interconnects on each conducting layer of the board by direct or indirect preparation of the artwork used to expose each of the photoresist-covered metal layers prior to a chemical etching step. If the fabrication technology employs either wire wrap interconnects or the newer "multi-wire" fabrication technology, the CAM module creates a magnetic or paper tape containing instructions which are employed directly by the automated wire wrap or the multiwire layout machines to generate each interconnect on a point-to-point or X-Y coordinate grid structure.

Many computer aided design software packages include an additional "computer-aided test" (CAT) module capable of generating test patterns which can be loaded into the new processor and executed to verify its correct operation. The test pattern generation module has access to the entire design data base of the new processor and, by implication, to all of the gate-by-gate observability criteria and all of the interconnect pathways in the system design; as a result, the CAT module can guarantee that the test patterns will be able to identify a majority of fabrication and manufacturing errors and a great majority of failed components.

Integrated CAD/CAM/CAT packages support in numerous ways the rapid design of new special purpose and general purpose processors. Perhaps of primary importance for the system designer, the CAD package reduces his paperwork burden, thereby allowing him to concentrate on creative design tasks. In general, there is an improvement of the efficiency and speed with which the logic design and layout processes proceed, since a well written CAD program provides intermediate design verification data which allow early detection of design and layout errors. In addition, the reliability of processors developed with CAD support can be higher than processors designed with traditional methods, because the new system designs are constrained to a carefully prepared, consistent, and comprehensive set of layout and interconnection protocols. When traditional design methods are employed, a large system developed by several engineers usually reflects several sets of ad hoc design and layout rules representing the unique experiences of the individual designers. The resulting hodgepodge of design approaches nearly always causes major problems which become painfully obvious when the individual subsystems must be connected to one another and tested. The reliability improvement achieved by adherence to a single set of design rules is recognized not only during initial system checkout, but also during normal maintenance of the operational system because the CAD program provides a uniform and comprehensive set of documentation and cross reference aids, a rarity when documentation is produced manually by several different designers.

The inclusion of computer aided manufacturing features in a comprehensive CAD package also aid in system stability by eliminating a large number of the mechanical errors which typically creep into a system during the conversion of the logical design into physical hardware. In a single recent example in which the design of a system containing 500 integrated circuits was carried out in our laboratory using computer aided design but traditional manufacturing methods, more than 90 percent of the errors detected during the initial system operational verification were traced to fabrication rather than design errors. Manufacturing errors frequently require more time to identify and correct than design errors, since the documentation associated with traditional manufacturing methods cannot assist the designers in their search for a fabrication error.

Although the value of CAD/CAM/CAT in the design of large computers is becoming more widely recognized and accepted, processors using conventional MOS or TTL technologies can be and are still being designed with traditional approaches. However, if high speed device technologies such as those described in the previous section are to be employed, the use of advanced CAD/CAM/ CAT packages becomes nearly mandatory, because the interconnects between logic components must all be created in a stringently controlled transmission line environment. It is, however, feasible to incorporate all of the requisite design rules for a transmission line logic family into a comprehensive computer aided design program which completely removes from the designer the entire burden of creating the transmission line environment. A greater number of designers can thus employ the highest speed logic families without first becoming experienced in transmission line theory.

VIII. ACKNOWLEDGMENTS

R. D. Beistad, K. M. Rice, D. J. Schwab, and S. V. Colvin for technical assistance; D. C. Darling, E. C. Quarve, M. C. Fynbo, D. F. Kasten, J. A. Lauer, S. J. Richardson, and E. M. Doherty, for assistance in preparation of text and figures.

VIIII. REFERENCES

Alexandridis, N. A. (1978). Bit sliced microprocessor architecture. IEEE Comput. Mag. 11, 56.

Atkins, D. E., and Rutenbar, R. B. (1980). Department of Electrical and Computer Engineering, private communications.

Control Data Corporation (1979). Advanced Flexible Processor Microcode Cross Assembler Reference Manual, Document 77900500.

Dias, D. M. and Jump, J. R. (1981). Analysis and simulation of buffered delta networks. IEEE Trans. Comput. C-30(4), 273.

Gilbert, B. K., Storma, M. T., James, C. E., Hobrock, L. W., Yang, E. S., Ballard, K. C., and Wood, E. H. (1976). A real-time hardware system for digital processing of wide-band video images. IEEE Trans. Comput. C-25, 1089.

Gilbert, B. K., Chu, A., Atkins, D. E., Swartzlander, E. E., Jr., and Ritman, E. L. (1979). Ultra high-speed transaxial image reconstruction of the heart, lungs, and circulation via numerical approximation methods and optimized processor architecture. Comput. Biomed. Res. 12, 17.

Gilbert, B. K., Krueger, L. M., and Beistad, R. D. (1980). Design of prototype digital processor employing subnanosecond emitter coupled logic and rapid fabrication techniques. IEEE Trans. Compon. Hybrids Manufact. Tech. CHMT-3, 125.

Gilbert, B. K., Kenue, S. K., Robb, R. A., Chu, A., Lent, A. H., and Swartzlander, Jr., E. E. (1981). Rapid execution of fan beam image reconstruction algorithms using efficient computational techniques and special-purpose processors. IEEE Trans. Biomed. Eng. BME-28, 98.

Kozdrowicki, E. W. and Theis, D. J. (1980). Second generation of vector supercomputers. Computer 13(11), 71.

LeLann, G. (1981). A distributed system for real-time transaction processing. Computer 14(2), 43.

Nahamoo, D., Crawford, C. R., and Kak, A. C. (1981). Design constraints and reconstruction algorithms for traverse-continuous-rotate CT scanners. IEEE Trans. Biomed. Eng. BME-28(2), 79.

Morse, S. P., Pohlman, W. B., and Revenel, B. W. (1978). A 16-bit evolution of the 8080. IEEE Comput. Mag. 11, 18.

Ornstein, S. M., Crowther, W. R., Kraley, M. F., Bressler, R. O., Michel, A., and Heart, F. E. (1975). Pluribus - a reliable multiprocessor. AFIPS Natl. Comput. Conf. 44, 551.

Ritman, E. L., Kinsey, J. H., Robb, R. A., Gilbert, B. K., Harris, L. D., and Wood, E. H. (1980). Three-dimensional imaging of the heart, lungs, and circulation. Science 210, 273.

Robb, R. A., Lent, A. H., Gilbert, B. K., and Chu, A. (1980). The Dynamic Spatial Reconstructor: A computed tomography system for high-speed simultaneous scanning of multiple cross sections of the heart. Journal of Medical Systems 4(2), 253.

S-1 Project Staff (1979). The S-1 Project, Volumes II and III. Lawrence Livermore Laboratory Report UCID 18619.

Siegel, H. J. (1979). A model of SIMD machines and a comparison of various interconnection networks. IEEE Trans. Comput. C-28(12), 907.

Siegel, H. J. (1980). The theory underlying the partitioning of permutation networks. IEEE Trans. Comput. C-30(4), 254.

Stritter, E. and Gunter, T. (1979). A microprocessor architecture for a changing world: The Motorola 68000. IEEE Comput. Mag. 12, 43.

Tasto, M. (1977). Parallel array processors for digital image processing. Optica Acta 24, 391.

Wittie, L. D. (1981). Communication structures for large networks of microcomputers. IEEE Trans. Comput. C-30(4), 264.

Wu, S. B. and Liu, M. T. (1981). A cluster structure as an interconnection network for large multimicrocomputer systems. IEEE Trans. Comput. C-30(4), 254.

Yamamoto, S. (1979). Hitachi Corporation, private communications.

Ziegler, S., Allegre, N., Johnson, R., Morris, J., and Burns, G. (1981). Ada for the Intel 432 microcomputer. Computer 14(6), 47.

PYRAMID ALGORITHMS AND MACHINES

Charles R. Dyer

Department of Information Engineering
University of Illinois at Chicago Circle
Chicago, Illinois

I. INTRODUCTION

Hierarchical structures are well suited for a) processing at
multiple levels of resolution ensuring the use of the most
appropriate resolution for the task and image at hand, and b)
representation at multiple levels of resolution so that
subsequent processes may make use of the most appropriate
(approximate) description. Pyramid data structures (also called
processing cones or layered recognition cones), for example, have
been used primarily for the purpose of efficiently computing
multi-resolution features (e.g. gray level, color, texture, edge)
for non-local neighborhoods in an image (typically nonoverlapping
squares).

In addition to their value as structures for efficiently
applying multi-size operators, pyramids also provide a regular,
hierarchical interconnection strategy from which various
hierarchical object representations can be derived. For example,
the quadtree (Dyer, Rosenfeld and Samet, 1980) representation of
regions can be constructed using a parallel bottom-up "pruning"
procedure which eliminates all subtrees which satisfy a given
homogeneity criterion. In general, given a segmented image and a
particular pyramid structure, multi-level representations can be
defined by marking those arcs which group pixels into regions,
and regions into relational object graphs.

Given such a representation, its computational efficiency
greatly depends on how well the pyramid's explicit
interconnections conform with the implicit links inherent in a
given description. For example, in early work with pyramid
structures for picture segmentation (Horowitz and Pavlidis,
1976), the lack of explicit pyramid links for efficiently
grouping the regions extracted by a pyramid split-and-merge

algorithm forced a final conversion to an entirely different data
structure (the region adjacency graph) which facilitated this
final grouping step. Thus the degree of match between a data
structure and the logical representation of the regions being
encoded, determines, to a great extent, the efficiency of the
algorithms which use that structure.

In this paper we develop an alternative, overlapped pyramid
structure which is similar to other proposed pyramids such as
(Hanson and Riseman, 1974; Tanimoto and Pavlidis, 1975; Burt,
Hong, and Rosenfeld, 1980). That is, it is a data structure
defined by stacking progressively smaller arrays on top of one
another, where the bottom level contains the image to be
processed and the top level contains a single node. Given such a
pyramid structure we then describe its advantages from two
divergent perspectives. First, we describe how such a structure
can be conveniently used for defining "overlapped quadtrees" for
region representation. That is, parallel propagation algorithms
are defined which link together all pixels within a region as a
subtree of the pyramid. The pyramid interconnection scheme used
here makes each such subtree have properties which are closely
related to the underlying region that it represents. Second, we
consider the feasibility of a hardware implementation of this and
related pyramid structures as the logical future extension of the
two-dimensional array machines currently being built. The
overall structure of the pyramid interconnection scheme is shown
to be ideally suited for VLSI technology and several possible
designs of such pyramid machines are outlined. See (Tanimoto,
1981) for other considerations related to the design of pyramid
machines.

The major difference between the pyramid proposed here and
previous versions is in the selection of a node's neighborhood
which explicitly links each node to others at its own level, the
level above it, and the level below it. Initially, pyramids were
defined such that each node was connected to four or eight
nearest neighbors at its own level, four children nodes in a 2 by
2 block at the level below, and a single parent node at the level
above (Tanimoto and Pavlidis, 1975). Hanson and Riseman (1974)
proposed structures with larger overlapping neighborhood sizes.
Recently, Burt et al. (1980) have considered pyramids in which
parent-child links are dynamically determined (within a 4 by 4
block at the level below and a 2 by 2 block at the level above),
varying from iteration to iteration of an algorithm.

The remainder of this paper defines our version of overlapped
pyramids in Section II, outlines parallel quadtree construction
algorithms using such a structure in Section III, and, in Section
IV, discusses design considerations for a pyramid machine for
VLSI.

II. OVERLAPPED PYRAMIDS

    The definition of overlapped pyramids that we propose can be
described loosely as the superposition of two nonoverlapped
pyramids in which at each level a node in one pyramid represents
a block of pixels in the original image which is shifted 50% with
respect to the corresponding node in the other pyramid. That is,
level k consists of an array of nodes such that one set of nodes
represents a partition of the base array into (2**k) by (2**k)
blocks (one node per block), and a second set of nodes at this
same level represents a second partition of the base array such
that each node's block is shifted by 2**(k-1) with respect to a
node in the other set. Fig. 1 illustrates the relationship
between a 3 by 3 block of nodes and the blocks of pixels at level
0 which are their pyramid descendants.
    Formally, assume we are given a (2**n) by (2**n) image. We
define an overlapped pyramid as an (n+1)-level structure, where
node (i,j,k) specifies the node at level k (level 0 is the base
array and level n is the root) and at coordinates (i,j) within
that array. Rows (columns) are indexed from the top (left) from
0 to (2**(n-k+1))-2, k > 0. Node (i,j,1) has four children
(i',j',0), where i' = i and i+1, and j' = j and j+1. Node
(i,j,k), k > 1, has five children (2i,2j,k-1), (2i,2j+2,k-1),
(2i+2,2j,k-1), (2i+2,2j+2,k-1), and (2i+1,2j+1,k-1).
Disregarding nodes on the border of each array level, each node
has four siblings and either one or four parents. Node (i,j,k)
has siblings (i-1,j,k), (i+1,j,k), (i,j-1,k), and (i,j+1,k).
Node (i,j,0) has four parents (i',j',1), where i' = i-1 and i,
and j' = j-1 and j. Node (i,j,k), 0 < k < n, has one parent
((i-1)/2,(j-1)/2,k+1) if i or j is odd; otherwise its four
parents are nodes (i',j',k+1), where i' = i/2 - 1 and i/2, and j'
= j/2 - 1 and j/2. Thus level 0 contains a (2**n) by (2**n)
array, level 1 a (2**n)-1 by (2**n)-1 array, level 2 a (2**(n-
1))-1 by (2**(n-1))-1 array, ..., and level k, 0 < k <= n, a
(2**(n-k+1))-1 by (2**(n-k+1))-1 array. Thus the pyramid
contains less than 7/3 as many nodes as in the original image
array.
    This pyramid definition differs from that of Burt et al.
(1980) in that here 4-neighbors at level k represent 50%
overlapping blocks at all levels k-1, ..., 0. In (Burt et al.,
1980) the degree of overlap varies with the height of a pair of
4-adjacent nodes. (For example, at level k-1 there is 50%
overlap, at level k-2 there is 60% overlap, etc. with the degree
of overlap approaching 66% as we descend the pyramid.) In our
case resolution is reduced at each level except the first by
sampling the level beneath it at alternate rows and columns. As
a result, about half of the nodes at a level have just one
parent. In this way four of a node's children represent
nonoverlapping blocks at each level k-1, ..., 0. The fifth child

FIGURE 1. A node's five children (shaded circles) and the blocks of pixels they represent.

represents the (2**(k-1)) by (2**(k-1)) block of pixels at level 0 which is centered beneath node (i,j,k).

The principal advantage of this pyramid definition is that it provides a reduced resolution partition of the original image at every level without incurring the disadvantages of nonoverlapped pyramids. This is due to the fact that a node's eight nearest neighbors at its own level represent blocks which overlap the given node's block by either 50% (its horizontal and vertical neighbors) or 25% (its diagonal neighbors). Furthermore, the structure guarantees that every pair of nodes at level 0 which are chessboard distance d apart, have a lowest common ancestor at level ⌈log d+1⌉. Thus algorithms involving data communication between nodes representing physically adjacent blocks require less data movement since these nodes are always at distance two apart rather than at a variable or image size dependent distance. Second, this structure provides for an alternative definition of the quadtree representation of regions which is less sensitive to object position and image size than is true of quadtrees. In particular, the root of the quadtree constructed from an overlapped pyramid will always be at a level proportional to the logarithm of the region diameter. This is discussed further in the next section. Finally, the pyramids defined here have nodes of degree considerably less than in other overlapped pyramids. This is important when we consider the feasibility of implementing pyramid machines in VLSI where chip layouts and area requirements are heavily dependent on the number and regularity of communication lines interconnecting processors.

III. REGION REPRESENTATION IN OVERLAPPED PYRAMIDS

Quadtrees (Dyer et al., 1980) are a variable-resolution data structure which have been shown to be an efficient representation for compact regions. Two main disadvantages of this representation have been observed, however. First, the size of the representation and the speed of processing a given object strongly depend on the position of this object in an image and the image's size. For example, in (Dyer, 1980) it was shown that

a square object of size 2**m in a 2**n by 2**n image may be
encoded in a quadtree ranging in size from O(n-m) nodes to
O(2**(m+2) + n - m) nodes. Second, representing multiple objects
within a single image often means that the quadtrees associated
with these objects are not disjoint. Thus an edge or node high
up in one quadtree may also be part of the quadtree
representations for an unbounded number of other objects.
Consequently, both for speed of processing a quadtree and for
maintaining disjoint quadtrees for disjoint objects, it would be
far more desirable if the quadtree representation of an object
always contained its root node at a level proportional to the
logarithm of the object's diameter.

In the remainder of this section we provide an alternative
definition of quadtrees based on the overlapped pyramid structure
defined above. For ease of exposition we first describe the
one-dimensional case.

A. One-Dimensional Overlapped Quadtrees

Given a size 2**n by 2**n binary image such that black pixels
represent object points and white pixels background points, we
construct the overlapped quadtree for each connected set of black
pixels using a parallel propagation algorithm in a one-
dimensional overlapped pyramid. More specifically, the (2**n)-1
nodes at level 1 each mark at time step 1 the type of size 2 base
which occurs below it. That is, each records whether it sees the
interior of a region (two black pixels), the background (two
white pixels), the left end of a region, or the right end of a
region. Beginning at the next step, a parallel bottom-up
propagation of these four label types occurs until the lowest
common ancestor node of each region's (left end, right end) pair
is marked. From the properties of the overlapped pyramid
structure it can be seen that the root node for a region of size
d always occurs at level $\lceil \log (d+3) \rceil$ and is centered above the
region.

As soon as the root node of a quadtree is marked, this node
can immediately initiate a downward signal to prune redundant
links and nodes in the tree as desired. In the worst case,
quadtree construction and pruning requires time proportional to
the logarithm of the region diameter. (Recall in nonoverlapped
quadtrees time proportional to the logarithm of the image
diameter was required.) The size of the overlapped quadtree
(i.e., number of nodes) is, in the worst case, proportional to
the region area. Furthermore, each quadtree is disjoint from all
others representing disjoint connected regions in the image.
Finally, if desired, each quadtree's root node could also
initiate a signal upwards in the pyramid in order to construct a
tree-structured relational description of all the regions in the
image.

B. Two-Dimensional Overlapped Quadtrees

In two dimensions an overlapped quadtree representation can be constructed by performing both horizontal and vertical propagation operations. In this case, however, when multiple objects are present in an image both the horizontal and vertical operations must be topology-preserving. In conventional nonoverlapped quadtrees, topology is preserved using the simple vertical shrinking operation which creates a black (white) node at level k iff its four children are black (white). While topology-preserving, this definition results in extremely deep quadtrees and, as a consequence when multiple objects are present, many common nodes and edges are used to represent several different objects.

In many circumstances where objects to be represented are numerous and relatively compact (e.g. in texture analysis), it may be preferable to obtain disjoint or almost disjoint quadtrees in which the root node of each is at a level proportional to the logarithm of the object's diameter, no matter what its position. To obtain such a representation and to preserve object topology, we will combine the vertical shrinking operation defined for nonoverlapped quadtrees with a horizontal parallel propagation operation which constructs an upright framing rectangle around an object in time proportional to the object's diameter (Dyer and Rosenfeld, 1980). This procedure is topology-preserving only when all rectangles are disjoint. For the remainder of this section we assume that all objects are sufficiently far apart so that their framing rectangles do not overlap.

The generalization of the one-dimensional algorithm consists of the following steps. The first step consists of a horizontal parallel propagation procedure which constructs the upright framing rectangle for each region using the transformation (and all 90 degree rotations of it):

$$\begin{matrix} 10 \\ 11 \end{matrix} \longrightarrow \begin{matrix} 11 \\ 11 \end{matrix}$$

As shown in (Dyer and Rosenfeld, 1980), this procedure successively fills in concavities in a given two-dimensional array, requiring at most object diameter parallel propagation steps to completely fill in the enclosing upright framing rectangle.

Next, each node at level 1 marks itself according to the contents of the 2 by 2 block of pixels below it. Since there exists a node at level 1 for every 2 by 2 block at level 0, the following features can be noted by the level 1 cells: interior, background, side (four types) and corner (four types).

At successive time steps each region's four corner marks propagate upward through the pyramid, always moving to the parent node which is in the direction of the center of the region's enclosing rectangle. Given an m by n rectangle (m < n), two pairs of corners will meet and merge at level $\lceil \log (m+3) \rceil$. These

marks continue propagating upward until the root of the quadtree
is marked.at level $\lceil \log (n+3) \rceil$. (This node is the lowest common
ancestor of the region's four corners marked at level 1.)
Finally, as in the one-dimensional case, each quadtree root node
can immediately initiate a downward propagating signal in order
to prune appropriate nodes from the tree, including those added
by the rectangle filling procedure.

Since the horizontal process is stable and requires at most
diameter time, in the worst case time proportional to a region's
diameter is required to construct its quadtree (the vertical
processes can effectively be ignored since the height of the
quadtree is proportional to the logarithm of the object's
chessboard diameter). When objects are relatively compact, the
speed of the horizontal process can be expected to be
significantly improved. It is an open problem, however, if this
construction can be performed in less than diameter time in the
worst case. In the general case where several regions' enclosing
rectangles overlap, the above algorithm is inappropriate. It is
also an open problem whether in this general case all quadtrees
can be constructed in parallel in time proportional to the
diameter of the largest region.

IV. A PYRAMID MACHINE

The design of special-purpose multiprocessor devices which
are suitable for VLSI implementation has recently received a
great deal of attention. Good characteristics of such devices
include a) the use of only a few, simple types of cells, b) data
and control flow which is simple and regular so that local and
regular interconnections of cells are possible, and c) algorithms
which use extensive pipelining and multiprocessing so that data
moves uniformly and synchronously through the device. With these
constraints in mind, the design task has become one of making
various data structures explicit in hardware in order to
efficiently solve a specific class of problems. Thus in addition
to conventional data structures such as one- and two-dimensional
arrays, chips are being designed to contain binary tree
structures, shuffle-exchange graphs, priority queues, and sorting
networks. In this way data paths which are determined by a
particular data structure and accompaning algorithms are
explicitly incorporated in the processors' organization.

The properties of pyramids which make them well-suited for
VLSI implementation include a) only a few simple cell types are
required, b) interconnections between cells are simple and
regular, c) operations can be pipelined when level-at-a-time
algorithms are specified, d) high cell-to-pin ratio means as chip
density improves larger packages won't be necessary, and e)
image-oriented representations and algorithms are more regular

and easier to implement in VLSI than object-oriented
representations such as a chain code, region adjacency graph or
strip tree.

For ease of exposition, we first describe the design of a
nonoverlapped pyramid. At the end of this section we address the
modifications necessary to implement overlapped pyramids. Each
node in the pyramid represents an identical, simple processing
element containing a few registers, buffers, words of memory, and
arithmetic and control circuitry. Specific details of this
circuitry depend on the class of operations to be performed and
will be omitted here due to lack of space. To determine chip
area requirements, we will assume each processor is square in
aspect with side length proportional to the number of wires
connecting it with its neighbors. This assumption is a common
one used in other VLSI models (Thompson, 1979). For the
overlapped case, each node can route messages directly to each of
thirteen neighbors -- five children, four brothers, and four
fathers. In order to provide adequate bandwidth for
communication, bit-parallel routing between processors will be
used. Assuming 8-bit pixel values, this means an 8-wire bus
connects adjacent nodes in the pyramid. Hence the degree of each
node is about 100.

A. Design

In this section we outline chip designs and layouts for a
nonoverlapped pyramid machine in which each node is connected to
one parent and four children. First, we consider the important
problem of determining a space-economical chip layout strategy.
The problem of producing area-efficient layouts for complete
binary trees was first considered by Mead and Rem (1979). They
showed that an "H" layout requires only O(n) area, where n is the
number of nodes in the tree. A simple modification of this H
layout scheme also provides a linear area layout for a 4-ary tree
of nodes.

Since only a fixed number of processors will fit on a single
chip, we next consider the problem of processor packaging and
chip interconnection. The strategy we will use is the same as
that proposed by Bentley and Kung (1979). That is, we will
define two types of chips: leaf chips and nonleaf chips. A leaf
chip will contain processors comprising a subtree of nodes at the
bottommost levels of the pyramid. A nonleaf chip will contain
processors corresponding to nodes higher in the pyramid. Fig. 2
illustrates this decomposition for a pyramid with four levels.

The advantage of this type of decomposition is that a leaf
chip requires only eight pins for interprocessor communication
(i.e., a single connection between a pair of adjacent processors)
regardless of the size of the subtree stored on the chip. Hence
as chip density improves with technological advances, there will

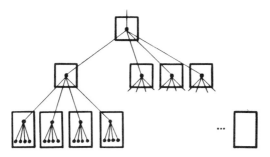

FIGURE 2. Decomposition of a pyramid into leaf chips
(rectangles) and nonleaf chips (squares).

be no pin limitations which will restrict the corresponding
increase in the subtree size contained on a single leaf chip.

On the other hand, the number of nodes on a nonleaf chip will
be restricted by pin limitations since pins are required for the
single top level node on the chip plus all of the nodes at the
lowest level of the subtree. Since packaging technology is not
expected to improve pin bounds significantly in the next decade,
nonleaf chips will be constrained by pin requirements and will
not be able to take advantage of improving component density.

B. A 512 by 512 Pyramid Machine

We now consider the possibility of fabricating in the near
future a pyramid machine which will process a 512 by 512 image.
This image size represents a realistic dimension in many
applications such as remote sensing and medicine for which a
pyramid machine could have significant impact in providing real-
time processing power. The number of nodes in such a 10-level
tree is 1 + 4 + 16 + ... + (4 ** 9) = 349,525.

Nonleaf chips can be implemented using current technology as
a 1-level subtree containing a single node. In this case eight
links (i.e. pins) will be used to support bidirectional data flow
between the node and its parent, and 32 (= 4*8) links will be
used to connect the node with its four children. With the
presence of a few additional pins for power, ground, and timing
and link synchronization, about a 45-50 pin package results.
Since pin bounds are expected to be limited to about 100 by the
end of the decade, the subtree size in nonleaf chips is not
likely to be improved significantly as VLSI technology advances.

Using current technology and the processing element area
estimates specified above, approximately ten nodes (i.e., a
pyramid with two or three levels) could be put on a single leaf
chip. Assuming component density continues to double every two
years, however, means that by 1985 approximately 100 nodes could

be put on a single leaf chip and by 1990 approximately 1000 nodes
per chip will be possible.

By the end of the decade we can therefore expect a leaf chip
to contain a 6-level subtree on a single chip (i.e., a total of 1
+ 4 + 16 + 64 + 256 + 1024 = 1365 nodes). In this case each leaf
chip stores a 32 by 32 subarray of the original image. A 16 by
16 array of leaf chips (= 256 chips) is then used to define the
bottom six levels of the pyramid machine. The top four levels
can be implemented using four layers containing 8 by 8, 4 by 4, 2
by 2, and 1 by 1 arrays of nonleaf chips (= 85 chips).

C. Overlapped Pyramid Machine

We now describe a possible VLSI implementation of overlapped
pyramids by augmenting the nonoverlapped pyramid machine. First
we consider the addition of sibling links, and then we add more
parent-child links.

Adding sibling links for every node would mean adding a
number of pins proportional to the perimeter of the base array of
nodes. Therefore we propose the following partial sibling
connection topology. For each node on a nonleaf chip, add pins
for connecting this node to each of its siblings. Each such node
is now connected to nine nodes (chips) -- its parent, four
siblings and four children -- via 8-bit data paths, requiring a
total of 72 pins.

Leaf chips are modified to incorporate a sibling link
whenever a node has its sibling present on the same chip. To
accomplish this requires a new layout with the additional wires
included. The VLSI model we will use is the same as above,
except we now allow two perpendicular wires to cross at a point.
(Using this scheme we can safely place all horizontally running
wires on one chip layer and all vertically running wires on a
second layer.) This model is similar to that used in (Thompson,
1980). Fig. 3 shows a linear area layout of the sibling-
augmented nodes on a leaf chip.

Those nodes which are on the perimeter of any one of the
sub-arrays in the six levels of a leaf chip's subtree will have
some of their sibling links missing. For these nodes to
communicate with their missing siblings, a message must be routed
to the parent of the root node on the given leaf chip. This node
then passes the message to one of its siblings which in turn
passes the message down to the appropriate node on the adjacent
leaf chip. While this is a relatively slow operation, messages
from all of the border nodes can be pipelined to the appropriate
destination nodes in a number of clock cycles which is
proportional to the perimeter of the base array on a leaf chip
(i.e., 128 cycles). Consequently, a parallel "find-neighbor"
operation in this augmented pyramid machine will take time which
is independent of the image size.

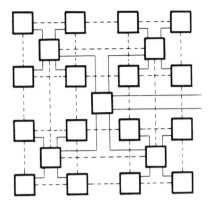

FIGURE 3. Augmented leaf chip layout. Solid lines indicate
parent-child links and dashed lines indicate sibling links.

Only minor modifications are needed to insert the additional
parent-child links to implement overlapped pyramids. Since
nonleaf chips contain one node each, no modification is needed
except to provide off-chip connections to four parents and an
additional child. Hence each node on a nonleaf chip must now
have pins for connections to thirteen neighbors, requiring a
total of about 100 pins.

Leaf chips must be modified since a leaf chip which stores a
32 by 32 subarray of the original image and a 6-level subtree
above this subarray must hold 1 + 9 + 49 + 225 + 961 + 1024 =
2269 nodes or almost twice as many as in nonoverlapped pyramids.
To add these extra nodes to our original layout in Fig. 3
requires only a small increase in total chip area, however, since
much of the area used in the original layout was occupied by data
lines. That is, the new nodes can be placed "in between" the
original nodes with little additional real estate needed.
Finally, a few additional outside connections are needed to
provide links to the four parents of the root node, i.e., a total
of 32 pins.

As in the case of adding sibling links, with this
implementation nodes which are on the border of a subarray in a
leaf chip do not have direct connections to all of their four
parent nodes (which are placed on separate chips). Again,
separate data routing would be required for these nodes to
communicate with all of their parents.

V. CONCLUDING REMARKS

An alternative pyramid data structure has been proposed which
has several desirable properties which make it more suitable for

region representation than other pyramid neighborhood
definitions. In particular, adjacent nodes at level k represent
50% overlapping blocks at all levels k-1, ..., 0. Also, a pair
of nodes at level 0 which are distance d apart have a lowest
common ancestor at level ⌈log d+1⌉. With these properties it
becomes straightforward to define hierarchical parallel
propagation algorithms within this structure. For example, we
have outlined algorithms for constructing "overlapped quadtrees"
using both horizontal and vertical propagation steps. Finally,
we have described a VLSI implementation of a pyramid machine,
taking into account the regularity of the structure (to obtain a
linear area layout), estimated component density improvements,
and packaging limitations (i.e., pin bounds).

REFERENCES

Bentley, J., and Kung, H. (1979). A tree machine for searching
    problems. Proc. IEEE International Conf. on Parallel
    Processing, 257-266.
Burt, P., Hong, T., and Rosenfeld, A. (1980). Segmentation and
    estimation of image region properties through cooperative
    hierarchical computation. Computer Science Center Technical
    Report TR-927, University of Maryland.
Dyer, C. (1980). Space efficiency of region representation by
    quadtrees. Proc. Workshop on Picture Data Description and
    Management, 31-36.
Dyer, C., and Rosenfeld, A. (1980). Propagation algorithms for
    framing rectangle construction. Pattern Recognition 12,
    211-215.
Dyer, C., Rosenfeld, A., and Samet, H. (1980). Region
    representation: boundary codes from quadtrees. Comm. ACM 23,
    171-179.
Hanson, A., and Riseman, E. (1974). Preprocessing cones: a
    computational structure for scene analysis. COINS Technical
    Report 74C-7, University of Massachusetts.
Horowitz, S., and Pavlidis, T. (1976). Picture segmentation by a
    tree traversal algorithm. J. ACM 23, 368-388.
Mead, C., and Rem, M. (1979). Cost and performance of VLSI
    computing structures. IEEE J. Solid-State Circuits SC-14,
    455-462.
Tanimoto, S. (1981). Towards hierarchical cellular logic: design
    considerations for pyramid machines. Department of Computer
    Science Technical Report 81-02-01, University of Washington.
Tanimoto, S., and Pavlidis, T. (1975). A hierarchical data
    structure for picture processing. Computer Graphics and
    Image Processing 4, 104-119.
Thompson, C. (1980). A complexity theory for VLSI. Ph.D.
    dissertation, Carnegie-Mellon University.

PROGRAMMING TECHNIQUES FOR
HIERARCHICAL PARALLEL IMAGE PROCESSORS[1]

Steven L. Tanimoto

Department of Computer Science
University of Washington
Seattle, Washington

I. INTRODUCTION

Human vision takes place across a series of levels beginning
with the retina and moving back via the optic nerve to the lateral geniculate and then to the visual cortex, and the visual cortex is itself stratified. Computer vision systems have been
structured after this observation, and in some, hardware for a
number of separate layers has been specified. Such systems have
been called recognition cones, processing cones, quadtree machines and pyramid machines. In other systems, hardware for a
single layer has been specified along with provisions for simulating a sequence of layers. These systems have been designed
to be general-purpose within image processing. Hence, common to
all of them are problems of programming the layers in a coordinated fashion to perform useful picture processing.
    Using a pyramid machine called PCLIP as an example of a
hierarchical parallel machine, we illustrate problems and solutions in programming for this class of processors. We also
suggest how some missing hardware features can be achieved in
software. We list techniques applicable in computer graphics as
well as image analysis; most of them illustrate particular problems and capabilities that arise from the use of hierarchical
structures.

[1] *Research supported in part by National Science Foundation
Grant ENG79-09246 and the University of Washington Graduate
School Research Fund.*

II. MACHINES AND TECHNIQUES

In the early 1970's recognition cone structures were introduced (Uhr, 1972) as a means of describing and recognizing two-
dimensional patterns. A series of two-dimensional layers of data
are interleaved with transforms to form such a cone. The transforms provide some specification for how the data is operated
upon and moved from lower levels to higher. A cone converges at
the apex and it is here that the results are output. It is intended that at any one time, some transform $T1$ may be operating from
some layer to the next, while others, $T2$, $T3$, etc., are operating
between other pairs of layers. The overall structure of Uhr's
cones as multiprocessing systems is a sequence of two-dimensional
processors, each of which applies a certain transform to all of
the points in its input layer and passes its array-valued result
to the next layer. The architecture is a highly-constrained MIMD
one, yet more flexible then SIMD ones.

A variation of Uhr's cone was subsequently developed into the
"processing cone" by Hanson and Riseman (1980). The transforms
are generalized to permit data movement laterally and down as well
as up through the levels. The computations of edges and texture
on a (simulated) cone have suggested that an SIMD configuration
of simple processors arranged in the conical structure is an appropriate hardware architecture.

Recently, a "quadtree machine" has been proposed by Dyer
(1981) and a similar "pyramid machine" is under development by
the author (Tanimoto, 1981) based upon the pyramid data structure
(Tanimoto and Pavlidis, 1975; Tanimoto, 1976). These, like the
cones, also call for a hierarchical network of simple processors,
but restrict the local interconnections of processors to either
four or eight lateral neighbors and four sons and a father
(rather than 25 or more sons and more than one father). The pro-
ponents of this restriction believe that it presents a more economically feasible structure without sacrificing any computational
generality or significant efficiency. A pyramid machine architecture called PCLIP is described in more detail in Section III.

A number of parallel processors for image processing including the CYTOCOMPUTER, the Massively Parallel Processor, and others
have been recently surveyed (Davis, 1980), and will not be discussed here as they do not directly make use of a hierarchical
structure. It should be mentioned that some researchers who use
parallel processors are applying them in a layered, hierarchical
fashion (e.g. Granlund, 1980) for which the discussion here may
be relevant.

Programming techniques for parallel image processors have
been described in terms of algorithms (Preston et al, 1979;

Toriwaki *et al*, 1980), and languages (Preston, 1980; Levialdi *et al*, 1980). General programming considerations for pattern recognition and image processing have been surveyed (Tanimoto, 1980).

III. A PYRAMID MACHINE

We have specified a pyramid machine called PCLIP (using the terminology of Duff, 1976) that consists of a set of cellular processors, and a single controller. The cells, arranged in a pyramid each accept inputs directly from thirteen neighbors and compute values that may be sorted in local memories or passed to other cells. The cellular operations provided in hardware consist of boolean pattern matching and transferring data from local memory to local registers.

PCLIP has three 1-bit registers per cell (in addition to local memories of 128 bits per cell). The first of the three is the "propagation" register, whose value is accessible to neighbors. The second, "target", register serves both to receive the result of a match instruction and as an additional input to it. The "condition" register disables its cell when set, causing it to ignore all instructions except a transfer of data into the condition register. These registers we identify by the letters P, T, and C, and local memory by $M1$, $M2$, ..., Mk.

To execute a match instruction, each processor receives two vectors of fifteen 0's and 1's from the controller; the first is the pattern to be compared with the neighborhood inputs (and the local P and T values), and the second is the "mask" which indicates which positions are "don't cares".

A simulator has been written for PCLIP which executes assembly language programs, processing data at a 32x32 base resolution. A VAX-11/780 mainframe and Deanza Systems display are used.

IV. PYRAMID MACHINE PROGRAMMING

Because its cellular processors perform boolean operations in one step, many of the bit-serial arithmetic techniques of CLIP are of use on PCLIP. Also, propagation labelling techniques are very appropriate. However, several programming methods (schema for parts of algorithms) seem particular to hierarchical parallel machines, and the following discussion concentrates on them.

Pyramid structures facilitate the design and executions of algorithms that make use of any of the following strategies:

(1) Planning: One creates a reduced "homomorphic" image (of
the original), analyzes it and uses the results to guide analysis
of the original.

(2) Divide and conquer: Dyer (1980) has demonstrated how the
closest pair problem (and several others) can be efficiently solved on a pyramid processor. (Dyer, 1979; Rosenfeld, 1980).

(3) Tree search: The locations of particular feature points
(e.g. northwest corner of a rectangle, or other extremes of objects in an image can be located in $O(\log N)$ time where N is the
number of rows in the image (Tanimoto, 1980a).

(4) Local/global analysis: Point neighborhood operations can
efficiently use both close-neighboring and more distant information in such tasks as adaptive thresholding and edge detection
and region analysis (Levine, 1980).

There are a number of relatively specific operations that
either recur in the above techniques or are otherwise of interest.
We examine these in more detail.

(1) Building pyramids: Starting with an image at the bottom
level of a pyramid data structure, pixel values may be assigned
at the upper levels by any of the following methods: logical operations AND, OR, XOR (assuming the image is binary); arithmetic
operations AVERAGE, MIN, MAX; local feature detection with template matching, etc. Also, one may first perform a transformation
on the image (e.g. for edges or texture) and then apply any of
these operators to fill in the upper levels. Here is an example
of assembly language code for building a pyramid using the AND
operation. This operation says to set a cell's value to 1 if and
only if all its four sons have value 1.

```
    LOAD                        ; get binary picture into P
    UNNCH P                     ; loop until no change in P
      ANDNB  DDDDDDDDDD1111D    ; set T to 1 iff all sons = 1
      ORPX                      ; OR old pyramid with new
    ENDUN                       ; end of loop
```

This operation could also be programmed without the loop by
repeating the body of the loop L times ($L+1$ is the number of
pyramid levels) or the loop could be controlled by a counter
rather than the detection of no change in P. The version presented illustrates our pyramid-data-dependent looping construct. Pyr-
amid-building is a preliminary step in many of the operations that
are listed below.

(2) Data-dependent selection: Certain kinds of feature points
can be efficiently located using bottom-up and top-down search.
For example, the location of the maximum intensity in an image is
found by first building a pyramid using MAX, and then tracing the
path of the maximum element from the root to its original location. The entire operation is performed in $O(L)$ time. Other

selection operations include finding the "pyramid maximum" of the
image (build pyramid using AVERAGE, then starting at the room
follow the father-son chain that moves to the largest son at each
step). Hierarchical template matching is a selection technique
that several researchers have investigated and is surveyed in
(Tanimoto, 1980a).

(3) Segmentation: Edge detection (Tanimoto and Pavlidis, 1975;
Hanson and Riseman, 1980), thresholding and region growing may be
performed in a manner which combines local and global information.
A variety of thresholding techniques are simple on pyramid machines that are otherwise awkward. Whereas commonly the average intensity of an image is chosen as a global threshold, in a pyramid
machine a separate threshold for each cell may be computed as a
function of the values of the cells' ancestors in a pyramid built
with AVERAGE. For example, one may let the threshold be the
average of the ancestors, thus making the threshold adapt to local
as well as global intensity variation. Cleaning of segmentations
may use a hierarchical form of relaxation labelling (Rosenfeld
et al, 1976). We demonstrate edge detection with noise removal
with a short PCLIP algorithm; it is given below.

Step 1. Input binary picture of noisy blob(s) to be
segmented (also save in TEMP1)
Step 2. Form AND pyramid (one additional level is enough)
Step 3. Identify edge cells as those with value 1 having
at least one lateral neighbor with value 0. Save results in
TEMP2.
Step 4. Project the level $L-1$ edges to level L by letting
all sons inherit the values of their fathers.
Step 5. Take intersection of projected edges with those in
TEMP2. These are the "externally cleaned" edges; save in EXTERN.
Step 6. Negate original data from TEMP1.
Step 7. Form "internally cleaned" edges by following steps 2
through 5 above, but not saving in EXTERN.
Step 8. Dilate the internally cleaned edges once.
Step 9. Take the intersection of these edges with those in
EXTERN, then stop. The cleaned edges are those in level L of the
pyramid.

(4) Graphics operations based on dilation: By successively
marking the neighbors of marked cells, a wide variety of tasks
can be accomplished. Region coloring is affected by starting the
marking from seed points within regions to be colored, and dis-
abling all cells not in any region. Marking may proceed vertically in pyramids as well as within levels to obtain acceleration
or special effects. By simultaneously labelling from two seed
points, the approximate midpoint may be found and recursively a

path between the points constructed. We now describe two labelling algorithms for region coloring. The first operates on a flat
cellular SIMD processor such as CLIP4; the second operates on
PCLIP. A Pascal-like language is used here, where $A[.,.]$ denotes
a single element of an array and $A[.,.+1]$ (for example) signifies
its neighbor to the east. We assume below that $A[0..N-1; 0..N-1]$
initially contains the input binary map and that $B[0..N-1; 0..N-1]$
contains a 1 in the seed position with zeros elsewhere.

```
procedure Flat-Color;
   iterate until nothing changes in B
      at each cell do
         if A[.,.] = 1 and at any neighbor B[.',.'] = 1
         then B[.,.] :=  1:
      end at each cell;
   end iterate;
end procedure;
```

In the second algorithm, the binary picture is initially in the
finest level of pyramid A; it's in $A[L,*,*]$. The seed pixel is
in $B[L,*,*]$ (somewhere in level L).

```
procedure Pyramid-Color;
   comment build pyramid using AND rule;
   iterate until nothing changes in A
      at each cell do
         A[.,.,.] := A[.,.,.] OR
            ( AND (i:=0,1; j:=0,1) A[.+1, 2 . +j,2 . +i];
      end at each cell;
   end iterate;
   comment now color as in flat case but with pyramid
            neighbors;
   iterate until nothing changes in B
      at each cell do
         if A[.,.,.] = 1 and
            at any neighbor B[.',.',.'] = 1
         then B[.,.,.] := 1;
      end at each cell;
   end iterate;
end procedure;
```

The algorithm Pyramid-Color can color a relatively convex blob in
$O(\log D)$ steps whereas Flat-Color always requires $O(D)$ where D
is the diameter of the blob in pixels.

(5) Nearest-neighbor operations: Given a set of "landmark" points on a discrete grid, and an arbitrary point Q on the grid, finding the closest landmark point to Q can be efficiently computed on a pyramid machine. "Iconic indexing" is a pictorial database access method that uses this operation (Tanimoto, 1976a). The problem of determining the closest pair of points in an input map can be solved efficiently on a pyramid automaton (Dyer, 1979).

(6) Accessing one cell: If no hardware instruction is provided for setting a particular pyramid cell (e.g. $A[5, 13, 25] := 1$), the cell can be selected by moving up to the root and output; or if the cell has been selected by marking it, it may communicate with the controller by placing bits into its P register (while all other cells have zero in theirs). Note that by marking several cells at a time, values may be moved and copied within the pyramid to create various geometric patterns rapidly.

(7) Global feature extraction: Besides the global average, maximum or minimum, there are other features that are easily provided with a pyramid machine. Reeves has found bit-counting to be important in threshold selection (Reeves, 1980) and has recently suggested a PCLIP algorithm for it (based on an idea of Rosenfeld) that uses bit-serial arithmetic. It requires that each node compute a sum over its four sons and pass the value up.

These seven kinds of operations augment the capabilities of flat cellular processors significantly, we believe.

V. CONCLUDING REMARKS

The design of appropriate programming languages for hierarchical parallel machines is a concern not discussed yet. It is not difficult to design extensions of Pascal, FORTRAN, LISP, etc., although constructing good compilers for such extensions by require some development effort. Perhaps of more importance in the long run will be development of interactive programming systems that use graphics not only to design image processing operators but also to display system diagrams that the user may annotate and/or modify to effect programming.

We see pyramid machines as important components of image processing systems but not usually as complete systems in themselves. The other components include host computer, graphic display, and imaging and storage elements. Special-purpose processors (e.g. for Fourier transforms) could also be combined with a system based around a pyramid machine.

Techniques have been described here for building pyramids, selecting pixels from images, segmentation, region coloring, and nearest-neighbor operations. We have not discussed some other aspects of programming these hierarchical structures such as controlling the level at which computations are performed or the details of bit-serial arithmetic (needed for multi-bit addition, etc. on PCLIP). Alternative architectures from PCLIP for pyramid machines may give rise to other programming techniques. For example, if the pyramid levels are perceived as independent processors, coroutine or pipeline paradigms are appropriate. VLSI circuitry is changing not only hardware but also the ways we think about software.

ACKNOWLEDGEMENT

Anil Gangolli made a significant contribution to our PCLIP simulator and programming experiments. Joseph Pfeiffer has helped with the design of PCLIP.

REFERENCES

Davis, L.S. (1980). Computer architectures for image processing. *In Proc. Workshop on Picture Data Desc. and Management,* IEEE Comput. Soc., Asilomar Conf. Grounds, CA, pp. 249-254.

Duff, M.J.B. (1976) CLIP 4: A large scale integrated circuit array parallel processor. *In Proc. 3rd Int. Joint Conf. on Pattern Recog.* Coronado, CA, pp. 728-733.

Dyer, C.R. (1981) A quadtree machine for parallel image processing. Tech. Rept. KSL 51, Univ. of Illinois at Chicago Circle.

Dyer, C.R. (1979) Augmented Cellular Automata for Image Analysis. Ph.D. Th., University of Maryland.

Granlund, G.H. (1980) GOP, a fast and flexible processor for image analysis. *In Proc. 5th Inst. Conf. on Pattern Recog.,* Miami Beach, FL, pp. 489-492.

Hanson, A.R., and Riseman, E.M. (1980) Processing cones: A computational structure for image analysis. *In* (Tanimoto and Klinger, 1980) pp. 101-131.

Levialdi, S., Isoldi, M., and Uccella, G. (1980) Programming in PIXAL. *In Proc. Workshop on Picture Data Desc. and Management,* IEEE Comput. Soc., Asilomar Conf. Grounds, CA, pp. 74-79.

Levine, M.D. (1980) Region analysis with a pyramid data structure. *In* (Tanimoto and Klinger, 1980) pp. 57-100.

Preston, K., Jr., Duff, M.J.B., Levialdi, S., Norgren, P.E., and Toriwaki, J.-I. (1979) Basics of cellular logic with some applications in medical image processing. *Proc. IEEE 67*, pp. 826–856.

Preston, K., Jr. (1980) Image manipulative languages: A Preliminary survey. *In* "Pattern Recognition in Practice" (E.S. Gelsema and L.N. Kaval, eds.) North–Holland Publishing Amsterdam.

Reeves, A.P. (1980) On efficient global feature extraction methods for parallel processing. *Computer Graphics and Image Proc.14*, pp. 159–169.

Rosenfeld, A. (1980) Academic Press, New York. "Picture Languages".

Rosenfeld, A., Hummel, R., and Zucker, S. (1976) Scene labelling by relaxation operations. *IEEE Trans. Systems, Man and Cybernetics 6*, pp. 420–433.

Tanimoto, S.L. (1976a) An iconic/symbolic data structuring scheme. *In* "Pattern Recognition and Artificial Intelligence" (C.H. Chen, ed.) Academic Press, New York, pp. 452–471.

Tanimoto, S.L. (1976b) Pictorial feature distortion in a pyramid. *Computer Graphics and Image Processing 5*, pp. 333–352.

Tanimoto, S.L. (1979) Image transmission with gross information first. *Computer Graphics and Image Processing 9*, pp. 72–76.

Tanimoto, S.L. (1980a) Image data structures. *In* (Tanimoto and Klinger, 1980), pp. 31–55.

Tanimoto, S.L. (1980b) Advances in software engineering and their relations to pattern recognition and image processing. *In Proc. 5th Inst. Conf. on Pattern Recog.*, Miami Beach, FL, pp. 734–741.

Tanimoto, S.L. and Klinger, A. (eds.) (1980) "Structured Computer Vision: Machine Perception Through Hierarchical Computation Structures." Academic Press, New York.

Tanimoto, S.L., and Pavlidis, T. (1975) A hierarchical data structure for picture processing. *Computer Graphics and Image Proc. 4*, pp. 104–119.

Uhr, L. (1972) Layered 'recognition cone' networks that pre-process, classify, and describe. *IEEE Trans. Computers C-21*, pp. 758–768.

MULTI- COMPUTER PARALLEL ARCHITECTURES FOR SOLVING COMBINATORIAL PROBLEMS [1]

W. M. McCormack [2]
F. Gail Gray
Joseph G. Tront
Robert M. Haralick
Glenn S. Fowler

Department of Computer Science
and
Department of Electrical Engineering
Virginia Polytechnic Institute and State University
Blacksburg, Virginia

I. INTRODUCTION

Combinatorial problem solving underlies numerous important problems in areas such as operations research, non-parametric statistics, graph theory, computer science, and artificial intelligence. Examples of specific combinatorial problems include, but are not limited to, various resource allocation problems, the traveling salesman problem, the relation homomorphism problem, the graph clique problem, the graph vertex cover problem, the graph independent set problem, the consistent labeling problem, and propositional logic problems [Hillier & Lieberman, 1979; Knuth, 1973; Kung, 1980; Lee, 1980]. These problems have the common feature that all known algorithms to solve them take, in the worst case, exponential time as problem size increases.

----

[1] This work was supported in part by the Office of Naval Research under Grant N00014-80-C-0689.

[2] The order of the authors was randomly chosen.

431

They belong to the problem class NP.

This paper describes a technique for the design of parallel computer architectures which most efficiently, or for the least cost, or for the smallest time to completion, execute parallel algorithms for solving these problems. The techniques we examine take into account the interaction between each specific algorithm and the parallel computer architecture. The class of architectures we consider are those which have inherent distributed control and whose connection structure is regular.

Combinatorial problems require solutions which do searching. In a very natural way, the algorithm for searching keeps track of what part of the search space has been examined so far and what part of the search is yet to be examined. The mechanism which represents the division between that which has been searched so far and that which is yet to be searched can also be used to partition the space which is yet to be searched into two or more mutually exclusive pieces. This is precisely the mechanism which can let a combinatorial problem be solved in an asynchronous parallel computer.

To help in describing the parallel combinatorial search, we associate with the space yet to be searched the term "the current problem". The representation mechanism which can represent a partition of the space yet to be searched can, therefore, divide the current problem into mutually exclusive subproblems.

Now suppose that one processor in a parallel computer is given a combinatorial problem. In order to get other processors involved, the processor divides the problem into mutually exclusive subproblems and gives one subproblem to each of the neighboring processors, keeping one subproblem itself. At any moment in time each of the processors in the parallel computer network may be busy solving a subproblem or may be idle after having finished the subproblem on which it was working. At suitable occasions in the processing, a busy processor may notice that one of its neighbors is idle. On such an occasion the busy processor divides its current problem into two subproblems, hands one off to the idle neighbor and keeps one itself.

The key points of this description are

(1)   the capability of problem division

(2)   the ability of every processor to solve the entire
      problem alone, if it had to

(3)   the capability of a busy processor to transfer a
      subproblem to an idle neighbor.

The parallel computer architecture research issue is:
to determine that way of problem subdivision which maximizes
computation efficiency for each way of arranging a given
number of processors and their bus communication links.

To precisely define this research issue requires

(1)   that we have a systematic parametric way of des-
      cribing processor/bus arrangements and

(2)   that we have alternative problem subdivision tech-
      niques.

For the purpose of describing processor/bus arrangements, we
use a labeled bipartite graph. The nodes are either labeled
as being a processor or as being a bus.   A link between a
pair of nodes means that the processor node is connected to
the bus node.   We do not consider all possible such graphs
but restrict our attention to regular ones.   Regular means
that the local neighborhood of any processor node is the
same as that of any other processor node and the local
neighborhood of any bus node is the same as that of any
other bus node.   As a consequence, each processor is connected to the same number of buses and each bus is connected
to the same number of processors.

It may not be readily apparent why different problem
subdivision techniques would influence computational efficiency.   After all, the entire space needs to be searched one
way or another.   However, subdivision has an integral relation to efficiency.   Processors which are not busy problem
solving can be either idle or transferring subproblems. Too
much time spent transferring subproblems will negatively
affect efficiency.   Excessive transferring of subproblems
can occur because the subproblems chosen for transfer are
too small.   A good problem subdivision mechanism transfers
large enough problems to minimize the number of times

subproblems are transferred, but transfers enough subproblems to minimize the number of idle processors.   The key
variable of problem subdivision is, therefore, the expected
number of operations it takes to solve the subproblem.
This, of course,   is a direct function of the size of the
search space for the subproblem, the basic search algorithm,
and the type of combinatorial problem being solved.

This paper addresses the interaction between the pro-
cessor/bus graph and problem size subdivision transfer
mechanism.   Once the relationships are determined and
expressed mathematically, the parallel computer architecture
design problem becomes less of an art and more of a mathematical optimization.   In addition, this paper examines the
effects of interconnection graph regularity on the physical
implementation of the system.   The problem of finding a
mathematical basis for a system partitioning which produces
a cost-effective VLSI implementation is examined.

Our ultimate goal is to allow computer engineers to
begin with the combinatorial problems of interest and determine via a mathematical optimization, the optimal parallel
computer architecture to solve the problems assuming that
the associated combinatorial algorithms, number of processors, number of buses, and costs are given.

II. PROCESSOR-BUS MODEL

In this section we discuss a processor-bus model which
can be used to model all known regular parallel architectures [Anderson & Jensen, 1975; Benes, 1964;Batcher, 1968;
Despain & Patterson, 1979; Finkel & Solomon, 1980; Goke &
Gipouski, 1974; Rogerson, 1979; Siegel & McMillan & Mueller,
1979; Stone, 1971; Sullivan and Bashkow, 1977; Thompson,
1978; Wulf & Bell, 1972].   The model does not currently
include the general interconnection and shuffle type networks.

The graphical basis for the model is a connected regular
bipartite graph.   A graph is bipartite if its nodes can be
partitioned into two disjoint subsets in such a way that all
edges connect a node in one subset with a node in the second
subset.   A graph is connected if there is a path between
every pair of nodes in the graph.   A bipartite graph is
regular if every node in the first set has the same degree
and every node in the second set has the same degree.   One
subset of nodes represents the processor nodes and one

subset represents the communication nodes in the parallel processing system. Every edge in the graph then connects a processing node to a communication node.

At this time we are not certain exactly how to compare the costs of various parallel architectures. Certainly the number of processors (n_p) and the number of communication nodes (n_c) will affect the costs. It is generally believed that design and manufacturing costs can be reduced by building the global architecture using a systematic interconnection of identical modules. If the modules must be identical, then each module must have the same number of neighboring modules. In graphical terms, this means that the bipartite graph must be regular. Let d_p be the degree of the processor nodes. This parameter defines the number of buses which the processor may directly access. Let d_c be the degree of the communication nodes (buses). This parameter defines the number of processors that a communication node must service. If $d_c > 2$, then either the communication nodes or the attached processors must possess arbitration logic to determine which processors have current access to the bus.

Any regular bipartite graph can be used to design a parallel computer structure by assigning the nodes in one set to be processors and the nodes in the other set to be communication links (or buses). Notice that theoretically either set of the bipartite graph could be the processor set. Therefore, each unlabeled bipartite graph would represent two distinctly different computer architectures depending upon which set is considered to be the processors and which set is considered to be the buses.

In systems for which $d_c = 2$, (i.e., each communiations link is reserved for transferring information between two specific processors), it is customary to model the system as a simple graph in which each processor is represented by a node and each communications link by an edge. For example, the Boolean n-cube (shown in Fig. 1) has been studied by many investigators.

Our model would require that each edge in Fig. 1 be replaced by a communication node and two edges as shown in Fig. 2. Figure 2 clearly conveys the alternative node assignments. We may make the dark nodes processors and the light nodes buses, producing the Boolean 3-cube or we may make the light nodes processors and the dark nodes buses [Armstrong & Gray, 1980]. In this case, each processor has access to two buses and each bus services three processors.

This second architecture cannot be adequately represented by
a graph in which each node is a processor and each edge a
communiations link.   If one considers the graph constructed
by replacing every edge of the standard drawing of the Boolean 3-cube, by a vertex and connecting two vertices if and
only if their corresponding edges were connected to a common
vertex in the original graph, then the fact that communication among the three processors is restricted to a common
bus is obscured by the resulting triangle structure which
implies three independent communication paths.

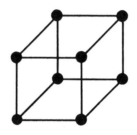

Figure 1
Common Representation
of Boolean 3-cube
processor Array

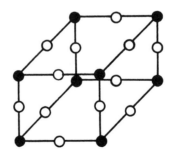

Figure 2
Bi-partite Graph
Representation of
Boolean 3-cube
Processor Array

The notation $B(n_p, d_p, n_c, d_c)$ will be used to denote a regular bipartite graph which represents an architecture with n_p processors (each connected to d_p communication nodes) and n_c communication nodes (each servicing d_c processors). The Boolean 3-cube will then be represented by a graph $B(8,3,12,2)$. In general, the Boolean n-cube will be represented by a graph $B(2^n, n, n2^{n-1}, 2)$. Reversing the assignment of nodes to processors and buses produces the $B(12, 2, 8, 3)$ graph of Fig. 2. This graph is called the p-cube by some investigators.

Other common architectures also have representations as bipartite graphs. For example, a planar array of size x^2 connected in the Von Neumann manner is modeled as a $B(x^2,4,2x^2,2)$ graph, the Moore connection results in a $B(x^2, 8, 4x^2, 2)$ graph, the common bus architecture (or star) with x processors is a $B(x,1,1,x)$ graph, and a ring architecture with x processors is a $B(x,2,x,2)$ graph. All existing architectures with regular local neighborhood interconnections can be modeled as a $B(n_p,d_p,n_c,d_c)$.

In addition to modeling existing architectures, we want to be able to generate new architectures. For example, all hardware architectures $B(n_p, d_p, n_c,d_c)$ with the same four parameters will probably have similar costs. Planarity considerations will have an effect of course, but we are not certain exactly how to quantify that cost. Neglecting planarity effects for now, we assume that two graphs with the same parameters will have similar costs. Several interesting questions arise immediately. How many bipartite graphs are there with a given set of parameters and how do we generate them? Assuming we can generate all or at least many graphs with a given set of parameters, are there other graph properties that will relate to performance evaluation?

We are currently doing simulations on similar graphs to determine whether any differences in performance occur and what the magnitude of these differences might be. The results to date are described in section IV. We plan to use simulation data to look for graphical properties that are useful in the prediction of performance. Certainly, graphs that are isomorphic must have similar cost and performance; however determining whether two graphs are isomorphic cannot be done in polynomial time. This problem is known to be equivalent to the problem of computing the automorphism partition for a given graph. We have been trying to develop efficient calculations for the automorphism partition of a graph because this partition has other performance implications.

For a given regular bipartite graph, all vertices will
not necessarily be equivalent relative to algorithm performance. In particular, nodes that are more central (have
more near neighbors) will perform differently than nodes
that are more remote (have fewer near neighbors). The automorphism partition classifies vertices into similar performing sets. We have some simulation results that indicate the
performance differences.

Another natural question asks whether a coarser partition than the automorphism partition might be used to classify performance. If so, this partition could be easier to
compute. We have been experimenting with several such classification schemes based on distance properties. Regular
bipartite graphs are said to be distance-1 regular since all
vertices have the same number of similar vertices distance-1
away. If in a distance-1 regular graph, vertices also have
the same number of vertices distance-2 away, the graph is
distance-2 regular. Similarly, a graph is distance-K regular if it is distance(K-1) regular and all vertices have the
same number of vertices distance -K away. We suspect that
the question of how to optimally overlay masks to mass produce large chips is related to distance-K regularity as
well. We are beginning to perform simulation experiments to
collect data in an attempt to relate system performance to
distance properties.

III. TREE SEARCHING IN MULTIPROCESSOR SYSTEMS

A. Introduction to Tree Searching

In order to make effective use of a multiple asynchronous processor system for any problem, a major concern is
how to distribute the work among the processors with a minimum of interprocessor communication. Kung [Kung,1980]
defines module granularity as the maximal amount of computational time a module can process without having to communicate. Large module granularity is better because it reduces
the contention for the buses and reduces the amount of time
a processor is either idle or sending or receiving work.
Also, large granularity is usually better because of the
typically fixed overhead associated with the synchronization
of the multiple processors.

In the combinatorial tree search problems we are considering, module granularity as defined by Kung is not as meaningful because each processor could in fact solve the entire

problem by itself without communicating to anybody. For our problem a more appropriate definition of module granularity might be the expected amount of processing time or the minimum amount of processing time before a processor splits its problem into two subproblems, one of which is given to an idle neighboring processor and one of which is kept itself.

When a processor has finished searching that portion of the tree required to solve its subproblem, it must wait for new work to be transferred from another processor. The amount of time a processor must wait before transmission begins and until transmission is completed is time wasted in the parallel environment that would not be lost in a single processor system. Thus, one must expect improvement in the time to completion to solve a problem in the multiple processor environment to be less than proportional to the number of processors. The factors that can affect the performance by either reducing the average transmission time or reducing the required number of transmissions include choice of algorithm, choice of search strategy, and choice of subproblems that busy processors transfer to idle processors.

B. Choice of Algorithm

In the single processor case, various algorithms have been proposed and studied to efficiently solve problems requiring tree searches. These usually involve investing an additional amount of computation at one node in the tree in order to prune the tree early and avoid needless backtracking. In work on constraint satisfaction [Haralicck & Elliott, 1980], the forward checking pruning algorithm was found to perform the best of the six tested and backtracking the worst.

For the same reasons, it seems clear that pruning the tree early should be carried over to a multiple processor system to reduce the amount of computation necessary to solve the problem. There are other reasons as well. Failure to prune the tree early may later result in transfers to idle processors of problems which will be very quickly completed. Since a transfer ties up, to some extent, both the sending and receiving processor, time is lost doing the communication and the processor receiving the problem would shortly become idle.

We would, therefore, expect that in the multiple processor environment the forward checking pruning algorithm for constraint satisfaction would work much better than

backtracking. However, in the uniprocessor environment Haralick and Elliott also showed that too much look ahead computation at a node in the search could actually increase the
problem completion time. It is also not clear why the best
algorithm for the single processor case would be the best
for the multiple processor system. Doing additional testing, as some of the other algorithms do, may be better in
the multiple processor case because it may eliminate more
nodes in the tree earlier and result in less communication
overhead and delay. Thus, it may be best to do as much
testing early in order to eliminate future transfers in contrast to the single processor case where only some extra
testing has been found to be worthwhile.

A second consideration in the selection of a search
algorithm is the amount of information that must be transferred to an idle processor to specify a subproblem and any
associated lookahead information already obtained pertinent
to the subproblem. In most cases this is proportional (or
inversely proportional) to the complexity of the problem
remaining to be solved. Thus the transmission time will be
a function of the problem complexity. Backtracking requires
very little information to be passed while, for forward
checking, a table of labels yet to be eliminated must be
sent.

C. Search Strategy

Search strategy is a second factor of importance to the
multiple processor environment. When a problem involves
finding all solutions, like the consistent labeling problem,
the entire tree must be searched. Thus, in a uniprocessor
system the particular order in which the search is conducted, i.e., depth first or breadth first, has no effect.
In a multiple processor system, however, this is a critical
factor because it directly affects the complexity of the
problems remaining in the tree to be solved and available to
be sent to idle processors from busy processors.

A depth first search will leave high complexity problems
to be solved later (that is, problems near the root of the
tree). This would seem to be desirable in the multiple processor environment because passing such a problem to an idle
processor would increase the length of time the processor
could work before going idle and thereby reduce the need for
communication. On the other hand, a breadth first search
would tend to produce problems of approximately the same
size. Since the problem is not completed until all

processors are finished, the breadth first strategy might be
preferable if it results in all processors finishing at
about the same time. It might be that the best approach
could be some combination of the two; for example, one might
follow a depth first strategy for a certain number of levels, then go breadth first to a certain depth, and then continue depth first again.

D. Problem Passing Strategy

A factor closely related to the search strategy occurs
when a processor has a number of problems of various complexities to send to an idle processor. The optimization
question is how many should be sent and of what complex-
ity(ies). Further complicating this is a situation where
the processor is aware of more than one idle processor. In
such a situation, how should the available work be divided
and still leave a significant amount for the sending processor?

Further complicating this question is the fact that the
overhead involved in synchronizing the various processors
and transmitting problems to idle ones will eventually reach
a point where it will be more than the amount of work left
be done. An analogous situation exists in sorting; fast
versions of QUICKSORT eventually resort to a simple sort
when the amount remaining to be sorted is small [Knuth,
1973].

In this case, it would appear that a point will eventually be reached where it is more effective for a processor
to simply complete the problem itself rather than transmit
parts of it to others. Determination of this point will
depend on the depth in the tree of the problem to be solved
and the amount of information that must be passed (which
depends on the lookahead algorithm being used).

E. Processor Intercommunication

One decision that has to be made is how the need to
transfer work is recognized. Specifically, does a processor
which has no further work interrupt a busy processor, or
does a processor with extra work poll its neighboring processors to see if they are idle.

The advantage of interrupts is that as soon as a processor needs work, it can notify another processor instead of

waiting to be polled. This assumes, however, that a processor would service the interrupt immediately instead of waiting until it had finished its current work. Furthermore,
when a processor goes idle, it cannot know which of its
neighbors to interrupt. This is related to handling multiple servers with a single queue which performs better than
using one queue per server. Using polling, an idle processor can be sent work by any available neighboring processor
instead of being forced to choose and interrupt one. In
addition, although an interrupted processor may be working
or transmitting (a logical and necessary condition) when
interrupted, it may not have a problem to pass when it is
time to pass work to the interrupting processor. In fact,
the interrupted processor could itself go idle. For these
reasons the simulation we discuss in section IV uses polling. Whenever a processor completes a node in the tree, and
as long as it has work it could transfer, it checks each
neighboring CPU and the connecting bus. If both are idle, a
transfer is made.

IV. SIMULATION EXPERIMENTS

In order to better understand the behavior of the
tightly coupled asynchronous parallel computer, we have
conducted simulation experiments using the consistent
labeling constraint satisfaction problem. The program
SIMULA [Birtwistle, Myhrhaug & Nygaard, 1973] was used to
perform these experiments. Let U and L be finite sets with
the same cardinality. Let $R \subseteq (U \times L)^2$ with $\#R/\#(U \times L)^2 =$
0.65. We use the simulated parallel computer to find all
functions $f: U \longrightarrow L$ satisfying that for all $(u,v) \in U \times U$,
$(u,f(u),v, f(v)) \in R$.

The goals of the experiments were to investigate the
effect of problem size (the cardinality of U and L), the
algorithm, the problem passing strategy, and the number of
processors. In the set of experiments we describe here,
each processor is connected via buses to six other processors in a regular manner. Specifically processor i is connected to processor i-1, i+1, i-7, i+7, i-11, and processor
i+11, all taken modulo the number of processors.

In the above regular architecture we varied the number
of processors and measured for each execution of the same
problem the average number of processors working. (where
working means working on the problem and not sending or
receiving problems or being idle). We compared the forward

checking and backtracking algorithms for various problem passing and tree searching strategies.

Our results indicate the following using the forward checking algorithm with #U=#L. As we increased problem complexity, by increasing the cardinality of U and L, and while keeping the number of processors constant, the average number of working processors got closer and closer to the number of processors. This indicated that for small problem sizes the problem at a depth of three or four in the tree soon became so easy to solve that it was not passed to processors far away from the processor having the original problem. A greater processor interconnection could solve this but at a greater parallel computer cost.

In [Haralick & Elliott, 1980], the superiority of forward checking to backtracking is clearly shown with respect to the number of checks needed to solve a problem. This superiority grows as the size of the problem increases. For example, when the problem size is #U=#L=12, then backtracking requires 7.5 times as many checks, but by size 24 it is over 14 times as many. Table I shows a comparison of these two algorithms for a small problem (size =12) and 25 processors. Because forward checking requires fewer tests, the time the problem was completed is, as expected, much less. The magnitude of the difference is reduced because, on the average, more processors are working with backtracking. The problem is too small for forward checking to keep all processors busy.

Also of significance is the disparity in the number of problems transferred. Since forward checking, by looking ahead, eliminates impossible solutions earlier, there are fewer problems to be transferred. This reduces the dependency of the problem on the processor interconnection scheme and reduces overhead associated with transferring work. Similar results were obtained in testing the two algorithms on other small problems. Tests were limited because the backtracking experiments took considerably longer to run; for example, in the case above, backtracking took over thirty times longer.

TABLE I. Comparison of the backtracking algorithm to the
forward checking algorithm with 25 processors
and #U=#L=12

	Time Done	#Problem Sent	Ave. # working
Backtracking	140298	1666	23.998
Forward checking	37710	252	19.058

In addition to the number of processors or algorithm
used, we tested other factors to study their effect. For
example, when a processor has a choice of subproblems to
transfer to another, intuitively the problem requiring the
most work to finish should be transferred because the second
processor will be longer which reduces overall communication
time. In one test to confirm this, the subproblem requiring
the least work instead of the most was always transferred.
It took 70% longer to complete the problem and there was a
250% increase in the number of subproblems transferred.
There was a corresponding drop in the average number of
working processors.

Given that it is better to pass more complex problems,
we conducted tests to compare doing the tree search breadth
first or depth first. The results clearly show the advantage of a depth first search. The average completion time
was typically at least 25% less and one fourth as many problems were transferred. Usually the differences were even
greater. This improvement is achieved because depth first
leaves larger problems available to transfer; thus processors sent work can work longer before becoming idle again.

A final issue in problem passing is whether or not very
low complexity subproblems should be transferred at all.
Table II shows the effect of not passing to other processors
problems at or below a particular depth in the tree, using
the forward checking algorithm.

TABLE II. The Effect of Not Sending Problems below depth d.

Depth d	Time done	#Prob. sent	Ave. working	Ave. sleeping	Expected # of consisdency checks to complete problem at this depth
3	144568	121	34.7	64.1	166.0
4	86059	671	56.7	32.3	81.7
5	85652	1730	56.3	16.5	39.6
6	89213	2238	53.3	13.8	22.2
7	93269	2426	51.4	15.0	15.0

There are 100 processors and #U=#L=16 with the forward checking algorithm, depth first search, and always passing the biggest problem.

For comparison the expected number of tests necessary to complete a problem at that depth, based on [Haralick & Elliott,1980] is included. The results clearly show that restricting problem passing for small problems can improve the performance of the system. Of course, too great a restriction will degrade performance because not all processors will be busy. It is not clear how the optimum value should be determined; although further comparisons or analytical and simulation results should provide insight.

V. EFFECTS OF REGULARITY ON SYSTEM IMPLEMENTATION

Advances in integrated circuit technology have led to VLSI circuits which have ten times the number of devices possible in LSI circuits [Foster & Kung, 1980; Mead & Conway, 1980]. This increase in the number of possible devices on a chip makes it more feasible to implement multiprocessor or multicomputer architectures [Sieworek & Thomas & Scharfetter, 1978]. The most significant of the recent advances in IC technology has been in the area of X-ray lithography. Laboratory circuits fabricated using X-ray lithography have been shown to be two orders of magnitude better than those built using optical techniques. X-ray lithography has also been shown to be ten times better than electron lithography and is in addition less expensive. Thus, depending on the

circuit complexity, it may be possible to put as many as ten
computing systems on a single chip.

Maintaining   high   speeds   in   a   multiproces-
sor/multicomputer system is one of the prime reasons that it
is necessay to use a higher density IC technology like VLSI
rather than using multiple LSI chips.   The inter-chip capa-
citance encountered in multiple LSI chip systems would prove
to be extremely detrimental in a system where the number of
processors is greater than ten.

In order to take the best advantage of VLSI technology
and in order to be cost effective, it is necessary to design
multiprocessor/multicomputer systems to be as geometrically
regular as possible [Mead & Rem, 1979]. Regularity improves
the ease with which a system may be designed and also leads
to higher chip yield. Additionally, regularity in the hardware makes the system easier to expand, allowing the basic
design to be useable in a broader set of applications.

Many different types of regularity can be incorporated
in multiprocessor/ multicomputer interconnection networks.
These types of regularity range from the global level, wherein the entire architechture is completely regular, (e.g.,
the Von Neumann or Moore array) to networks wherein the
regularity is quite local(e. g. snowflake architectures).
In a network which is globally regular, each element in the
network will have an identical structure in terms of its
capabilities and its interconnection facilities.   Local or
sub-global regularity implies that there may be more than
one type of network element, each having different capabilities or interconnection facilities.

The degree of regularity is linked to the robustness of
the interconnection network as well as to the overall system
performance and cost constraints.   Tradeoffs are therefore
possible, which will produce the design of a system that
meets the basic design objectives and is physically realiza-
ble.

Realization of a VLSI implementation of a large parallel
processing system mandates much interplay between the system
architects and the integrated circuit designers. A systematic approach to the problem is to divide the overall network up into small hardware subsections.   This subsection
division is best made at regularity boundaries. The ideal
situation would be to have each hardware subsection identical to every other hardware subsection.

In the case where the subsections are not identical, the implementation details of one subsection should not strongly influence the details of other subsections. Some interaction between subsections may occur, especially if an entire system is to be implemented on a single chip. Examples of typical interaction include layout area overlap, crossing wires, and power consumption restriction. To maintain the integrity of a modular approach to the implementation, these interactions should be minimized as much as possible.

Once the individual subsections have been identified and their implementation details specified, the IC designer must decide on the type and number of subsection which can be placed on a single chip. This decision is of course highly dependent upon the limitation of the IC technology being used. Not only must the needed chip area and power dissipation be considered, but also, and most importantly, the amount of information which will need to be conveyed off-chip must be taken into account. (The need for finding an efficient means of transferring sufficient information to/from a VLSI circuit is a difficult problem which will require solutions different from those used with today's LSI technology.)

As an example of the types of subsection placement that might be found in a system, consider the Von Neumann array of computers. One VLSI realization might be as a two chip-type set. The first chip-type might be one which has five computers and their four associated interconnecting bus controllers on a single chip. The second chip-type would then consist of a single computer and four bus controllers. The Von Neumann interconnection would be formed by appropriately wiring each of the two different chip-types to form the complete array. See Fig. 3

Certain problems are obvious in this example. Probably the foremost is that the first chip-type will need to have the facility for connecting to twelve off chip buses. Although this may seem to be an unrealistic burden to have been placed on a single chip, it may be that the design objectives necessitated a more powerful computer in the array positions occupied by the second chip-type. Assuming that the more powerful computer takes up more semiconductor real estate, the tradeoff by the IC designer may be justifiable.

It can be seen in the above example that the regularity
of the system contributes heavily to the ability to make
subsection realization positioning tradeoffs at IC design
time.   In addition, system regularity generally reduces the
problem of programming a large parallel processing system
and greatly improves the overall understandability of the
architecture.

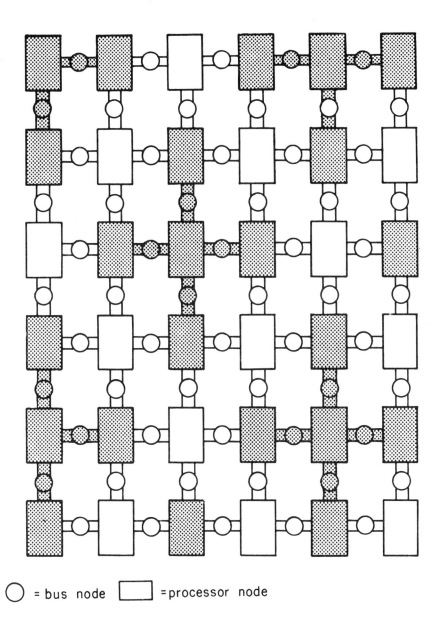

○ = bus node ▭ = processor node

Fig. 3.   Regular layout, VonNewmann Array.

REFERENCES

Anderson, G. A., and E. D. Jensen, "Computer Interconnection
    Structures: Taxonomy, Characteristics, and Examples,"
    Computing Surveys, Vol. 7, Dec. 1975, pp. 197-213.

Armstrong, J. R. and F. G. Gray, "Some Fault Tolerant Pro-
    perties of the Boolean n-Cube," Proceedings of the
    1980 Conference on Information Sciences and Systems,
    Princeton, NJ, March 26-28, 1980, pp. 541-544.

Benes, V. E., "Optimal Rearrangeable Multistage Connecting
    Networks," Bell System Technical Journal, July 1964,
    pp. 1641-1656.

Batcher, K. E., "Sorting Networks and Their Applications,"
    Spring Joint Computer Conference, 1968, pp. 307-314.

Birtwistle, G. M., Dahl, O. J., B. Myhrhaug, and K. Nygaard,
    SIMULA Begin, Auerbach Publishers Inc. Philadelphia,
    PA, 1973.

Despain, A. M. and D. A. Patterson, "X-Tree: A Tree Struc-
    tured Multi-processor Computer Architecture," 5th
    Annual Symposium on Computer Architetura,architecture,
    1978, pp. 144-151.

Finkel, R. A. and M. A. Solomon, "Processor Interconnection
    Strategies," IEEE Transactions on Computers, Vol.
    C-29, May 1980, pp. 360-370.

Foster, M.J. and H.T. Kung, "The Design of Special Purpose
    VLSI Chips", Computer, Jan. 1980.

Goke, R. L. and B. S. Lipovski, "Banyan Networks for Parti-
    tioning Multiprocessor Systems," Proceedings of First
    Conference on Computer Architecture, 1974, pp. 21-28.

Haralick, Robert M. and G. Elliott, "Increasing Tree Search
    Efficiency for Constraint Satisfaction Problems",
    Artificial Intelligence, Vol. 14, 1980, pp. 263-313

Hillier, F. S. and G. S. Lieberman, Operations Research,
    Holden Day, Inc., San Francisco, 1979.

Knuth, D. E. The Art of Computer Programming, Sorting and
    Searching, Addison-Wesley Publishing, Reading, MA,
    1973.

Kung, H. T., "The Structure of Parallel Algorithms," in Advances in Computers, Vol. 19, edited by M. D. Yovits, Academic Press, 1980.

Lee, R. B., "Empirical Results on the Speed, Redundancy, and and Quality of Parallel Computations," Proceedings of 1980 International Conference on Parallel Processing, 1980.

Mead, C. A. and M. Rem, "Cost and Performance of VLSI Computing Structures," IEEE J. Solid State Circuits, sc-14(2), pp. 455-462, 1979.

Mead, C. A. and L. A. Conway, Introduction to VLSI Systems, Addison-Wesley, Reading, Mass. 1980.

Mirza, J. H., "Performance Evaluation of Pipeline Architectures," Proceedings of 1980 International Conference on Parallel Processing, 1980.

Rogerson, P.C. Fault Tolerant Networks of Microprocessors, Master Thesis, Virginia Polytechnic Institute and State University, March 1979.

Siegel, H. J., R. J. McMillan and P. T. Mueller, Jr., "A Survey of Interconnection Methods for Reconfigurable Parallel Processing Systems," National Computer Conference, June 1979. Siewiorek, D. P., D. E. Thomas and D. L. Scharfetter, "The Use of LSI Modular in jComputer Structures: Trends and Limitations", Computer, July 1978.

Stone, H. S., "Parallel Processing with the Perfect Shuffle," IEEE Transactions on Computers, Vol. C-20, Feb. 1971, pp. 153-161.

Sullivan and Bashkov, "A Large Scale, Homogenous, Fully Distributed Parallel Machine," Proceedings of IEEE Computer Architecture Conference, 1977, pp. 105-124.

Thompson, C. O., "Generalized Connection Networks for Parallel Processor Intercommunication," IEEE Transactions on Computers, Vol. C-27, Dec. 1978, pp. 1119-1125.

Wulf W. A., and C. G. Bell, "C.mmp- A multi-mini-processor," Fall Joint Computer Conference, 1972, pp. 765-777.

A VISUAL ANALYZER FOR REAL-TIME INTERPRETATION OF TIME-VARYING IMAGERY

Bruce H. McCormick
Ernest Kent
Charles R. Dyer

Integrated Systems Laboratory
University of Illinois at Chicago Circle
Chicago, Illinois

A design strategy utilizing highly parallel processing for real-time analysis of time-varying imagery is proposed. This strategy focuses on identifying models and algorithms employed in biological visual analysis, mapping their correspondence with existing and new algorithms for computer-based visual analysis, and simulating biological visual mechanisms with VLSI arrays of parallel processors.

Our objective is to design a visual analyzer for spatially and temporally distributed image processing, employing a three-dimensional architecture derived from central nervous system concepts. Output of the visual analyzer will be high-level real-time image constructs, suitable either for guidance of immediate action or for symbolic description of images.

I. INTRODUCTION

A. Levels of Analysis

Within this general design strategy, four hierarchical levels of analysis can be distinguished:

Sensory and Perceptual Function. Identification of those sensory and perceptual functions of mammalian visual processing which would be most profitably modeled in an artificial visual system to be used in a variety of three-dimensional real-time environments;

Neurophysiological Expression. Specification of the neuro-physiological expressions of these sensory and perceptual functions, and the design of parallel processing algorithms for their implementation;

Array Simulation. Computer simulation of multiple array structures designed to emulate the parallel processing employed at the neurophysiological level in the mammalian visual system;

Realization in VLSI. VLSI implementation studies for assessing low-level design procedures to optimize speed/hardware tradeoffs and minimize interconnection problems.

B. Unique Aspects of the Proposed Visual Analyzer

The unique aspects of the proposed visual analyzer are:

Events in Time Are the Units of Analysis. The current action occurring over time, not a hypothetical frozen instant of a visual field, is the element of visual analysis.

Real-Time Analysis of Time-Varying Imagery Is a Design Goal. The ability of the visual analyzer to process dynamic imagery in real-time is a necessity for advanced applications, and is a design goal from the start.

Image-Flow Architecture Is Employed. A three-dimensional array of functional units is employed to provide adequate computational power for real-time analysis of events in four dimensions (three-space plus time). Computational processes in the system may interact along either temporal or spatial dimensions of the input data.

Biologically-Based Design Is Compatible With Advances in VLSI Electronics and Neuroscience. The design uses VLSI technology to express neural architecture as currently understood. Further convergent developments in these areas may be incorporated into the system at a later time.

Technology Can Be Extended to Cognitive Machines. The initial system can later function as a preprocessing stage for subsequent advanced cognitive machinery; the linkage between the neuroconnective and symbolic models of visual description is built in. This extension is foreseen in the structure of its output, which is derived from studies of perceptual function and its computational requirements, as observed in human cognitive processes. On a more theoretical level, the studies proposed here may provide insight into the general problem of applying parallel processing techniques to the development of machine intelligence.

C. Significance

Achieving these goals will have both practical and theoretical benefits. Man-made visual systems of the type described here may be employed in such applications as visual guidance of high-speed low-altitude flight of missiles and aircraft over unfamiliar terrain; undersea exploration; remote sensing and control of operations in hazardous and inaccessible environments; high-level monitoring of radar and other imaging devices; and in general, any application requiring real-time image processing in generalized three-dimensional time-varying environments.

II. PROJECT REALIZATION AND RATIONALE

Underlying the proposed design strategy are two assumptions which we explicitly adopt. The first is that functional modeling of selected aspects of biological image processing is a promising and largely unexplored approach for further development of machine intelligence. The second is that parallel processing computers of the sort described below offer the machinery most likely to realize a real-time visual analyzer successfully. In light of previous attempts at biologically-oriented modeling, and the existence of other possible design approaches, these assumptions require comment. In this section we shall address these issues in an interconnected fashion, describing the design of the device and -- from the perspective of principles which we perceive to be embodied in the results of a half-billion years of visual experimentation -- its rationale. We will adopt the order of topics adopted in Section IB to describe the unique aspects of the visual analyzer.

A. *Events in Time are the Units of Analysis*

The natural unit for visual information processing, in the machine as in the animal, is the current action, not the current instant. We might capture this distinction by the analogy that in the real world we look at a movie, not at sequences of individual static pictures. The direct perception of reality as an ongoing process is fundamental to our ability to interact with it. Even if the human observer is artificially constrained to look at a single static picture, we know from observation of his eye movements that the scene is rapidly decomposed into a time-sequential montage of sub-scenes, with some sub-images repeatedly and redundantly reexamined; he imposes the movie organization on the static picture. Our everyday visual "image" of the world is just such a montage, sequentially built up from views through the tunnel of clear foveal vision. Thus, the ability to analyze time-sequences of images into coherent perceptual scene-models is fundamental to our perception of events in both space and time.

For the purposes of computer vision, we find it useful to generalize the concept of movie to include: time-sequential images (robotics); spatial sampling of an extended high resolution image (cartography); multispectral imagery (remote sensing of the environment); and consecutive serial sections of a three-dimensional object (spatial reconstruction in microscopy; computer tomography). Having made this generalization, we must also point out that, in reality, except for the engineering subdiscipline of multispectral image analysis, there is today only a very limited corpus of experimentation in any of these areas. Furthermore, we find essential a careful separation between the two concepts of image management and image computation. Utilizing this distinction we have attempted to generalize from prior experience with spatial data management and multispectral image analysis to a projected time-varying image-flow architecture for computer vision.

B. Real-time Analysis of Time-varying Imagery is a Design Goal

Advanced computer architecture is still wedded to habits of thought which have evolved from the analysis of static imagery. While such architecture might be employed to analyze dynamic imagery, it cannot cope with it in real-time, and thus loses the principal advantage to be derived from such analysis. It is for this reason that we place such emphasis on learning from biological models of vision: biological vision developed explicitly for real-time processing of dynamic imagery. From dynamic biological models, correspondences in principle can inspire appropriate computer architecture.

The natural evolution of action-oriented architectures for visual information processing has been little studied by computer scientists and engineers. Before characterizing what a biological visual system does, it may be useful first to discuss what such a visual system need not do. Consider a chipmunk running across a field. Its visual system certainly does not need to resolve and name each individual blade of grass in its visual field, perhaps some 10-20 times per second. By overloading its visual system, this inessential processing would inhibit attention to the potential presence of a hawk circling overhead. It would suffice for the animal's visual system to discern grass as a textural element, crudely segmented through consecutive frames. A similar argument applies equally well in the case of a biological pilot in a low flying aircraft or a machine-guided low flying missile. The system must optimize its use of time by performing urgent analyses early at low levels of sophistication, and sophisticated analyses later as the image flows through the system -- while simultaneously performing a new set of low level analyses on the next moment of the image as it enters the system. Clearly a visual system with this property -- an action-oriented architecture -- would have an evolutionary advantage.

Cyclostomes (e.g., lampreys), which first appeared in Ordovician times 300 million years ago, already had well-developed eyes and visual brain centers. Insects took a different but remarkably parallel evolutionary course. Animals of different classes look upon the same natural scene, but "see" and respond differently. The lamprey and the house fly share few common visual interests; the eyes and visual pathways of these two animals reflect their disparate concerns.

By analogy, we have no reason to anticipate that the rapid industrial evolution of vision-equipped missiles and robots will give rise to one universal architecture for computer vision. Rather, the design of individual computer vision systems must be predicated upon the actions and responses required in their circumscribed visual environments, and diverse evolutionary paths can be expected. The architecture of our device should lend itself to flexible implementation.

The artificial intelligence view of vision does not seem to account very well for the lamprey's excellent visual apparatus. We have evolution quite twisted around. An unfortunate paradigm for vision has sprung up which views it as the ability, when shown an object, to respond by naming it. In the extreme view, vision is conceived of as a restricted question-answering system: "A word is worth a thousand pictures." But this point of view is counter to natural evolution, in which the contemporary pastiche of intelligence and symbol manipulation is a very late arrival on the world scene. The principles of vision appear to be far older and more fundamental than language, with survival value that will undoubtedly outlive man. We do not denigrate the importance of the naming function of vision, but we feel that it must arise as a development out of an already sophisticated device for real-time navigation, and in higher animals, the construction of dynamic perceptual scene-models.

C. *Image-Flow Architecture is Employed*

We view the real-time image processor as being implemented by a three-dimensional stack of two-dimensional processing ARRAY ELEMENTS interconnected in the third dimension by intervening INTERCONNECTION ELEMENTS (see Fig. 1). In physical structure, the real-time image processor would resemble the multiple elements of a compound lens (or a club sandwich): with alternating array and interconnection elements.

The image data from a particular "moment" of image-time will be contained in, and processed by, an array element at each stage of its analysis, and then passed to the next array element in sequence. Meanwhile, new image data is entering the device continually from the front, so that successive moments of the image flow through the device in a pipelined fashion. Thus, image-time is represented linearly in the third dimension of the device (perpendicular to the elements), while the three dimensions of the visual scene at any given moment of image-time are represented by data residing on the two-dimensional surface of an array element.

    1. Realization of Array Elements. Each array element is composed of a mosaic of VLSI chips, much as the ITEK high resolution CCD array for the space telescope is built from a mosaic of CCD chips (see Fig. 2). Processing arrays at different levels of the device are anticipated to be functionally quite distinct, and to parallel the comparable processing schemas at different functional levels of biological visual systems. In a number of ways, the constraints on the design of these proposed processing arrays are quite unlike those of other existing image processors, in that:

    a. Each array element consists of a regular 2-dimensional array (rectangular or hexagonal) of processing elements (PEs). The PEs provide distributed feature extraction and database storage for several different properties, with some PEs specialized for one set, and others for another.

    b. The array elements are computationally bound to complete their processing of a time-segment of the image in a typical frame time of 33 msec, for use with commercial United States television equipment. Nonetheless individual PEs may encode for time-varying imagery over several frame intervals.

    c. Since dynamic imagery flows through the system -- moving successively from one processor array to the next, it becomes necessary to devote special attention to the interconnection between processing elements. Such a connectionistic framework for real-time image processing places severe constraints on the space-time trade-offs for any one processing element and on the level of data distribution to neighboring PEs, both within the same array and between adjacent arrays. We consider that an algorithmic study of these trade-offs is of fundamental importance, with direct bearing on the appropriate design of semiconductor array circuitry and the third-dimension interconnection elements. Judging from physiological evidence, interesting visual processing should begin after three levels of processing, and probably will require at least five levels to reach higher levels of abstraction.

    2. Interconnection Elements. One potential interconnection technology available today would use optical linkages between the integrated circuit array elements. In this model the connection elements can be viewed as generalized fiber optic bundles allowing the fan-in and fan-out of signals as image information is passed from one array element to the next.

    For devices intended for the rugged battlefield environment or for commercial robotics, we are attracted to an optical/fiber-optics technology in which the individual processing arrays are readily replaceable, much like individual lens elements of a compound optical system. This would also permit easy customization of systems for varied tasks. That the interconnection elements are inert fiber optic networks embedded in a supporting matrix is also attractive, as these components of the design would be hardy and should require virtually no maintenance. Interconnection elements also serve to conduct heat from the adjacent array elements.

These remarks are intended only to illustrate the intrinsic feasibility of implementing the device in the proposed 3-dimensional architecture. We are not prematurely wedded to any one interconnection technology.

D. Biologically-based Design is Compatible with Advances in VLSI Electronics and Neuroscience

The attempt to produce machines that perform cognitive functions by emulating biological systems is not new, and its motivation is quite straightforward. The brain is the only example we have of a working system which accomplishes the desired ends. Because the basic elements of the brain perform functions similar to those of elements of electronic systems, it is tempting to suppose that many operating principles of the brain will transfer easily to electronic systems.

1. Inappropriate Levels of Modeling. Despite these seemingly plausible arguments, attempts to base practical devices on biological models have not been highly successful. A principal reason appears to be that designers have focused on aspects of biological function unsuited to the available technological base. For example, there has been no shortage of designs for "electronic neurons," despite the total lack of an adequate technology for producing and interconnecting such devices in the quantities needed to simulate brain processes. Attempts at simulating this level of biological functioning often ignore the lack of understanding of detailed brain operation on which to base algorithms for these devices.

At the other end of the scale, there have been very many attempts to model the gross functional aspects of brain operations on traditional serial processors. While such experiments in artificial intelligence have produced many interesting insights, they have not been able to generate practical programs for real-time solutions on a useful scale. It seems clear that they are inherently limited by the size and speed of the machinery on which they run, and that no foreseeable improvement in this technology will enable them to compete in real time with a biological system whose architecture is adapted to the efficient solution of problems requiring large numbers of simultaneous operations. In this case as well, information is lacking about the physiological details of the processes being modeled. While such programs may attempt to simulate the overall functions of biological systems on the large scale, they are unable to draw upon any existing body of information concerning the actual underlying procedures of large-scale biological systems. They are thus denied any real advantage of having an existing system to emulate. It is an unfortunate truism that theories of cognitive psychology owe more to the field of artificial intelligence than vice versa.

2. *Proposed Level of Modeling.* We do not believe that these problems in previous research affect the validity of the biological modeling approach. Rather, such problems seem to reflect difficulties in selecting levels of approach appropriate for existing technology or existing biological knowledge. Both the approach detailed here and the functional level of the processes targeted for emulation have been carefully chosen to avoid these pitfalls.

The visual analyzer proposed here would model certain lower-level properties of the biological visual system in order to serve as an input preprocessor for advanced visual systems. At this level, existing physiological models offer sufficient detail for substantial guidance. In addition, a technological base adequate to realize the system in VLSI hardware either exists or is under development. The functional level represented by these stages of visual input processing is sufficiently advanced to make the development of such a device attractive, while not being so specialized as to preclude its subsequent use as a building block in generalized cognitive machinery.

At the level of image processing represented by the proposed device, numerous theoretical models of biological visual operations exist which provide detailed insight into a variety of important and fundamental perceptual processes. For example, detailed theoretical models describe perceptual information "property channels" in the visual system which extract information on luminosity, color, texture, motion, depth, location and other elementary properties. Other models specify mechanisms for boundary detection and scene segmentation, form, the deletion of redundant information and the creation of location-free statements of scene content for use in categorization and naming.

E. *Technology can be Extended to Cognitive Machines*

We shall define the output properties of this system to fit the input requirements expected for higher-level generalized systems. We propose to anticipate these requirements by study of the functional properties of human higher perceptual processes. In this way, we will be able to draw upon existing knowledge of human cognitive/perceptual processes in defining the proposed system, even in the absence of specific knowledge of the underlying physiology of these higher processes. So defined, the

system should provide a significant improvement in the real-time
performance of advanced computer cognitive or perceptual systems,
whether such systems are implemented in software as artificial
intelligence algorithms, or ultimately as hardware devices. It
will do this by freeing such systems from the costs of data
reduction from raw input to higher level perceptual constructs.
A particular advantage of the proposed approach is that, by
designing the output to conform to the kinds and organizations of
data employed in human higher perceptual processes, further
research on simulating higher processes will be facilitated.
"Bootstrapping" the development of successively more advanced
systems based on biological models thus becomes possible.

F. *Summary*

     The application of models of biological visual processes to
the design of artificial systems as proposed here is made
particularly attractive by the fact that biological vision is
fundamentally parallel in its organization. As a result, we may
employ these biological models not only as paradigms of useful
and perhaps necessary processes in image recognition, but also as
demonstrably successful examples of such processes realized in a
parallel processing environment.
     The major task of the proposed program is to   model
biological techniques for mapping perceptual processes   onto
parallel arrays. Effort can then be directed to the   more
restricted problem of determining efficient methods   for
implementing these mappings in electronics-based as opposed to
neuronally-based architecture.
     The level of biological modeling to be employed in the
detailed design of the processing array elements will   be
intermediate between that of a "neuron by neuron" model of the
visual system, and that of a mere "black-box" functional
description of cortical operation. At this level we will be able
to obtain significant guidance from the biological model without
resorting to unrealistic technology or physiology.
     Our assumption that parallel processing arrays are the most
plausible approach to the problem of computer vision follows
immediately from an acceptance of the applicability of biological
models in this area. While it is true that neural systems are
neither strictly parallel nor serial in organization, it is clear
that at the level of the visual system, the biological machine is
overwhelmingly parallel in its approach. To model its operation
in any other fashion seems to waste the detailed insights gained
by study of its organization. Coupled with their obvious success,
this choice of organization on the part of natural systems serves
to reinforce the conclusion, from a priori logical premises, that
the problems of image recognition are best suited to parallel
approaches. This follows from the large number of similar
operations which must be performed, and the fact that the

necessary data for all of these operations is simultaneously available to the system.

A three-dimensional architecture with space-time interactive processing such as that described above most closely addresses the requirements of a system designed to emulate biological processing. This is because the biological system likewise employs a multi-stream pipelining architecture with strong interaction in the longitudinal (temporal) dimension as well as transverse (spatial) interactions. We feel that no other architecture is likely to realize our goal of real-time analysis of dynamic imagery.

APPENDIX: VLSI DESIGN CONSTRAINTS

A. *The Input Level*

The visual analyzer must accommodate a variety of imaging techniques and transducers customized to particular environments (low light levels, infrared, x-ray, telescopic, etc.). For example, the first array element can be a simple CCD imaging array. Alternatively, a remote television video signal could be sent to the device and buffered here. Both approaches have advantages. One benefit of the visual analyzer structure that we propose is that 'focal plane' array elements can be readily replaced by others with different characteristics or even different sensory modalities, without necessarily changing the remainder of the visual analyzer -- a significant advantage in using this modular architecture.

To facilitate presentation of essential concepts, we have shown the device in Figure 1 as having a single "retina" or input. However two side-by-side image areas are required to generate retinal disparity signals for depth channels. These in turn generate vergence and accommodation movements to define the fixation point in depth.

B. *Number of Processing Elements Required*

A rule of thumb of the image processing community has been that it takes on the order of 1000 instructions/pixel to process an image through to the intermediate level of vision. These instructions, of course, are on a contemporary serially-organized (von Neumann) computer. On this basis we can place a lower bound on the number of processing elements required to exhaustively process commercial television imagery.

The number of picture elements per second to be processed is

$$no.\ of\ pixels/sec = (480\ X\ 640\ pixels/frame)\ X\ 30\ frames/sec$$
$$= 9.2\ X\ 10^6\ pixels/sec.$$

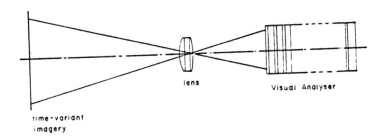

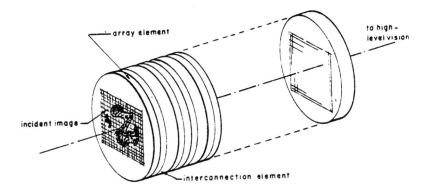

Fig. 1. *Visual analyzer as stack of array and interconnection elements.*

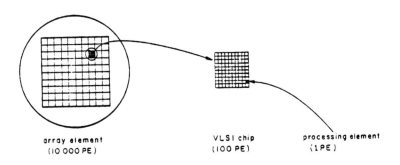

Fig. 2. *Mosaic construction of an array element.*

Allowing for line returns and interframe time cuts into the time available for video digitization. Undoubtedly similar limits on processing rate would constrain the operation of the visual analyzer. Accordingly we will assume

pixel sampling rate = 12.3 MHz (max).

We will assume 1000 instructions/pixel and a conservative $3\,\mu$ sec/instruction to derive

$$\text{no. of processing elements} = (12.3 \times 10^6 \text{ pixels/sec})$$
$$X \ (10^3 \text{ instr/pixel})$$
$$X \ (3 \times 10^{-6} \text{ sec/instr})$$

$$= 36,900 \text{ PE}$$

This estimate is quite conservative. It assumes a slow instruction time of $3\,\mu$ sec, whereas increased microminaturization may improve this situation. Nonetheless the contemporary estimates of instructions/pixel are for long wordlength machines. The architecture of the visual analyzer, which is designed to alleviate needless address computation and to exploit an estimated five-fold temporal/spatial redundancy in television imagery -- should compensate for this difference.

C. *Total Area of Silicon Required*

Assuming continuing advances in microlithography and that line widths continue to halve every 2.3 years, we anticipate that a contemporary modest microprocessor in 1985 will occupy approximately 1 (mm)^2 of silicon surface. On this basis, we require 4×10^4 (mm)^2 of silicon surface, or in the design proposed in Section IIC we require 4 *array elements,* each (100 mm X 100 mm).

In summary, the design proposed is conservative in its estimate, and continuing improvements in solid state technology should in time allow even less silicon/array element, possibly fewer array elements, but not otherwise fundamentally change the device.

Index